Phenomenology and The Ghost in the Machine

An Investigation Into Wilfrid Sellars, Jean-Paul Sartre, Gilbert Ryle and The Concept of Mind

Timb Hoswell

Phenomenology and The Ghost in the Machine

An Investigation Into Wilfrid Sellars, Jean-Paul Sartre, Gilbert Ryle and The Concept of Mind

ISBN 978-1-365-60331-0

2016
Written by Timb D. Hoswell
Published by
The 7th House of Cult Literature
Strathfield
Sydney, Australia
Tittle page by Roger Foley Fogg

ISBN 978-1-365-60331-0

90000

9 781365 603310

tist: Roger Foley Fogg

Ar

Explorers' Guide to Phenomenology and the Ghost in the Machine

Two Ways To Read the Work

I have always believed that philosophy should be both a source of inspiration and an experience. Thus I suggest there are two ways to read the work.

Inspiration.

If you want to use this work as a source of inspiration, I suggest start anywhere and read it, and if you don't like what you read, start somewhere else and read from there. Take inspiration from the ideas and concepts in this work.

The Experience

There is a specific set of experiences in this work. To get to them you have to go through stages. I want to show you an insight into 20th Century Philosophy of Mind that crosses the Analytic-Continental Divide and starts with Gilbert Ryle, Jean-Paul Sartre and Wilfrid Sellars. It is an insight that relates to ongoing research in Neuroscience and Psychiatry as well as the way we study and explore the mind. Like the Empiricists and Rationalists during the Enlightenment, the latter part of the 20th Century has come to a division in the Philosophy of Mind about the fundamental nature of mind and how to unlock its secrets. This division I should argue, has splintered Philosophy between Psychologism and Anti-Psychologism with The Ordinary Language Theorists and Neuro-philosophers on one side and the Phenomenologists and philosophers who advocate developing a 'first person philosophy of consciousness' on the other. The argument that splinters philosophy is about how do we progress in our understanding of the relationship between language and the mind in the wake of technological advances in neuroscience? How does the language we use to talk about the mind relate to the processes, states, mental phenomena and chemical imbalances that make up our newly emerging 21st Century understanding of the 'Scientific Image of Man'? What are terms like 'a flash of anger' or 'a man's motives' actually referring to?

This argument which pierces psychology, psychiatry, medicine, medical science and neuroscience through 21st Century philosophy has roots that go back in one way or another to three central philosophers in historical dialogue with each other. Wilfrid Sellars, Gilbert Ryle and Jean-Paul Sartre stand at the centre of the controversy and conflict between Psychologism and Anti-Psychologism that splinters philosophy today.

On one side of the divide we have the classical phenomenologists. These are people like Jean-Paul Sartre, Henri Bergson, Franz Brentano and Husserl who set out to study consciousness and give a disciplined account of it. They are part of a breed of philosophers who broke away from the Husserl-Frege debates on psychologism to follow the other path and form part of the Continental tradition. But alongside these we have a 'new breed' of analytic philosophers who we might see as having aims similar to the 'classical phenomenologists'. These are people involved in Analytic philosophy with Analytic philosophical training who are interested in a "first person science of the mind". People like David Chalmers or Frank Jackson and revisionists like Tim Crane.

These are philosophers who want to explore the conscious experience of the mind and who argue that there is knowledge to be from gained exploring our conscious experience of the world.

On the otherside we have philosophers who are Anti-Psychologistic. People like Robert Brandom, J. Stanley and T. Williamson, Gilbert Ryle himself, Michael Dummett and a whole generation of Ordinary Language philosophers like Paul Grice, John Searle and J. L. Austin who have their own positions on language and meaning. Indeed Michael Dummett, here, is pivotal for understanding Analytic Philosophy and its ongoing search for a theory of meaning in language and how this project relates to its Anit-Psychologistic agenda. These are philosophers who see the mind as either constrained by language in our talk about it or thought as analogous to language. Thus they see the way forward is to understand the nature of language and its relationship to the mind and knowledge about both by developing a theory of meaning in language.

Michael Dummett takes this view and divides language theorists into two schools on the basis of how we might go about finding a theory of meaning to explain language based on explicit and implicit formulations. The former are Pro-Fregeian and formal about language and its sentential structures while the latter are Dummettian and characteristic of the implicit Ordinary Language school. These distinctions will become clearer in the paper as I explain them at length.

These two vast trees of Psychologism and Anti-Psychologism that make up 20th Century Philosophy of Mind have branches that intersect in a historical link between Gilbert Ryle and Jean-Paul Sartre which grow fruitful in *The Concept Of Mind*. The link between Gilbert Ryle and Jean-Paul Sartre is at the centre of this division that runs through 20th Century Philosophy of Mind and their influence rebounds in philosophy today. This is no mere coincidence. Both philosophers were read on either side of the Analytic-Continental divide and both of them were influenced within key components of the framework making up 20th Century Philosophy and the Analytic-Continental divide.

We can see from Gilbert Ryle's writings that he was influenced by the writings of David Hume. Hume was adopted by early Analytic philosophers because of his relationship with Kant, Empiricism and the Logical Positivists' project for a criteria of meaning and epistemology. Ryle was also of a generation of philosophers keenly interested in and influenced by developments in Symbolic Logic. There are parts of *The Concept of Mind* where he engages in talk about the modalities of the ordinary user's understanding of 'can' and uses logical distinctions there. While original in its scope and project, *The Concept of Mind* shares with many other works of its period a keen interest in using language analysis as a philosophical tool and a method for solving philosophical problems.

Indeed the direct link between Jean Paul-Sartre and Gilbert Ryle is Sartre's argument on David Hume which Ryle lifts and later admits he lifts and is part of what he admits troubles him about his account of the imagination in *The Concept of Mind* in one of the last interviews of his life.

The original form of the argument is a phenomenological argument Sartre develops against David Hume in *The Imaginary*. In it Sartre compares properties of the percept and the imagined object to show that the two are two different types of things entirely, and one not merely a more lively version of the other. Gilbert Ryle reads this, then uses that same argument in his chapter on *The Imagination* in *The Concept Of Mind*. Thus the branches from the rival philosophical trees of Psychologism and Anti-Psychologism become entangled in the orchards of philosophy.

Sartre has been influenced by a strain of philosophy that comes down from the other side of the Anti-psychologism/Psychologism divide and a school of philosophy that has roots in the conflict between Husserl and Frege. Sartre is interested in Kant, but he is also keenly interested in Hegel's logic and the way that Hegel deals with Kant's unification of time and space under the apperception of the I. Hegel, of course, forms a fundamental difference to how Sartre and the Continental Philosophers read Kant. Indeed Sartre's work *Being and Nothingness* is deeply engaged in dialogue with Hegel's system of logic and contains chapters where Sartre quotes directly from *Hegel's Encyclopedia of Logic* and other works Hegel wrote on logic.

Thus when we come to Robert Brandom, for instance, we can see how the two schools have interpreted Kant differently. Brandom has insights from Frege into the way that propositions work while Sartre has developed a phenomenological doctrine about theitic consciousness and the way an "I" or "me" appears within our thoughts. We see on either side the emergence of Psychologistic and Anti-Psychologistic Themes with parentage that goes back to Husserl and Frege but which emerges between Sartre and Brandom in a new form, as a new interpretation, with a new rivalry in how to read back into Kant. Does one adopt a linguistic and propositional treatment of judgements, as Brandom does, where the minimum unit of thought that one is capable of is a proposition that is affirmed and prefixed by "I think" ? Or does one adopt, as Sartre does, an apperception of the I in which the necessity for the possibility of an "I" is always present, but de facto, it does not always accompany our thoughts?

If we return to Ryle and look to the dialogue between Wilfrid Sellars and Gilbert Ryle and the relationship that emerges from Wilfrid Sellars in some of his other papers we can also see a link to the emerging structure of neuroscience which revolutionizes psychiatry in the last decades of the 20th Century. Wilfrid Sellars was aware of the developing technologies of neuroscience and the way they could come to change how we understood the human being and our understanding of the nature of mind. Sellars more than most philosophers was ahead of his time in recognizing the impact of neuroscience on our understanding of the mind as the 20th Century drew to a close. His discussion of the Manifest Image and the Scientific Image of Man tell us about what these impacts would do to reshape the mind-body problem. Wilfrid

Sellars and his writings on the mind are deeply prophetic in this sense. He anticipates many of the ideas of Paul and Patricia Churchland and the intial scope of the arguments that would emerge between David Chalmers and Daniel Dennett. Thus the dialogue between Sellars and Ryle can tell us something about the emerging and changing concept of mind in the wake of technological change and innovation. When we reflect on this and what Dummett has to say about implicit and explicit forms of Anti-Psychologism we see that these relationships are part of a web of deep philosophical questions and meditations to arise out of philosophy, language and science on the nature of mind. The attractive insights this web, when systematically explored, can reveal are a vast tapestry. This tapestry is coloured by the reactions of the leading philosophers of today to the work of Gilbert Ryle, Wilfrid Sellars and Jean-Paul Sartre. These reactions are the threads that run throughout the Philosophy of Mind. Once we understand this we can step back and see the picture in the tapestry formed by the complexity of colour and thread.

Through exploring this relationship between Gilbert Ryle, Jean-Paul Sartre and Wilfrid Sellars, and following the steps in this book one can take a trip through 20th Century Philosophy and into the outer reaches of the mind.

The Outer Reaches of The Mind.

When I was originally planning out this work, I had the problem of how do I convey the deep insights in Gilbert Ryle, Wilfrid Sellars and Jean-Paul Sartre to an audience? The problem is that one must be very deeply entrenched in philosophy to understand concepts like "psychologism" and "anti-psychologism". Moreover the insights I wanted to share arose from a historical link between Jean-Paul Sartre and Gilbert Ryle that had been little explored.

Gilbert Ryle's *The Concept of Mind* was heavily influenced by Jean-Paul Sartre's work *The Imaginary* via an argument about David Hume which Ryle himself admits in a little known piece that was originally written in French. Once you know this you can begin to see how the influence on Ryle from Sartre's philosophy actually introduces a fault line into the arguments in *The Concept of Mind* and changes Ryle's original language based account of the mind.

Ryle has a revolutionary concept about how we can explore the mind based on the relationship between mind and language. Ryle's idea is that we can understand the mind by the sorts of things people say and the way they say them. He thinks that all of the important knowledge about the nature of our minds is available in language. Ryle's idea is that we already know about the nature of people's minds because they talk about it. For Ryle, the man who talks about going fishing on the weekend is having thoughts about going fishing on the weekend. The woman who worries about what to wear to the Christening on Monday is thinking about what to wear to the Christening on Monday. The couple who discuss what to have for dinner are thinking about what to have for dinner.

Ryle argues that the mind is not an mysterious otherworldly place full of Cartesian demons and Oedipus complexes connected to the ID via one's anus. Ryle argues that people think about going fishing on the weekend, what to have for dinner, and what type of clothes they're going to wear to work or social events because this is the sort of thing that people talk about. He thinks that the mind is accessible to us through language and thus for Ryle language is the tool one uses to understand the mind. Ryle thinks we can understand the mind through the every day language that people use to talk in, and which he maintains they think in. Ryle affirms that the mind is accessible through an 'internal log keeper' which he argues can tell us what someone is thinking if we ask them. Asking someone what they are thinking about allows them to phrase a 'special status report'. The internal log keeper presents the internal monologue as a 'special status report'. The internal log keeper of the mind, according to Ryle, can retrospect, but it can not 'reflect'. For Ryle 'retrospection' is the act of reciting aloud what the internal log keeper has been keeping track of in the monologue of internal thought for the purposes of one of these 'special status reports'.

From this 'log keeper' role Ryle develops a concept of mind based on Ordinary Language analysis. He thinks we can understand the mind by analysing the way people use language. The vast majority of the arguments in *The Concept of Mind* draw on this concept of an internal log keeper, retrospection and special status reports in one way or another. He formulates an analysis of the mind based on the 'linguistic behaviours' of words found in every day conversation and ordinary language used by people in non-theoretical contexts. The reason why he focuses on 'ordinary language' and rejects the use of theoretical vocabularies like we find in Cartesian analysis of the mind and psychological text books, is that Ryle thinks 'Ordinary Language' is the language that people think in.

Although Ryle doesn't admit it in *The Concept of Mind* many of the arguments in that work also have phenomenological roots. A few arguments are purely phenomenological and based on getting the person to imagine an object or trying to remember what a certain emotion felt like. Some arguments are both, they have a phenomenological side, and are also based on distinctions that arise from an analysis of ordinary language. Exploring the two-sided nature of these arguments reveals something very interesting about philosophy of mind which can open up the Psychologistic and Anti-Psychologoistic debate.

This is an incredibly interesting thing for anyone familiar with the history of 20th Century Philosophy of Mind. One of the claims that Ryle makes in *The Concept of Mind* is that consciousness is a myth that originates in the Protestant Reformation. For this reason Materialists like David Armstrong read him as a polemic against dualism and a possible way to cross over the mind-body problem in favour of Materialism. Ryle became included in much of the Pro-Science Philosophy strains at the time. Through readings of him by philosophers like David Armstrong, J.J.C Smart and U.T. Place. This lead to Ryle being classified by those who haven't read him as part of the 'jack-booted pro-science Analytic philosophical wave' because he was read and used by philosophers involved in that wave.

However, as you will see, Ryle himself was actually anti-scientific in his stance about understanding the nature of mind. Ryle thought that science and historical periods of science, what Thomas Kuhn would later famously refer to as 'scientific paradigms', were in fact the origins of problems and muddles about the mind. Ryle thought that periods like the Enlightenment had introduced mechanical and causal vocabularies that contaminated and caused riddles in our ordinary language understandings about the mind, and thus muddled up the ways that we talk about, and think about the mind. Ryle has a whole chapter dealing with these contaminations which he calls "the Bogey of Mechanisms" in *The Concept of Mind* and repeats this assertion through-out the work.

This, of course, must come as a shock to those who haven't actually read Ryle, but have only read about him by those who classified him in the Pro-Science 'jack-booted' analytical schools, because he was read and associated with those schools.

To also find that Ryle himself had phenomenological roots and had lifted some of his arguments out of Jean-Paul Sartre, the same Gilbert Ryle who had been read by Materialists and erroneously classed as "Pro-Science" must also come as a shock to many who haven't read Ryle, but simply taken his classification as a jack-booted pro-science Analytic philosopher on face value.

But I argue it is a mistake to dismiss the real Ryle. To find that these phenomenological arguments hidden in Ryle's work and the roots of some of his arguments could open up and account for the inconsistencies in *The Concept of Mind* is an exciting thing! It offers us a historical nexus between Analytic and Continental Philosophy with rich material to be explored and opens us up to the rich originality of Ryle's thought.

Few people, without having read the sources I present in this work, would connect Gilbert Ryle with the 'wild and woolly' French philosopher of Existentialism, Jean-Paul Sartre. The two, on first appearance, are worlds apart.

Indeed no two philosophers, prima facie, could seem further apart than Gilbert Ryle and Jean-Paul Sartre. One is connected to an introspective discipline of mental analysis similar to various strains of Eastern meditation, wrote about freedom, anguish and left-wing-political discourse. Sartre was influential in popular culture of the 60s, and his Existentialist themes echo through the music of that period in the works of John Lennon and Jim Morrison. You also find Sartrean themes in California Psychotherapy and his ideas about Existential Life projects are in tune with the work of psychologists like Erich Fromm and Carl Rogers.

Ryle's influence, as mentioned, goes in the other direction. He was read by hard-nosed 'science in jackboots' philosophers who grew up on the establishment bloc reading papers on Logical Positivism and Falsification and were excited by advances in physics. The two were received in entirely different wave lengths. One was read in the counter culture while the other in the Analytic philosophical establishment.

However, by careful analysis of the arguments in *The Concept of Mind,* one can start to piece together a picture that allows one to reach an understanding that not only shows how the two are connected through Sartre's influence on Ryle, but why this connection is philosophically rich and offers us a way to bring together 20th Century Philosophy of Mind in an over-arching thesis about Psychologism and Anti-Psychologism. The insights offered in this work are attractive because they also show why the issues raised by Frege are still relevant, and why Dummett's critical distinction between implicit and explicit theorists about the meaning of language has important insights into the philosophy of mind.

The link between Gilbert Ryle and Jean-Paul Sartre allows us a way in to understand the rivalry between arguments that draw on phenomenological methodologies and linguistic analysis. This link allows us to systematically explore the way that different types of arguments about the mind work, and to compare them against each other to unlock deeper insights about the nature of arguments about the mind that continue on into the important work done by philosophers today.

Because Gilbert Ryle was so influential everybody read him. It was not just the Materialists. In fact most of the Analytic Philosophy of the 60s and 70s is immersed in his work. This impact means many of the themes and components of his work still run though Analytic Philosophy today since most of the Analytic Philosophers active today either read him or their own Professors did. Because Gilbert Ryle read Sartre, there are phenomenological arguments that run through *The Concept of Mind* and which can offer insights into developments in Analytic Philosophy today and where it has come since then. Unlocking these insights teaches us something about the philosophy of mind and the way that certain claims work. Thus the inconsistencies in *The Concept of Mind* sit at the centre of both historical 20th Century Discourse, and what happens now in modern philosophical discourse. The relationship between Gilbert Ryle and Jean-Paul Sartre is not just historically interesting it is also philosophically interesting.

However this interesting relationship and rich seam of philosophy does not end with Gilbert Ryle and Jean-Paul Sartre because one of the most important philosophers of the last part of the 20th Century read Ryle and wrote about him in two of the most important papers published last century.

Ryle read Sartre.

But Willfrid Sellars read Ryle.

Wilfrid Sellars is important for a large number of reasons, and one of those reasons is he recognized the use of EEG brain scans as a new source of technological information about the human mind well before it became a topic espoused by Daniel Dennett, David Chalmers and the Churchlands. Wilfrid Sellars brought together advances in neuroscience with deep philosophical insights into language and the mind, and combined this with a revolutionary insight into the way we learn to see the world as part of linguistic communities.

Ryle's "log keeper" of the mind re-emerges as Sellars "In forro interno" strain of arguments and he constructs a 'Rylean language' from the ground up. Wilfrid Sellar's philosophy, like Gilbert Ryle's was, and is vastly influential today. Indeed it is hard to find many philosophers who have not been influenced by Wilfrid Sellars work in the journals and books that dominate philosophical publications in the Anglo-Analytic world at the time of writing this Explorers' Guide.

I argue that work presented in this book should excite any philosopher involved in 20th Century, or 21st Century thought because we have a line of arguments about the mind that cross the Analytic-Continental divide and involve three of the most influential philosophers last century; Jean-Paul Sartre, Gilbert Ryle and Wilfrid Sellars, who argue about some of the biggest issues in Analytic and Continental Philosophy of Mind. Indeed the work of these philosophers comes to characterize and play a role in defining these issues and discovering a line of nexus between them opens up vast and original philosophical vistas for what happens next in philosophy of mind when we move past the Psychologistic and Anti-Psychologistic debates.

The insights on offer in this work are attractive because they concern themes and tensions that run through modern Analytic Philosophy today. For anyone who has studied philosophy long enough the insight offered by this work is something of an epiphany. However, because of the complexity of the arguments leading up to that epiphany there are stages and deeply complex steps one should go through to arrive at an understanding of philosophy strong enough to comprehend the attractive insights on offer.

This was my problem.

How might the general reader arrive at the insight on offer without having read a library full of modern philosophy books?

One way I thought of introducing this epiphany was a spiral.

The reader starts in a small circle, and then as the spiral expands the insights become deeper and deeper.

In this way, I could trace out some of the threads from this spiral and how they impact on philosophy today from the small spiral of influence three philosophers had on each other last century.

To get the "experience" of *Phenomenology and The Ghost in the Machine*, particularly if you have never studied philosophy before I suggest you begin with the essay at the end of this book (1) *Fun and Phenomenology Guest Starring Jean-Paul Sartre and Gilbert Ryle*. After reading the essay, and having a firm grasp of the Ordinary Language and Phenomenological link between Gilbert Ryle and Jean Paul Sartre via David Hume, (my first discovery), you then progress to (2) *The Synopsis*. You may find you need to draw diagrams and sketch out the way the concepts connect with each other as you read these two documents. I encourage you to do so that you do not become lost in the complexity of the stages in the overall argument of the work. The better your understanding of the link in Ryle and Sartre the better you will be able to understand the taxonomy I develop to break up Ryle's arguments in *The Concept of Mind*. Understanding the relationship between (i) Ordinary Language arguments, (ii) Linguistic Behavioural arguments and (iii) phenomenological arguments is critical to understanding the subsequent stages of the thesis. If you do not understand these three types of argument you will not understand the rest of the paper. The origin to understanding these three types of arguments comes from understanding why Ryle and Jean-Paul Sartre overlap in their argument on Hume. (3) Once you understand the structure of the arguments in the paper and have a fair idea of how they link together I suggest you read the central body of the work without referring to the footnotes from start to finish following the notes you made from the *Synopsis*. Refer to the footnotes if you need clarifications or you need to look up a source, but if possible try to concentrate on the body of the document. There is much scholarship in the footnotes that can distract you from the central argument while you are trying to understand it. (4) Re-read the work with the footnotes. The footnotes contain information and deeper distinctions that elucidate some of the arguments and their implications across philosophy today.

The following are some 'concept maps' which you might like to study, use and refer back to in your travels. The progression of the chapters of the main body of this work follow the breakdown of Ryle's terminology leading up to the final insight that this paper has to offer. A familiarity with Ryle's terminology will help the reader to follow much of the detail of the arguments presented in this book, and provide them with much of the insight that Ryle's work offers. It is in this spirit that I present them.

Best wishes

T.

Understanding Linguistic Behavioural and Phenomenological arguments in this paper. Part 1

Ryle's uses a specific type of argument I call in the paper a **"Linguistic Behavioural argument"**. This type of argument bases a claim about the mind on the behaviour of the language. For instance Ryle separates dispositions based on the behaviour of sentential subclauses by separating verbs that respond to a "how/that" distinction in order to uphold a distinction between knowledge and belief based on the behaviour of the words themselves. He does similar things with Dispositional and Episodic terms.

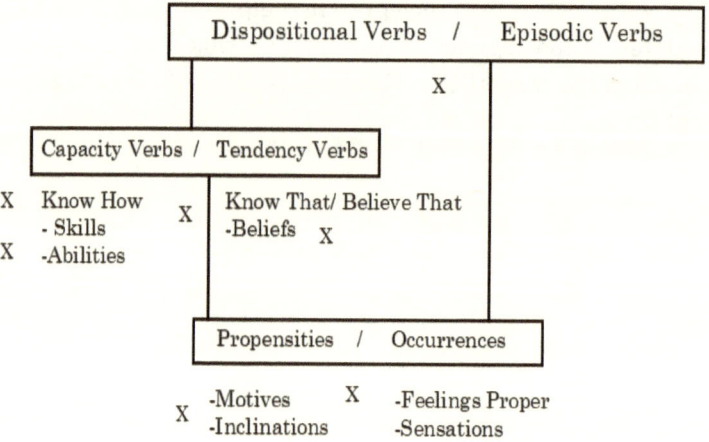

X : Distinction Based On A Linguistic Behavioural Argument

Professor Weitz's is the earliest example of someone who uses the term "Logical Behaviourism. He tries to model Ryle's arguments into three model sentences that work within a 'Fregeian propositional framework', and what Dummett would call an "explicit" framework. Weitz's mistake was to try and classify all of Ryle's dispositions with three model sentences. This would have seemed like a good idea at the time because whole sentences with propositional contents can have truth values. Thus it would become possible to map Ryle's terms in a Logical Framework. Robert Brandom and A. J. Ayer are other examples of this Pro-Fregeian "Logical Behaviourism".

Brandom, via Kant, argues that Judgements are propositional and must be either affirmed or denied by the presence of the I. The minimum unit of thought is thus a propositional which contains an I. A. J. Ayer argues that language must be present in order for thought to be true or false based on the propositional content of a thought. Without language Ayer thinks it is impossible to form sentences and only sentences can be true or false. This is the Logical Behavioiristics thread of the paper which I refer to as the explicit theorists basing the distinction on Dummet's 'implicit/explicit' position. .

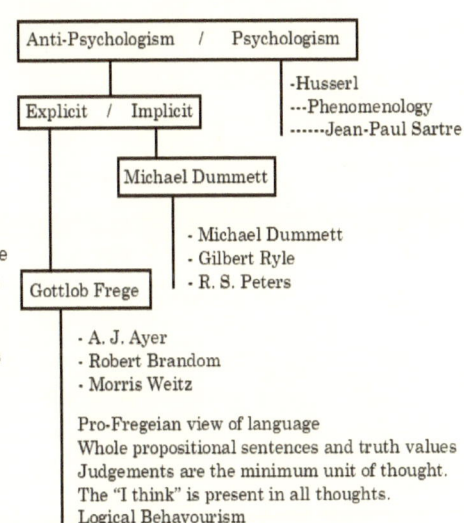

Anti-Psychologism / Psychologism

Explicit / Implicit

-Husserl
---Phenomenology
------Jean-Paul Sartre

Michael Dummett

Gottlob Frege

- Michael Dummett
- Gilbert Ryle
- R. S. Peters

- A. J. Ayer
- Robert Brandom
- Morris Weitz

Pro-Fregeian view of language
Whole propositional sentences and truth values
Judgements are the minimum unit of thought.
The "I think" is present in all thoughts.
Logical Behavourism

If we return to Weitz, the cost of a sentential propositional models is it ignores the behaviour of the dispositional verbs themselves at the level of the configurations of the verbs, nouns and adverbs that Ryle himself is interested in and hence Weitz doesn't have the resources to distinguish between Capacity and Tendency verbs which Ryle does. This begins the "Linguistic Behavioural Strain" of arguments in this paper.

Ryle's Ordinary Language Map of Linguistic Behaviours

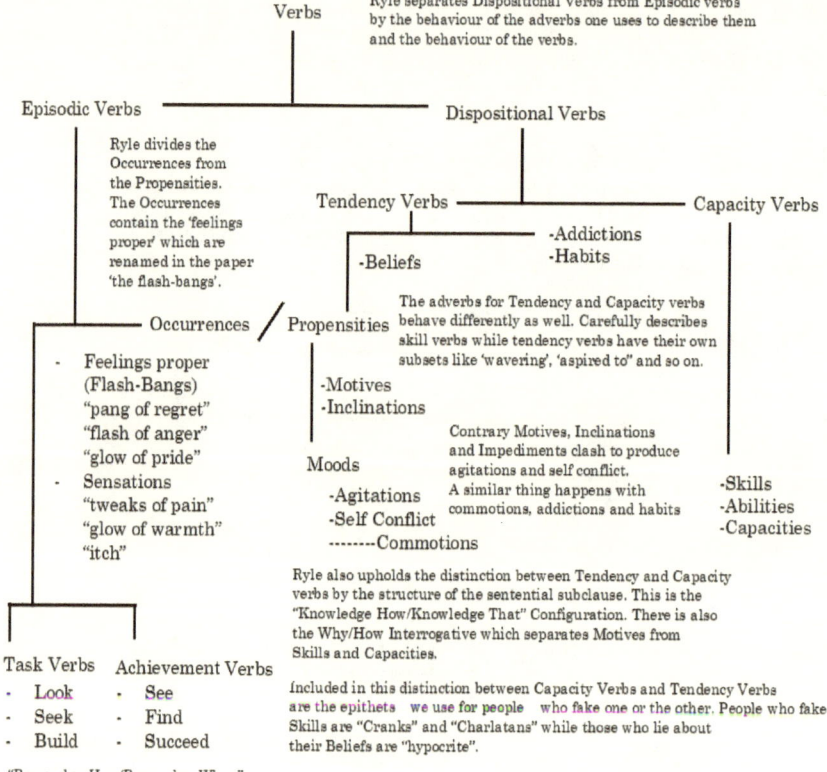

Verbs

Ryle separates Dispositional Verbs from Episodic verbs by the behaviour of the adverbs one uses to describe them and the behaviour of the verbs.

Episodic Verbs

Dispositional Verbs

Ryle divides the Occurrences from the Propensities. The Occurrences contain the 'feelings proper' which are renamed in the paper 'the flash-bangs'.

Tendency Verbs

Capacity Verbs

-Addictions
-Habits

-Beliefs

Occurrences / Propensities

The adverbs for Tendency and Capacity verbs behave differently as well. Carefully describes skill verbs while tendency verbs have their own subsets like 'wavering', 'aspired to' and so on.

- Feelings proper (Flash-Bangs)
  "pang of regret"
  "flash of anger"
  "glow of pride"
- Sensations
  "tweaks of pain"
  "glow of warmth"
  "itch"

-Motives
-Inclinations

Moods

-Agitations
-Self Conflict
-------Commotions

Contrary Motives, Inclinations and Impediments clash to produce agitations and self conflict.
A similar thing happens with commotions, addictions and habits

-Skills
-Abilities
-Capacities

Ryle also upholds the distinction between Tendency and Capacity verbs by the structure of the sentential subclause. This is the "Knowledge How/Knowledge That" Configuration. There is also the Why/How Interrogative which separates Motives from Skills and Capacities.

Task Verbs   Achievement Verbs
- Look   - See
- Seek   - Find
- Build   - Succeed

Included in this distinction between Capacity Verbs and Tendency Verbs are the epithets we use for people who fake one or the other. People who fake Skills are "Cranks" and "Charlatans" while those who lie about their Beliefs are "hypocrite".

"Remember How/Remember When".

Ryle uses this configuration for differentiating between dispositions and episodic verbs. Remember when utilizes sets of verbs listed under 'Task verbs' and 'Achievement verbs' in the subclauses of 'remember when' statements. 'Remember how' uses the sets of dispositional verbs under capacities and their various adverbs of manner.

One of the most important things to watch in the paper are the "flash-bangs".
"Flash Bang" is the term I give to a subset of the Linguistic Distinctions in Ryle's
Occurrences which he calls "feelings proper". These unlike the other distinctions,
marked with an X, lack the linguistic resources for a Linguistic Behavioural
Distinction. I have put an arrow where Ryle has this problem.

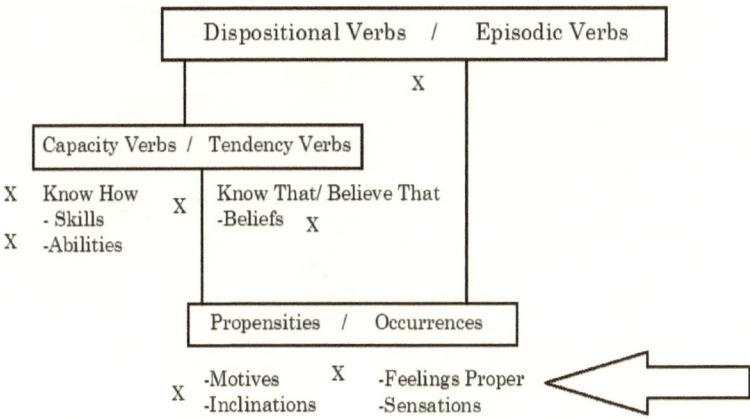

X : Distinction Based On A Linguistic Behavioural Argument

"Feelings Proper" are things like a "glow of pride" or a "flash of anger", "a pang of
regret". The problem Ryle has with these is he needs a way of telling the
difference between a 'glow of warmth' and a 'glow of pride' apart on a linguistic
basis without recourse to a theory of consciousness or phenomenological insight.
Indeed there are several arguments where he relies on an 'occult' or hidden
phenomenology in order to make his arguments.

This distinction between 'glows of pride' and 'glows of warmth' becomes the
phenomenological thread of the paper and is important to the overall thrust of
the argument for a return to a Pre-Fregeian Psychologism. If the difference
between a 'glow of pride' and a 'glow of warmth' is sublinguistic, or pre-linguistic,
that is the meaning of the terms relies on knowledge outside of a domain of
linguistic meaning, which is to say, if meaning in talk about glows of pride and
glows of warmth has foundations in non-linguistic awareness i.e. consciousness
of what a glow of pride and a glow of warmth feel like, then this gives us a view
of language as something like a code that we compound our thoughts into and
meaning in language has foundations in a theory of consciousness. This is what
Anti-Psychologism as elucidated by Michael Dummett rejects.

Understanding Linguistic Behavioural and Phenomenological arguments in this paper. Part 2

The relationship between Ordinary Language arguments and Linguistic Behavioural arguments is very similar to the relationship between a Lingua and Parole in semiotics. The Ordinary Language argument acts like the Lingua and draws on the body of knowledge the natural language speaker has in order to make claims about the mind. These claims are based on terms like 'it makes sense to say this' or 'it does not make sense to say that', or 'it makes sense to talk about this' to affirm or deny a claim about the mind. Ordinary Language arguments draw on the normative source of the body of knowledge shared between speakers of the same language to make the listener agree with the claim forwarded by the argument.

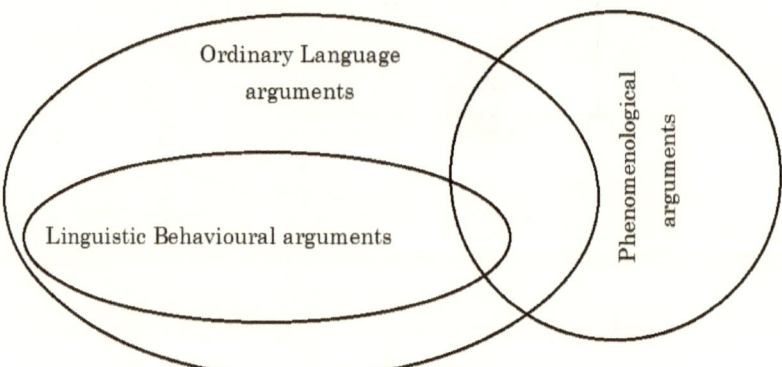

Linguistic Behavioural arguments act more like a parole. They are specific configurations of language, manifestations in the form of verbs, nouns, adverbs, adjectives, slang and so on. For instance Ryle offers one in the way that episodic and dispositional verbs behave.

Now all Linguistic Behavioural arguments are Ordinary Language arguments because they rely on what it makes sense to talk about. But not all Ordinary Language arguments are Linguistic Behavioural arguments. There are some arguments in Ryle which are Ordinary Language arguments but which do not have Linguistic Behaviours to uphold them. Ryle's argument about 'Feelings Proper' aka 'flash bangs and the difference between a 'glow of warmth' and a 'glow of pride' is an example of an Ordinary Language argument without a Linguistic Behavioural distinction. The English natural language speaker knows the difference between 'glows of warmth' and 'glows of pride' but there is no 'Linguistic Behavioural' difference to base this on. Ryle has a number of arguments like this.

This is the taxonomy that is utilized for sorting through Ryle's arguments showing the relationships between the different taxa.

All Linguistic Behavioural arguments are Ordinary Language arguments

Linguistic Behavioural arguments

These focus on the behaviour of bits of ordinary language to make claims about the mind. For instance Dispositional Verbs like 'believe', 'aspire' and 'posses' do not behave like 'wake up', 'jog', 'shout' which belong to Episodic Verbs.

Ordinary Language arguments

These affirm a claim about the mind based on "it makes sense to say" or "it does not make sense to talk about" and other similar turns of phrases.

But some Ordinary Language arguments have hidden analogical structures and phenomenological and insights.

Some Linguistic Behavioural arguments have hidden Phenomenological sides and possible introspective forms of interpretation.

Phenomenological arguments

These make some sort of claim based on getting the person to imagine or introspect or perform an exercise of consciousness. Sometimes these are hidden in Ordinary Language arguments like Ryle's claim about a difference between a 'glow of pride' and 'a glow of warmth' which ordinary language speakers know but which are not based on specific linguistic behaviours.

Occult Phenomenological arguments.

These are arguments that have both a Linguistic Behavioural and a Phenomenological side like the "remember when/remember how" configuration. We can distinguish episodic and capacity dispositional verbs from each other by the language and the linguistic behaviour of the sentential subclauses one attaches to 'remember when' and 'remember how', but one can also think about the difference between these two types of claims. Memories are different sorts of things to skills and there is a phenomenological difference between these things when we reflect on them. We think and experience them organized in different ways within our thoughts.

There are also "Pure" and "Naked" forms of phenomenological argument inside of Gilbert Ryle's *The Concept of Mind*.

These Pure Phenomenological arguments do not have Linguistic Behavioural distinctions or Ordinary Language justifications and cause problems for Ryle's polemic against consciousness.

Pure Phenomenological arguments.

There are arguments like the Reader/Witness argument that are purely phenomenological and do not have any direct relationship with linguistic behaviours or ordinary language claims. One has to try the argument out to see the point of it.

This is a map showing the complex historical interaction and influence between Ryle, Sartre and Sellars against the back drop of the Heterophenomenological and Autophenomenlogical distinction.

Heterophenomenology

Autophenomenology

Neurophysiology

Sartre:
Imagined Object
Percept

Wilfrid Sellars

Philosophy and the Scientific Image of Man

Sellars tries to link these two together.

Neurophysiology involves devices like EEG, Polygraphs, etc.

'continuity'
'irreality'
'plenitude'

Ryle admits he shared Sartre's argument against David Hume

Treating thought as 'language'.
(In Forro Interno)

He gets this idea from Gilbert Ryle.

(Ryle's Log Keeper)
Status Reports

Phenomenology

Ryle also has phenomenological arguments hidden in *The Concept of Mind.*

(Mindologue)

Ryle
Linguistic Behavioural
arguments

Ordinary Language

This is the authority for the arguments that Ryle makes.

These are claims about the mind based on the way people use ordinary language.

Dummett's
Anti-Psychologistic
position

Gottlob Frege

There are two strains of Anti-Psychologism which Dummett identifies. Explicit Anti-Psychologism and Implicit Anti-Psychologism both strains are post-Fregeian reactions.

Brandom's Explicit
Anti-Psychologistic position

Phenomenology and The Ghost in the Machine

An Investigation in to Wilfrid Sellars, Jean-Paul Sartre, Gilbert Ryle and The Concept of Mind

By

Timb D. Hoswell

Contents

Chapter One Introduction: The Search for a methodology

1.1 The Historical Thesis

I think originally, before Ryle wrote *The Concept of Mind*, he had a unique concept of mind that involved an argument about the mind in dialogue with itself. His central argument against a hidden Cartesian theatre of the mind may have started out with the argument that writing something down is no different to thinking it or saying it. Furthermore, using this idea of the mind in dialogue, he could explain causal dilemmas about the mind and body interactions with the world by examining the way actions are talked about in philosophy and the mechanical sciences and focusing on the ordinary language use of verbal descriptions in adverbial clauses and sub-sentential structures that place the mind "inside the adverb" describing the action[1]. For Ryle consciousness is a myth and introspection isn't introspection at all. It is retrospection: a form of "speaking in the mind" or internal narration. He could rest these claims on two central assumptions. Firstly, that the place the mind happens is in the world. Secondly that people already know how to talk about the mind in the world. These presented him with a normative source for arguments in the Philosophy of Mind that he could use to affirm his own arguments or negate those of other philosophers. For his insight was that he could refer to the domain of knowledge possessed by the natural language speaker in order to craft arguments to work against those of other philosophers, or as a source of authority to support his own arguments. These are "Ordinary Language arguments", a special type of argument that draws its authority from

---

[1] Ryle, Gilbert. *The Concept of Mind*. Middlesex: Peregrine Books, Penguin, 1983 Pg 133.

appeals directed at the knowledge people have of ordinary language. It will be one of the central aims of the philosophical thesis of this paper to unravel some of the ambiguities in the ways in which Ryle's "Ordinary Language arguments" do this.

My historical conjecture is that Ryle then had a change of heart and this change of heart pushed him in another direction. It caused him to develop a different account of the imagination, one that involved a concept of consciousness and an account of "the mind's eye". Here the fault line opens up in *The Concept of Mind*. This provides the second piece of evidence for the historical thesis of this paper.

The historical thesis of this paper is that inconsistencies in Ryle's *The Concept of Mind* derive from a confusion between arguments about phenomenology and arguments about the behavior of language. This confusion arose from a little known or unacknowledged influence on Ryle's philosophy by the French philosopher Jean-Paul Sartre. Specifically I argue that Ryle's original concept of mind involved a linguistic treatment of the mind and that this treatment was different from the positive and negative structures of the account that I argue Ryle has appropriated from Jean-Paul Sartre's phenomenology. The confusion of these two threads is what lies at the heart of a number of inconsistences throughout *The Concept of Mind* and the confusion in the way that much of Ryle has since been read. We can see the start of these inconsistencies if we look at a few key pieces of evidence.

For the first piece of evidence, this paper looks at one particular strand of argumentation in *The Concept of Mind,* which is the 'contra David Hume' argument Ryle offers in his account of the imagination[2]. However it is a phenomenological argument. I will argue that this argument has a significant resemblance to the one that Sartre uses at the start of his treatise on *The Imaginary*[3]. In this regard I will suggest Ryle has a debt to Sartre.

Another piece of evidence for this historical thesis is a second account of the imagination which Ryle produced in the posthumous

---

[2] Ryle, *Concept of Mind,* 1983. Pp 236-237
[3] Sartre, Jean-Paul. *The Imaginary*. Translated by Jonathen Webber. Abington: Routledge, 2004 Pg 1 – 16.

work, *On Thinking*. I argue that this is closer to what his original account of the mind in *The Concept of Mind* may have once been[4].

My third piece of evidence is an interview with Bryan Magee published shortly before Ryle died. In this interview, Ryle admitted, he was troubled by the account he gave of the imagination in *The Concept of Mind*. He spoke briefly of a gap in the earlier work and problems related to a positive and negative account as one could read it. This seemed to bother Ryle in the interview. It is through this positive and negative account we will draw the 'gap' out[5].

Lastly, there is the confession. In an article Ryle wrote and originally published in French Ryle admits that the argument against Hume by Sartre in *The Imaginary* and that published in *The Concept of Mind*, are one and the very same argument[6]. Jean Paul Sartre is the missing philosophical influence that explains Ryle's phenomenological style of argument in *The Concept of Mind*.

1.2 The Philosophical Thesis

The philosophical contention of this thesis is that we need a move towards a reconsideration of a Pre-Fregean Psychologism. Specifically I argue that at least some parts of language are merely codes for at least some of our thoughts. The position I am arguing for is that some pre-linguistic knowledge is necessary for a theory of meaning and that this insight undermines the authority of Ordinary Language arguments in the philosophy of mind. My final position will have three central psychologistic theses to support this. Firstly there is the Short Narrow Thesis, the 'Sharp Thesis' of the paper, which is the claim that an appeal to autophenomenology is a stronger basis for an argument that involves individuation on the nature of mind than heterophenomenological sources such as an appeal to the analysis of linguistic behaviour or physiologistic symptoms. This is the essential

---

[4] Ryle, Gilbert. *On Thinking*. London: Basil Blackwell, 1979. Pp 51 - 64
[5] Magee, Bryan. *Modern British Philosophy* Oxford: Oxford University Press, 1986. Pg 130
[6] See Ryle, Gilbert "Phenomenology Vs the Concept of Mind." In *Critical Essays*, edited by Julia Tanney. Oxon: Routledge, 2009. Pg 32.

thesis of the paper and it can only be established once the Strong Narrow Thesis has been argued.

Secondly, the Strong Narrow Thesis concerns the emotions and involves an attack on their articulation as dispositions and occurrences as defined by Ryle, concentrated in several key directions. The key line of argument here is to show that the emotions are irreducible to linguistic behaviour, and that some sort of non-linguistic knowledge is necessarily prior to and fundamental in establishing meaning in talk about them. Since some non-linguistic knowledge of the mind is necessary for a theory of meaning, and a theory of meaning is essential for a theory of language, language can not be the whole story when it comes to investigating the nature of mind.

What I am attacking is the anti-psychologistic position of implicit meaning. This position is championed chiefly by Ryle and Dummett who both hold an implicit theory of meaning. An 'implicit theory of meaning' takes it that meaning is 'implicit' and 'practical' in competent language use and can only be explored, investigated, or mapped out in some manner whether in or through its uses. This position takes it that meaning in language is fundamentally implicit and cannot be fully stated explicitly. As such an implicit theory of meaning advances arguments against an 'explicit' theory of meaning such as that offered by Frege[7] and typical in much of the Analytic tradition of philosophy. However what both the explicit and implicit theories of meaning share in common is that they are both anti-psychologistic. This is important, for in arguing for the Strong Narrow Thesis, I will be arguing that some non-linguistic psychological knowledge is prior and fundamental in relation to a theory of linguistic meaning.

This Strong Narrow Thesis brings us to the Broad General Thesis which puts it much in the camp of David Chalmers[8]. The Broad General

---

[7] This paper follows the dominant view and scholarship espoused by the Analytic School and sides with Michael Dummett on interpretation of Frege as an anti-psychologist. However recent revisionist scholarship by Tim Crane argues that Frege was not entierly an anti-psychologist. Crane argues that there are psychologistic strains in Frege if we look closely enough with the right reading. This paper will deal with the dominant view and scholarship, rather than Tim Crane's controversial papers on the topic. For Tim Crane's interesting and thought provoking work see Crane, Tim. *Aspects of Psychologism*. Massachusetts: Harvard University Press, 2014.

[8] And also much of the work of Tim Crane who argues for a Psychologistic turn in the philosophy of mind. However, I have refrained from using much of Crane's work and his arguments on

Thesis of this paper has it that some mental states, operations or acts are simply irreducible to a Cognitive[9] or Functionalist[10] account, and that we need a theory of consciousness to give an account of these. I want briefly to elaborate on this.

1.3 Arriving at the Philosophical Theses

There is an important psychologistic insight on offer here that I want to share. It arises from several arguments that point to non-linguistic meaning in people's thoughts, and to an insight into how we think about emotions and what their relationship to language is. But that insight itself arises from reflection on the clash between Ordinary Language Philosophy and the influence on Ryle of Sartre's phenomenology in some of the arguments in *The Concept of Mind* and which can reveal something interesting about the way some of his arguments work.

To reveal this insight and arrive at the above three theses, I shall put forward the view that carefully distinguishing between different types of argument that Ryle uses brings out inconsistencies in *The Concept of Mind* that can offer us a way of looking at Ordinary Language style arguments in the philosophy of mind, and the relationships between them, their linguistic structural analysis, and the normative form of Ryle's claim. The normative form of Ryle's claim is

---

Psychologism, insightful and interesting as they are, because I side with Dummett's reading of Frege as an anti-psychologist, at least for the history of 20th Century Analytic philosophy. There may be some room for some argument about whether Frege had certain psychologistic leanings, but for the purposes of his general influence on Analytic Philosophy and the history of Analytic Philosophy Frege has been read as presenting very strong grounds for anti-psychologism.

[9] See Sousa, Roland de. "Emotion." In *The Stanford Encyclopedia of Philosophy*, edited by Edward Zalta, Spring 2013 Edition. http://plato.stanford.edu/archives/spr2013/entries/emotion/ 2013. Section 5 on cognitive theories for where I am drawing my definition from. According to De Sousa what characterizes cognitive approaches is the claim that emotions are propositions.

[10] See the section on Chalmers later in this paper, in 'Chapter Three : Towards a Methodology', the subsection '3.1 The difference between Linguistic Behaviorism and Logical Behaviorism' and '3.1.1 David Chalmers'.

important because it allows one to reveal underlying phenomenological strains hidden inside the argument itself.

Indeed wrapped inside the puzzle of the inconsistencies in *The Concept of Mind* is a mystery about the appeal that sources of authority and language make on us as arguments. Solving that mystery gives us an attractive insight into how language works. To assist in directly addressing this part of the paper the content is divided into the historical thesis and the philosophical thesis. The historical is the first, the philosophical is the second. The second part[11] of this paper is the part that unlocks this mystery. I shall now explain the methodology I use to unlock this mystery behind the arguments in *The Concept of Mind*.

By developing a detailed and principled taxonomy of Ryle's arguments I can lead the reader to an attractive insight into the ways that a single argument might have several different levels of appeal. I do this, firstly, by drawing attention to a gap between two types of argument that Ryle uses. The first is what I call a 'Linguistic Behaviourist[12]' argument, which relies upon analysis of the behaviour of language. The Linguistic Behavioural argument, the first type of argument, and the claims about the mind Ryle makes based on it, ultimately rest upon a normative source of knowledge possessed by the natural language speaker. This normative source gives us the second type of argument. This second type of argument is an Ordinary Language argument. Not all Ordinary Language arguments are Linguistic Behavioural arguments. There are some Ordinary Language arguments that do not have Linguistic Behavioural parts. Instead they rely upon the reader using forms of reflection that one might associate with phenomenological style arguments and in particular Jean-Paul Sartre's phenomenology. The historical thesis gives us grounds for thinking that Sartre's influence on Ryle created the phenomenological style arguments that fill this gap between Linguistic Behavioural arguments and Ordinary Language claims.

---

[11] See 'Chapter Three : Towards a Methodology' and subsections 3.1.1, 3.1.2, and 3.2' in this paper for the beginning of the philosophical thread

[12] Where I use the capitals I refer to my original taxonomy.

We can sum up the argument this introduction has laid out so far thus: Ryle originally had a concept of mind that involved a linguistic analysis that treated thought like language. I then posited that Ryle was influenced by Sartre's phenomenology and this created the confusions and inconsistencies that riddle *The Concept of Mind*. These inconsistencies can be put more precisely thus: that Ryle's particular style of argument which is making claims about the nature of mind based on analysis of the behaviour of language, and which draw their authority for making claims about the nature of mind from the knowledge of the 'ordinary language speaker', was contaminated by Sartre's influence. I argue that this influence on Ryle by Sartre created a subclass of arguments in Ryle that are phenomenological but which Ryle treats as Ordinary Language arguments. Ryle treats them as Ordinary Language arguments because he claims the source of the authority he gives for these arguments comes from the knowledge the 'ordinary language speaker' possesses about what it makes sense to say.

In the next section of this introduction I will reveal the methodology I developed for 'processing' Ryle's arguments to draw these underlying inconsistencies out. I will explain how I classify the normative claim of an 'Ordinary Language argument', the nature of a 'Linguistic Behavioural argument' and how to locate the underlying phenomenological thread hidden in *The Concept of Mind*. The rivalry between the phenomenological arguments in Ryle and his Linguistic Behaviourism will in turn reveal something interesting about language and lead us to the three philosophical theses about psychologism.

Thus so far we have considered the two central threads of this paper, firstly the historical thesis which sets the scene and argues that an inconsistency in *The Concept of Mind* arises from a missing historical influence and that this influence is Sartre. Secondly we looked at the philosophical thread, which argues that distinguishing between the types of appeals in *The Concept of Mind* can reveal something interesting about Ordinary Language arguments. This philosophical thread will conclude with the three psychologistic theses that were outlined. The first psychologistic thesis was the Short Narrow Thesis which claims that auto-phenomenological sources are the strongest source of appeal for an argument individuating on the nature of the mind. The second psychologistic thesis, and necessary for establishing the first, was the Strong Narrow Thread about the emotions and an argument that non-linguistic knowledge is prior and fundamental to a theory of linguistic meaning. The third psychologistic thesis was the Broad General Thesis that some mental phenomena are irreducible to language or reductive forms of functionalism and we need a theory of consciousness to give an account of them. I will now turn to a brief discussion of the methodology I use to reveal the different levels of appeal in Ryle's arguments. This will be followed by sketching out the arguments I use to advance the three psychologistic thesis claims of the paper which will occupy the remainder of this introduction.

1.4 The Methodology

Basically, what really interests me is the different types of arguments Ryle constructs and the sorts of normativity they appeal to. Ordinary Language arguments are one species that appeal to a normative force to persuade the reader on the basis of the authority of shared knowledge constituted by competence in ordinary language and what it 'makes sense to say'. However Ryle also uses another type of argument where he examines specific configurations and structures in the use of verbs, nouns, adjectives, epithets and so on, revealing the intricate patterns of linguistic behaviour connecting the mind and our behaviour in the world. Not all of these second species of argument map on to the first. There's a space left between them. In that space I will argue Ryle has phenomenological arguments. These are appeals to bits of experience, but in order to perform them, one needs to perform what Sartre would call a 'theitic' act. A theitic act utilizes the maxim that 'consciousness is always conscious of something' by making consciousness itself its own object in an act of consciousness. It is usually done in order to demonstrate or to learn something interesting about consciousness and conscious acts.

Ryle uses these sorts of arguments but often doesn't acknowledge that he is doing so. They can usually be found in Ryle where he makes an Ordinary Language argument with some sort of distinction or feature which doesn't offer fruit on the level of a language description in the form of an analysis into the way that bits of natural language like verbs, nouns, subclauses and so on behave. In some cases it does not even emerge on the level of an appeal to common linguistic intuition in the competent language user as a distinction that can be put in the way that the ordinary language user might talk about something.

For an example consider the argumentative moves that Ryle makes to distinguish imagination and perception. On page 237[13] of *The Concept of Mind* Ryle points out that an imagined shriek is not ear splitting but a real one is. Thus an imagined shriek and a real shriek are categorically different sorts of things. To see his point, we must try to 'imagine a shriek so loud it hurts our ears?' Similarly, to grasp the difference between see and 'see' we might try to imagine a light so bright it hurts our eyes[14]. These are appeals to intuitive acts or exercises that involve introspective scrutiny. One needs to perform the act for one's self in order to see the point of it. The official use of such moves is to prop up the argument that imagination is a different sort of thing from perception, but they also invite the question: at what level does the argument make its appeal? Does the difference between imagined and perceived light and sound offer fruit merely at the introspective level of scrutiny? Or does it bottom out in a genuine *linguistic difference* in the way the relevant words behave? Or does it make sense on the level of an appeal to the intuition of the natural language user that it *doesn't make sense* for one to talk about imaginary lights that hurt the eyes? The latter phrase I just put in italics will be highly important later in the thesis because it usually signals the move to advance an Ordinary Language claim.

To work up to what will be a major move in the paper[15] imagine these different types of appeal in the same argument as having a polarizing effect in the way you might feel the force to agree with the claims in an argument. At one end of the spectrum of arguments we have linguistic appeals to the behaviour of words, at the other end, we have appeals to the actual exercises of imagining bright lights or

[13]Ryle, *Concept of Mind*, 1983 Pg 237
[14]Ryle, *Concept of Mind*, 1983 Pp 232 - 237
[15]See 'Chapter Four : Theitic Consciousness and the Interlude' and its various subsections as well as 'Chapter Five : Ordinary Language arguments' both in this paper.

recalling a time of conflict in one's mental life when two contrary motives or impulses clashed. These two different arguments appeal to different things, one linguistic, one phenomenological, and form different strains of argument in Ryle. Using those two strands, the Phenomenological and Linguistic Behaviourist strands, I locate a contradiction in Ordinary Language arguments to show why appeals to an audience, based on the assumption that ordinary language is a unified and authoritative source of appeal, are not very good types of argument to use. I do this by first pointing to the possibility for conflicts to arise between Linguistic Behavioural and Phenomenological strands of Ordinary Language arguments, that present rival claims, such that if the rival claims were taken together they would result in a contradiction. Then I present one.

The strategy behind this move is to break up the authority of Ordinary Language arguments as a platform for either lodging a negation of another philosopher's argument, like Ryle does with Augustine's 'volitions'[16] and David Hume's 'passions'[17], or for an affirmation of his own bit of analysis like his doctrine about 'inclinations', 'moods' and 'agitations'.

The breakup of Ordinary Language arguments as a unified source of argumentative appeal is designed to reveal that the normative force in a piece of analysis can be attributed to the analysis itself. The result of this argument is to show that the overall appeal of the Ordinary Language arguments Ryle makes can be divided by the type of analysis they make on the basis of the appeal in the analysis itself into the two forms I've been laying out. These of course are the Linguistic Behavioural arguments and an occult[18] phenomenological kind of

---

[16]Ryle, *Concept of Mind,* 1983 Pg 62

[17]Ryle, *Concept of Mind,* 1983 Pg 91

[18] Where I use the term 'occult' I use it to mean 'hidden'. An 'occult phenomenological argument' is an argument that gets you to perform a phenomenological act without acknowledging that you are doing phenomenology or engaging in introspective acts of consciousness. By the term 'phenomenology' I mean specifically Sartre's phenomenological egology or reflective consciousness, wherein one turns consciousness upon itself to learn something about consciousness. I will explain this in detail later in the thesis. See 'Chapter Four. Theitic Consciousness and the Interlude', specifically '4.2.4 Phenomenological Style arguments and acts of Introspective Scrutiny' and '4.2.5 Theitic Consciousness and the Apperception of the I at the Reflective Level' for an explanation of Satre's phenomenology. Phenomenological argumentation is left largely undeveloped until these points where I will adopt technicalities that will clarify much of what this earlier use generalizes or leaves obscure.

argumentation Ryle makes but doesn't acknowledge. This division of what is presented as a unified normative appeal into two rival sources (and both, I argue, turn out to be derived from different types of argumentative appeal), is an important stage of my overall argument. It is an important stage because from it I then launch what I argue is the strongest move in the paper towards arguing for a reconsideration of Pre-Fregean psychologism. On the strength of the normative basis of what I shall call the 'first personal direct appeal' I argue that phenomenology is a better candidate for individuating motives than a bit of Linguistic Behavioural analysis. The 'first personal direct appeal' which is characterized by phenomenological arguments becomes important here. This is the basis for the Short Narrow auto-phenomenological Thesis of the paper. This line of thought, in turn, is backed by two arguments, one which draws on a dilemma in the general shape of a syllogistic disjunction between dispositions and the other which collects a developing thread of arguments about the emotions that bottoms out in structural aspects of consciousness.

The first argument is a dilemma drawn from a case study of a patient of Pierre Janet where the subject cries each time she comes to see the psychologist. Sartre, analyzing this case, asks whether the girl cries because she is dispositionally overwhelmed by the task of confessing her secrets to the psychologist or if she cries in order not to confess her secrets. The answer relates directly back to a thread I start with Robert Brandom about reason-explanation frameworks. The argument is simply this: that motives are unjustifiable within a game of giving and asking for reasons. They relate to non-linguistic knowledge that must be set up prior to the game and the person must arrive with the knowledge in order to play it.

The second argument involves the thread of arguments I develop in Sartre's shadow and arises in the paper as a distinct possibility following from the breakup of Ryle's attempt at a unified Ordinary Language platform for justifying phenomenological and Linguistic Behavioural style arguments. This breakup sets the scene and develops the normative strength of the two rival sources of analysis. The insight offered here is that the emotions aren't 'states' or bits of 'linguistic behaviour'. What they are, instead, is raw and real, and mixed up with the way we view the world. They manifest specific aspects of consciousness because they are a type of consciousness. They need consciousness. Structurally, they fixate on an object, as a way of seeing that object, which, in turn, implies a form of consciousness that can illuminate the object. The way I make that argument is complex, so I'll only give you the jist of it here. Basically, emotions are ways we feel about things, which implies that emotions are, at base, a consciousness of those things.

It follows from that argument that first and foremost, we need a theory of consciousness and not a linguistic theory of meaning in order to understand the mind. Emotions are not abstract states that people float around in or to which they are related linguistically by descriptions of the emotional states or propositions[19]. Rather they are embedded in the world and their structure involves fixation on an object, situation or scenario. I argue that, with a few limited exceptions, an emotional state involves fixation on something as an exercise of consciousness. A fear is not constituted by the proposition 'I am afraid' or 'that tiger is fearsome' nor by a linguistic description of the fear such as a "chill" or "shiver of fear" but rather involves fear of something. Most commonly it involves fear as an act of consciousness in apprehending that object. The apprehension of that "something" in turn, implies a view that is not neutral or impartial, but an act of consciousness involved in the world. That fear implicates consciousness. The fear is not purely linguistic, propositional or sentential in nature, but rather I suppose, phosphorescent[20] a bit like sea water, to use Ryle's terminology, as a way of apprehending the object. This is to say the fear is intrinsic and fundamental to the act of consciously apprehending the object of the fear. The apprehension of the object is not a linguistic behaviour. The apprehension of the object is the fear itself. It is charged up you might say.

---

[19] See Sousa, Roland de. *Emotion*. 2013. Section 5. http://plato.stanford.edu/entries/emotion/. De Sousa argues what characterizes cognitive theories about the emotions is the view that they are propositional in form and affirmed. In this regard they tend to follow the stoic notion of judgments.
[20] Ryle, *Concept of Mind*, 1983 See Pg 153 for Ryle's 'self luminous', 'refulgent', 'phosphorescent' terminology. Where I use 'phosphorescent' in this paper, in using it, I refer to consciousness in a specifically technical sense I draw out of Ryle. Ryle contrasts 'phosphorescent' against 'retrospective'. The latter, retrospective, is where an internal voice narrates events when a person is 'alive to what they are doing' and which has a 'log keeper' function that can play this back if asked for a special status report. Where I use 'phosphorescent' I use it in the technical sense of referring to consciousness without this internally narrated aspect.

Take anger as another example. If something has made you angry, then you are angry about whatever that something is. Being angry does not involve the steps 1) apprehending the object 2) adopting an angry attitude or proposition towards it. Rather anger is one's way of apprehending the object. What one apprehends is the anger-making object. That is to say, if emotions are ways of seeing the world, then they imply a way of seeing the world that imbues them with a 'charge' of meaning. This involves a theory of consciousness and not a theory about meaning in language. Later, when we reflect back on emotions and talk about them or try to describe them we codify them into a language as a separate linguistic act to the anger or the fear itself. Later we say 'I was so angry' but in the moment we do not apprehend the anger linguistically as a proposition about how we feel or the expression of a linguistic faculty. The anger, rather, is tied directly into our consciousness of the world. We need a theory about *that* if we are to explain the nature of mind. Essentially what I am arguing is that the moment of anger and its linguistic description, which may come later or be given as a report at the time, are two different things. The former involves a theory of consciousness while the later involves a theory of meaning. Attempting to understand the latter without the former creates the confused illusion in which Ryle's linguistic behaviourist arguments are caught up.

Indeed this illusion is what drives what I term Linguistic Behaviourism. We mistake the codification for a manifestation of consciousness, because we are looking at the way we use the words, not at how we apprehend the world. The illusion is a powerful one. The move of resisting this illusion is significant because it means returning to a Pre-Fregan Psychologism and the view that language is a code for at least some of our thoughts. A theory of consciousness must come before and be fundamental in a theory of linguistic meaning in building a theory of mind. That is to say we need a return to a theory of psychologism before we can begin the task of constructing a theory of meaning in language and ultimately, understanding the nature of mind. Such is the philosophical thesis of the paper.

Chapter Two: The Historical Thesis

The historical thesis sets the groundwork for the paper. The historical thesis posits that the inconsistencies in *The Concept of Mind* arise from an influence on Gilbert Ryle by Jean-Paul Sarte's phenomenology. This influence of Sartre's philosophy on Ryle has been largely unrecognized if it has been recognized at all. The job of the historical thesis, thus, is to make credible this influence of Sartre's phenomenology on Ryle's linguistic concept of mind. It does this by presenting several pieces of evidence that reveal Sartre's influence on Ryle.

Later in the paper we will return to Sartre in order to construct a chief arm of our methodology which will give us a way of drawing out the phenomenological content in Ryle's arguments to give us his phenomenological patterns of arguments. Some of these arguments have both Phenomenological and Linguistic Behavioural sides. Some have Ordinary Language justifications and phenomenological elements, and some have all three. By this later stage of the thesis we will have developed a precise classification of Ordinary Language arguments and Linguistic Behavioural arguments and the relationships between them.

The importance of the historical thesis presented in this section is this it establishes the historical foundations for the inconsistencies in *The Concept of Mind* and identifies the relationship with Jean-Paul Sartre.

This section will begin with Ryle's negative and positive account, shift focus to his negative account, discuss his view of Hume, look at Sartre's irreal and lastly show how all of these tie together with Ryle's confession.

2.1 The Two Accounts of The Imagination

In *On Thinking*[21] Ryle produces an account of the imagination that is significantly different to that in *The Concept of Mind*. In the former account Ryle concentrates on examples such as a boy finding a plank of wood and imaginatively using it for a horse, or a bridge, and how his behaviour differs in its description from that of a man who discovers the same plank of wood and would use it for a panel of his chook pen. He examines such cases as why 'imaginative' is a bad or negative adverbial phrase to use in describing a boy's mathematical work and considers in what other ways our use of the word 'imagination' and its corresponding adverb 'imaginative' differ in our everyday use. There is something very different about this version of the imagination in *On Thinking* when compared to the more vivid account in *The Concept of Mind*.

The account in *On Thinking* feels almost 'exorcised'.

What is missing from the chapter on the imagination in *On Thinking* is an account of "the Mind's eye" which his earlier account in *The Concept of Mind* has. This is interesting because while talking to Bryan Magee, shortly before he died, Ryle admitted to being haunted by something.

He says

> I'm boastful enough to think and to say the general theme of the *(Concept of Mind)* is dead right. But, of course, I committed some howlers in the writing of the book, and there are certain pages of the book which I

---

[21] Ryle, Gilbert. *On Thinking*. London: Basil Blackwell, 1979. Pp 51 – 64.

can't bear to read now because it reminds
me of how stupid or hasty I was in those
days. I'll just mention two of them[22].

And indeed, of the two, the one that seems to really be bothering him is
the account he gives of the imagination in *The Concept of Mind*. In
describing it he suddenly brings up the notion of a 'gap' in the positive
account of what the imagination is, and this, he confesses, is what
troubles him.

> Then (there is a chapter about) the
> imagination where the negative things that
> I was saying, were dead right. . . (T)here's a
> positive part of the story, which I didn't
> capture, and haven't captured yet. But here
> I don't feel so very guilty, because nobody
> else that I have seen, that has chanced his
> arm on this particular thorny topic, has
> made very much more headway than I did.
> Somebody's going to fill the gap someday – I
> don't expect it will be me – but whatever
> filling of the gap is produced it isn't going to
> do much damage to the book, or the
> negative theme of this chapter[23].

I should like to bring this out and demonstrate the way I think this
negative account works, against the background of the positive account,
before extracting the phenomenological content.

There are two tiers to the account of the imagination in *The
Concept of Mind*. The first is Ryle's argument that there is no
'imagination' as such because there is no one kernel function, faculty or
– we can't say consciousness, can we?[24] – process, which embodies all

---

[22]Magee, Bryan. *Modern British Philosophy* Oxford: Oxford University Press, 1986. Pg 130.
[23] Magee, *Modern British Philosophy,* 1986. Pg 130.

the different ways that we use "imagination" as an everyday term in our ordinary language descriptions. Indeed this negative type of account which focuses on the way people use the term 'imagination' in their everyday use is somewhat similar to his account in *On Thinking*. Notice also why this is an interesting claim. Here he is using the authority of ordinary language to negate the notion there is a single 'faculty' of the imagination. This is a type of argument Ryle uses where he negates a view or a claim about the mind on the basis of the authority of ordinary language usage. He does this to Augustine's account of human will and 'volitions' as we will see later on. However, regarding 'imagination', he is doing it on the basis of scattering or dividing the word into different ordinary language uses.

> There is no special Faculty of Imagination, occupying itself
> single-mindedly in fancied viewings and hearings. On the contrary, 'seeing' things is one exercise of imagination, growling somewhat like a bear is another; smelling things in the mind's nose is an uncommon act of fancy, malingering is a very common one, and so forth[25].

This account of growling like a bear and 'playing bears' is much of what occupies his negative account in *On Thinking*, along with examination of how the adverbial term ' imaginative' is applied to activities. He focuses on the adverbial descriptive content in what I have been calling 'take care' forms and Ryle calls in his earlier work *The Concept of Mind* 'heed concepts', 'mongrel categoricals', and 'semi-episodic forms[26]'. These allow him to look at and make arguments about the imagination on the basis of the way we use the term in ordinary language discourse, without engaging ourselves in actual acts of the imagination.

---

[24]See the section titled '4.1 Introspection' and the subsection titled '4.1.1 Ryle's Use of the Terms Consciousness and Introspection' in this paper.
[25]Ryle, *Concept of Mind*, 1983 Pg 244.
[26]Ryle, *On Thinking*, 1979, see the first chapter for his discussion on this point.

That given what of this second account and the omission of 'seeing' things in the mind's eye, and 'smelling' things in the mind's nose? To grasp the full scope of this argument it is necessary to go briefly into the oddities of Ryle's treatment of Hume's causal theory of consciousness.

We will now do this.

Ryle has it, that Hume thinks objects impress themselves on to us with a certain vivacity to create impressions, and these impressions fade with less vivacity into ideas. This is more or less typical of a standard treatment of Hume. Where Ryle differs is that he thinks Hume's account is part of a background story that develops into a myth of consciousness. This myth, Ryle argues, began with the Protestant reformation and ended in what Ryle describes as 'para-optics'[27]. Where Ryle's analysis of Hume differs in this myth, from say his reading of John Locke's theory of consciousness is that Ryle thinks Hume has picked up a sophistication from another myth that starts with the mechanical theory of mind[28]. This sophistication has it that ideas get their reality from impressions and impressions get their reality from the world. Ryle's view is that Hume has developed a causal hypothesis about the natural structure of consciousness.

> We are now in a position to locate and correct an error made by Hume. Supposing, wrongly, that to 'see' or 'hear' is to have a shadow-sensation, (which involves the further error of supposing that there could be shadow-sensations), he put forward the causal theory that one could not have a particular 'idea' without having previously had a corresponding sensation, somewhat as having an angular bruise involves having been previously struck by an angular object. The colours that I see in my mind's eye are,

---

[27] Ryle, *Concept of Mind*, 1983 Pp 152 – 154.
[28] Ryle, *Concept of Mind*, 1983 Pp 74 – 80 for the Bogey of Mechanism, see 236 -242 for Ryle's application of the 'para-mechanical theory of mind' to Hume's impressions and ideas, and 256 – 257, for Ryle's discussion of Hume treating ideas as 'shadow sensations' left over from impressions.

he seems to have thought, traces somehow left by the colours previously seen by me with my eyes open.[29]

That is, Ryle fancies, that Hume fancies, or rather theories with causal strains of consciousness attached to them like Hume's, posit that there is something in these impressions that then crosses over into ideas and that the ideas are 'entities' or shadow-sensations left behind.

Hume notoriously thought that there exist both 'impressions' and 'ideas', that is, both sensations and images; and he looked in vain for a clear boundary between the two sorts of 'perceptions'. Ideas, he thought, tend to be fainter than impressions, and in their genesis they are later than impressions, since they are traces, copies or reproductions of impressions. Yet he recognised that impressions can be of any degree of faintness, and that though every idea is a copy, it does not arrive marked 'copy' or 'likeness', any more than impressions arrive marked 'original' or 'sitter'. So, on Hume's showing, simple inspection cannot decide whether a perception is an impression or an idea[30].

Now what is so striking about Ryle's treatment is that it classes Hume in a special type of category, a *causal theory of consciousness*. That is we are *put in mind* of the world by the object impressing on to us and producing ideas. Now it is debatable how far Hume might actually go along this line. I should argue that Hume does not think that knowledge of causation, in the sense Ryle means to treat Hume's theory as a casual theory of consciousness, is actually possible[31]. Rather I

---

[29]Ryle, *Concept of Mind,* 1983 Pg 257

[30]Ryle, *Concept of Mind,* 1983. Pp 236-237

[31]Ryle, *Concept of Mind,* 1983 Pg 236, Ryle in *The Concept of Mind* actually thinks this originates

should argue that all we can have knowledge of on Hume's view is association[32].

That is to say, original 'Historical Hume', or Hume as I read him, states, that all we can really know about causation is the conjunction of events which produces a bond between them which we think of as knowledge. The idea of causation is built up from repetition of one idea conjoined with another which we might take to be an implication of one object to another, and so on, but is really, nothing more than the *sentiment of belief* attaching two impressions together. For Hume this *belief* is all that really separates out the conjunctions of ideas, produced from impressions, from the sorts of conjunctions produced by fiction and fancy in the imagination[33]. To step out of this conjunction, and assume that there are objects causing these impressions, 'somewhat as having an angular bruise involves having been previously struck by an angular object'[34] is more than Hume's skeptical resources are willing to admit. This is a tension in Hume and it is that same tension that Kant inherits and is criticized for by Sartre[35].

---

from a confused use of distance and tactile senses that get treated as the one type of sense see pg 36. While physical senses require some sort of tactile element which implies an object pressing upon the person in a causal chain, Ryle argues, that the visual and distance senses don't. Hence why Hume's position, Ryle thinks, originates from the confusion of 'impressions' as visual and auditory events that 'impress' upon the senses. This combined with what Ryle refers to as the 'Bogy of Mechanism', see Pg 79, *The Concept of Mind,* creates the 'Causal Theory of Consciousnesses'.

[32] See Hume, David. *Enquiries Concerning Human Understanding and Concerning the Principles of Morals.* Reprint from the 1777 edition ed. New York: Oxford, 2005. Pg 57, Sect VI, which reads "But finding a greater number of sides concur in the one event than in the other, the mind is carried more frequently to that event, and meets it oftener, in revolving the various possibilities or chances, on which the ultimate result depends. This concurrence of several views in one particular event begets immediately, by an inexplicable contrivance of nature, the sentiment of belief, and gives that event the advantage over its antagonist, which is supported by a smaller number of views, and recurs less frequently to the mind. If we allow, that belief is nothing but a firmer and stronger conception of an object than what attends the mere fictions of the imagination, this operation may, perhaps, in some measure, be accounted for."

[33] Hume, *Enquiries*, 2005, Pg 46, Section V, Part II reads, "It follows, therefore, that the difference between *fiction* and *belief* lies in some sentiment or feeling, which is annexed to the latter, not to the former, and which depends not on the will, nor can be commanded at pleasure. It must be excited by nature, like all other sentiments; and must arise from the particular situation, in which the mind is placed at any particular juncture. Whenever any object is presented to the memory or senses, it immediately, by the force of custom, carries the imagination to conceive that object, which is usually conjoined to it; and this conception is attended with a feeling or sentiment, different from the loose reveries of the fancy. In this consists the whole nature of belief."

[34] Ryle, *Concept of Mind,* 1983. Pg 257

[35] Jean-Paul Sartre. *Being and Nothingness.* Translated by Hazel E. Barnes. Abington: Routledge, 2003. Pp 2 - 4

That is, I should like to argue that 'Historical Hume' argues even though the impressions produce their ideas by conjunction which in turn grounds our knowledge of causation in belief – that is the association between impressions and their repetition in the conjunction of ideas gives rise to the concept of causation – this concept of causation is merely that, for Hume: a human concept. 'Historical Hume' argues that this concept of causation arises as a 'sentiment of belief' caused by the conjunction of ideas first experienced as impressions. Given that this is the case in Hume there is none or very little grounds in 'Historical Hume' for moving from conjunction of ideas to causation and into a causal theory of consciousness as Ryle's Hume would have it. Even in the case of the impressions themselves to posit that they have causal powers beyond the impressions themselves is to go beyond the resources of Hume's epistemology[36]. To press this point, however, is of course, to tread the line between Hume and Berkley. Nonetheless for both Historical Hume and in Ryle's reading of Hume, Hume presents ideas and impressions *as the same sort of thing*. The difference between the ideas and the impressions for Hume in both cases is, of course, the level or degree of vividness[37].

From Ryle's analysis of Hume's causal theories of consciousness; memories, images, thought-images, ideas, impressions, and so on, get their reality, or rather the ontological existential status that makes them eligible for 'reality', from the causal process that puts them there. This is what I will call in Ryle an inheritance condition[38] with the fading properties of 'vivacity' in Ryle's Hume's case. In Original 'Historical Hume', as opposed to Ryle's Hume, this inheritance condition would be analogous with the ideas themselves and the sentiment of 'belief' that gets attached to them through repetition of experiences and their conjunction in experience.

There is an obscurity in what Ryle is trying to do that I am at pains to bring out. He is trying to turn an ontological argument aimed against the causal strain of the theory of consciousness into a larger

---

[36]Hume, *Enquiries*, 2005 Pg 48, Section V, Part II "And in philosophy we can go no farther, than assert, that it is something felt by the mind, which distinguishes the ideas of the judgment from the fictions of the imagination. It gives them more force and influence; makes them appear of greater importance; infixes them in the mind; and renders them the governing principles of all our actions."
[37]Ryle, *Concept of Mind*, 1983 Pg 236, hereafter, we will refer to Ryle's treatment of Hume, rather than the historical Hume and fine line distinctions one might draw.
[38] See 'Linguistic Behaviorism: Episodes and Dispositions. What exemplifies the linguistic Behaviorists claims' in this paper.

psychological one in two ways. Firstly he has expanded the causal theory of consciousness, in particular the products thereof, to include 'entities' in some 'mental' sense. This is the Cartesian quasi-theatre strain of argument in Ryle which is based on the analogy of a stage where the stage murder is not a copy of the murder in the real world but a different sort of thing[39]. The reason why he has done this is he is going to argue that the actors on that mental stage as entities do not exist. We shall call this a combined psychological and ontological argument. Secondly, he is going to argue that the objects of the imagination don't exist and thus *there is nothing to be conscious of*. These two moves are linked by an analogy. He's going to treat that analogy as part of an Ordinary Language Therapy for the causal theory of consciousness based on a discrepancy in Hume's use of the term 'vivid' (which is of course 'phenomenologically loaded' with what I will describe later as the 'direct first personal perspective'). It is a phenomenological style argument that works by getting the audience to participate in the argument through some exercise to illustrate a point.

First, however, let us look closely at the ontological part of this argumentation against Hume.

Ryle writes

> The crucial problem is that of describing what is 'seen in the mind's eye' and what is 'heard in one's head'. What are spoken of as 'visual images', 'mental pictures', 'auditory images' and, in one use, 'ideas' are commonly taken to be entities which are genuinely found existing and found existing elsewhere than in the external world[40].

Now the consciousness part of the argument.

---

[39]Negri, Antonio. *Political Descartes: Reason, Ideology and the Bourgeois Project*. Translated by Alberto Toscano Matteo Mandarini Radical Thinkers New York: Verso, 2006. This metaphorical use and analogy with a 'theatre of the mind' is actually an Anglo misreading of Descartes. The metaphor of a 'theatre of mind' actually originates with the Lullists who Descartes was in opposition to. See Antonio Negri Political Descartes Pp 56 – 62, particularly regarding Schenkel, Pg 59.
[40]Ryle, *Concept of Mind,* 1983 Pg 236

So minds are nominated for their theatres. But, as I shall try to show, the familiar truth that people are constantly seeing things in their minds' eyes and hearing things in their heads is no proof that there exist things which they see and hear, or that the people are seeing or hearing. Much as stage-murders do not have victims and are not murders, so seeing things in one's mind's eye does not involve either the existence of things seen or the occurrence of acts of seeing them. So no asylum is required for them to exist or occur in[41].

This denial is where Ryle runs into a problem because his positive account of the imagination has a theory of consciousness in it. That is to say, in order to differentiate between the objects of the mind and the objects of perception, before he can claim the objects of the imagination do not exist, he needs a theory of consciousness to distinguish them. That is, to claim the objects seen with the "mind's eye" are a 'different sort of thing to those seen with ordinary eyes' he needs a theory of consciousness, prior to any existential claim about whether they 'exist' or not.

He does this by putting quotation marks around the word 'see', like I've just done.

Ryle writes

One way in which people tend to express this difference is by writing that, whereas they see trees and hear music, they only 'see', in inverted commas, and 'hear' the objects of recollection and imagination. The victim of delirium tremors is described by others, not as seeing snakes, but as 'seeing' snakes. This difference of idiom is

[41]Ryle, *Concept of Mind,* 1983 Pg 236

reinforced by another. A person who says that he 'sees' the home of his childhood is often prepared to describe his vision as 'vivid', 'faithful' or 'lifelike', adjectives which he would never apply to his sight of what is in front of his nose. For while a doll can be called 'lifelike', a child cannot; or while a portrait of a face may be faithful, the face cannot be any such thing. In other words, when a person says that he 'sees' something which he is not seeing, he knows that what he is doing is something which is totally different in kind from seeing, just because the verb is inside inverted commas and the vision can be described as more or less faithful, or vivid. He may say 'I might be there now', but the word 'might' is suitable just because it declares that he is not there now[42].

Thus, if we return to the causal theory of consciousness in Ryle's Hume, the attack is on the 'vivacity' of the image, and a mixed sense, of which 'vivid' can mean in relation to these two different types of 'seeing'. Something can be vivid in the sense of being loud, or bright, at least enough to hurt the eyes or the ears. It can be vivid in the sense of ' 'see' ' in the sense that it seems real, it is a good imitation, or the claim that 'the hallucination was so real and life-like you almost believed it'. But it *doesn't make sense to talk about* a mixed sense of both like the 'deafening thunder of the mind'. No matter how loud someone imagines a sound, that is, no matter how vivid it is in the second sense, they can not deafen themselves with it. No matter how vividly bright the light they can imagine, they can not blind themselves with it.

Ryle writes

---

[42]Ryle, *Concept of Mind,* 1983 Pg 233

Hume's attempt to distinguish between ideas and impressions by saying that the latter tend to be more lively than the former was one of two bad mistakes. Suppose, first, that 'lively' means 'vivid'. A person may picture vividly, but he cannot see vividly. One 'idea' may be more vivid than another 'idea', but impressions cannot be described as vivid at all, just as one doll can be more lifelike than another, but a baby cannot be lifelike or un-life-like. To say that the difference between babies and dolls is that babies are more lifelike than dolls is an obvious absurdity. One actor may be more convincing than another actor; but a person who is not acting is neither convincing nor unconvincing, and cannot therefore be described as more convincing than an actor. Alternatively, if Hume was using 'vivid' to mean not 'lifelike' but 'intense', 'acute' or 'strong', then he was mistaken in the other direction; since, while sensations can be compared with other sensations as relatively intense, acute or strong, they cannot be so compared with images. When I fancy I am hearing a very loud noise, I am not really hearing cither a loud or a faint noise; I am not having a mild auditory sensation, as I am not having an auditory sensation at all, though I am fancying that I am having an intense one. An imagined shriek is not ear-splitting, nor yet is it a soothing murmur, and an imagined shriek is neither louder nor fainter than a heard murmur. It neither drowns it nor is drowned by it.

Now, I think Ryle is right here, in this account, that there is a radical difference between things that we can 'see' in the mind's eye, and things we normally see. But I should argue that trying to imagine the 'deafening thunder of the mind' is a part of that phenomenological distinction. Furthermore, I don't agree with Ryle that the ontological argument is one and the same with the 'psychological'. That is, I do not think he has enough resources to make that distinction and still maintain his position on consciousness like he argues for with the two forms of vivid without a contradiction. That is, Ryle has already boxed himself out of using consciousness and introspection in order to make a phenomenological claim that requires one to make an act of introspective scrutiny. Now as I point out later in the paper Ryle explicitly states that the mind can't introspect because "the myth of consciousness is a piece of para-optics[43]".

To be clear my argument is that he can't offer an account of the mind that rules out consciousness, introspective acts and reflection, and still offer an account of the imagination based on the differences between the uses of 'see' and see, like he introduces, as a base for his ontological argument about the objects of the imagination and do so without a theory of consciousness. There is a direct contradiction between 1) Ryle's attack on introspection, 2) Ryle's attack on consciousness and 3) Ryle's attack on reflection and 4) the evidence he offers for his ontological argument about imaginary objects in the account he puts forward in the chapter on the imagination in *The Concept of Mind*, 5) various exercises he gets his readers to perform. 1), 2) and 3) clash with 4) and 5).

This contradiction is the cause of the inconsistencies that zigzag through out *The Concept of Mind* like the border between New South Wales and Victoria.

---

[43]Ryle, *Concept of Mind,* 1983 Pg 153

2.3 The Mysterious Philosopher Hypothesis

My contention is that the original 'concept' of mind that Ryle had in writing *The Concept of Mind* is rather like that which Dennett identifies as the central thesis in *On Thinking*[44]. That is Ryle's original position was that of treating thought as though it consisted of different species of language. This dissolves the illusion that thought is inside the head, since language can take place in the everyday world. It can be written down, spoken and is publically accessible since it is a form of communication. The meaning must be publically accessible if we can communicate in it. This is a philosophically original and interesting hypothesis.

I think that Ryle's original hypothesis may have been something like what follows: by treating consciousness as a myth and introspection as a development of that myth, Ryle could develop a line of argument against the Cartesian and Empirical schools from the normative position of Ordinary Language[45]. Consciousness could be developed as a form of internal narration, and memory as 'retrospection', that is a form of speaking in the mind and what Wilfrid Sellars refers to as 'in forro interno' in his reading of Ryle and the body of Sellars' work[46]. Treating ordinary language as an authoritative normative source allowed Ryle to connect these pieces together with an argument that has the following steps. Firstly the world is the place the mind happens[47]. The mind as

---

[44]Dennett, Daniel C. *Consciousness Explained*. London: Penguin, 1993.Pp 223 – 225, specifically, in *The Concept of Mind* see pg 153 and 'retrospection'. Ryle has a log keeper function in memory and various types of thought speech or 'mindologue' I will draw out later in the thesis. See '4.1 Introspection' and '4.,1.1 Ryle's Use of the Terms 'Consciousness' and 'introspection', and '4.1.2 The Species of Mindologue' in this paper. See also Dennett's discussion of Ryle's concept of thinking in Dennett, Daniel. "Re-Introducing the Concept of Mind." *The Electronic Journal of Analytic Philosophy*, no. 7 (2002): http://ejap.louisiana.edu/archives.html.
[45]Ryle, *Concept of Mind,* 1983 Pp 157
[46]Ryle, *Concept of Mind,* 1983 Pp 153, see Sellars, *Empiricism & the Philosophy of Mind*, 1997 Section XII Our Rylean Ancestors, Subsection 48, for Sellars 'in forro interno' version of Ryle's log keeper.
[47]Ryle, *Concept of Mind,* 1983 Pp 56 – 57 NB see the "historian's argument". See also Parkinson, G. H. R. "Translation Theory of Meaning." In *Communication and Understanding*, edited by Godfrey Vesey, 1-19. New Jersey: The Royal Institute of Philosophy; Humanities Press, 1977. Pg 1-19 for his

Ryle sees it isn't occupied in a detached esoteric Cartesian realm, but rather the mind is occupied with deliveries and driving lorry trucks, or making passes on football fields, or driving forklifts and stacking pallets. The realm of the mind is in the everyday world around us and not some abstract place inside the head. Secondly, that people already know how to talk about the mind in the world[48]. He could back this up with an argument against the domain of science, by claiming that overzealous influences from science contaminate our ordinary language conceptions of mind[49] and thus he could mark out the domain of the philosophy of mind as the area in which we talk in ordinary ways of day to day discourse about the things people do and the way they do them. This would allow Ryle to argue that we can think about 'mental operations' as simply various species of 'mind talk'[50]. This, of course, is what he does. This was a good theory.

I suggest that while putting this theory together he either had a conversation with a colleague or he sat down and read another philosopher. This confused Ryle, and at first, he thought he could solve the problem by re-writing the chapter on the imagination in *The Concept of Mind* by focusing on the linguistic structures that demarcate an account of "the mind's eye". But this new account clashed with an earlier section, namely his polemic against consciousness, thus resulting in the puzzle of the fault line running through *The Concept of Mind*, and the rich seam of Ordinary Language Philosophy and Phenomenology we find in the work as we have it today. All we need to make this bit of detective work plausible is the missing philosopher or colleague and a link between the two. To this we now turn.

---

discussion on the importance of the 'historian argument' in Ryle in the relation between the mind and the world and other minds. See also Ryle, *Concept of Mind,* 1983. Pp 44 – 50 for Ryle's discussion on the exercise of intelligence.

[48]Ryle, *Concept of Mind,* 1983 Pp 9 -11

[49]Ryle, *Concept of Mind,* 1983 Pp 74, see the "Zealot's Argument" where Ryle claims that 'zealots' trying to force scientific paradigms onto our concept of mind create ordinary language muddles.

[50]Ryle, *Concept of Mind,* 1983 Pg 151 for the 'log keeper' function, and Pg 153 for 'retrospection'.

2.4 Sartre's Irreal

In Sartre's early work, *The Imaginary*, he sets out an investigation into the differences between perception, conception and the imagination at the level of our conscious experience. The transition which occurs between *The Imaginary* and *Being and Nothingness* is based on an extended refutation of Bishop George Berkley. The structure of that argument depends upon differentiating properties of consciousness at the level of the mental image from those of perception.

For example, Sartre distinguishes one of the properties of the 'irreal' as 'detachment'. A centaur, were one to imagine it, would appear without a past or a future in a sort of 'timelessness' and with a lack of spatial properties. That timelessness and lack of spatial qualities is different to the lamp which if we picked it up would have a desk under it and which if we moved the desk would have a wall hidden behind it. If we examined the wall we might be surprised to find marks from the furniture. This 'surprise' is an important feature of Sartre's philosophy. It indicates that perception is passive, susceptible, predisposed and open to discovery which is unlike the imagination where we have to actively construct the objects. In his ontology this power of 'surprise' is one of the key features of perception, and points towards a *'transphenomena'* of being.

Continuity likewise is another feature of objects found in perception. The lamp is continuous with the desk and the wall in its reality. It is spatially attached in a way that the imaginary centaur is not. Likewise, I know I am here, because I got up this morning and my immediate presence is connected to a sequence of events I can recall which attach themselves to a rolling out of bed where I fell asleep the night before. The centaur however appears without a history or a past in a sort of 'abstracted space' where we imagine it. We could give the centaur a continuity of events which starts when we say he was born in a magical barn, but then, this raises the question 'what is the barn attached to?' And also 'Where did the barn come from and what's behind it?'

Sartre thinks that Berkley's universe lacks the sorts of traits that would make it either a percept or an imagined part of consciousness. Berkley's universe, Sartre reasons, must either be irreal or posses a transphenomena. The act of apprehending it would either be a percept like the desk or an imagined consciousness like the centaur. Objects in Berkley's universe would either be perceptual and contain a plenitude of experiences such as angles and ways we can look at the desk, details in the wood grain of the desk that increase each time we look at it, and a continuity with the rest of the perceptible world. The imagined object would be detached, have only what we actively construct in the image and have limitations in how we might view and experience it.

Sartre begins his extended argument against Berkley off with a refutation of Hume's notion of 'vividness'[51]. For instance, if one were to examine a piece of paper and then to imagine it one would find that no matter how carefully one has examined the piece of paper, that upon returning to the original it would contain a plenitude of detail that the imagined copy would not possess.

Sartre writes

> Let us consider this sheet of paper on the table. The more we look at it, the more it reveals its characteristics to us. . . Each new orientation of my attention, of my analysis, reveals to me a new detail; the upper edge of the sheet is slightly warped. The end of the third line is dotted, etc. But I can keep an image in view as long as I want: I will

---

[51]Sartre. *The Imaginary*, 2004. Pg 1 - 16

never find anything there but what I put there. This remark is of the utmost importance in distinguishing the image from perception[52].

Sartre's central argument against Hume is that consciousness can distinguish between an 'imaged' and a 'perceived' act, based on certain intrinsic properties that are specific to each. The perceptual content can be distinguished by an "over-filling" of consciousness. There is more detail, or resolution in the percept than there is in the imaged. This comes out at the practical level of, for instance, the fine lines of the wood grain in a desk, along with the termites that might surprise us and be hiding inside it. The desk of the mind has none of these traits unless we put them there as we imagine it. That element of surprise is missing here in the mentally imaged accounts. Thus, one of the central differences between imaged and perceptual forms of consciousness is that perceived consciousness over-flows with perceptual content while the imaged content has a certain poverty about it. The imaged content will never startle us with termites or bits of gum, it will never overwhelm us with a seemingly infinite level of detail in wood grain, painful brightness, tactile coarseness and so forth. These are all qualities of the percept, which Sartre argues in *The Imaginary,* point us towards perceptual consciousness, and in the opening chapters of the investigation in *Being and Nothingness,* point us further towards what he refers to as the 'trans-phenomena' of being. Consciousness for Sartre is the space where we can make this qualitative distinction.

Sartre writes

In a word the object of perception constantly overflows consciousness; (whereas) the object of an image is never more than the content one has of it; it is defined by that consciousness. One can never learn from it what one does not know[53].

---
[52]Sartre. *The Imaginary,* 2004 Pg 9

Sartre maintains that the two acts even annihilate each other if one tries to bring them together at the same time in the same act.

> As long as I look at this table, I can not form an image of Pierre; but if all at once the irreal Pierre surges up before me, the table that is under my eyes vanishes, leaves the scene. So these two objects, the real table and the irreal Pierre can only alternate as correlates of radically distinct consciousness: how could the image, under these conditions, contribute to forming the perception?[54]

Likewise the imaged consciousness can be distinguished and detached, Sartre thinks, from conceptual consciousness by the fact we can reduce the imaging consciousness to certain facts that do not require us to think about the imaged consciousness. Take his cube example. Consider the number of right angles or how to work out the surface area of the cube or the number of squares in a cubed cube. None of these require us to revert back to either the imaged form or even the percept of the cube to work out. These functions can be developed in pure abstraction without referring back to either the mental image or visual percept of the cube itself.

We might put the three levels of distinction in this way with the example of, say, buying a refrigerator. The refrigerator in the store contains a level of detail that the one depicted in the catalogue and which we imagined do not have. We may, critically, notice dents at a level of detail in the floor room model that were not available for assessment in the level of resolution in the catalogue. It might contain extra draws. Certain compartments may swing out a different way to

---

[53]Sartre. *The Imaginary*, 2004 Pp 9 NB This overflowing becomes the trans-phenomena in *Being and Nothingness* and the core of his refutation of Kant.
[54]Sartre. *The Imaginary*, 2004 Pg 121

how we imagined when we were looking at the catalogue. The refrigerator in our minds assumed a blue back, like the sides and front while the back of the one in the store is painted red irrespective of what we thought or imagined. These make up the perceptual and imaged differences. Now thinking about the internal size capacity of the refrigerator, how much it costs, whether there are separate compartments for milk and meat, what it might do to our bank balance, and getting it home: these are all conceptual elements of our thinking with their own specific qualities.

If we think back to Ryle's distinction between 'see' and see, between 'hear' and hear, and the refutation he develops, in great complexity, of Hume's degrees of 'vividness', we can see, that in fact, it has a startling resemblance to the argument that Sartre makes in *The Imaginary*. Moreover it reveals, in essence, a possible explanation for why the chapter on imagination Ryle presents in *The Concept of Mind* sits at the centre of a fault line that runs between his Ordinary Language argumentation and Phenomenological style arguments,

If this were a detective's case all that one might need now would be a confession. And indeed it is to this we now turn.

2.5 Ryle's Confession

In an obscure 1959 article originally published in *La Philosophie Analytique* and written in French[55] Ryle admitted that he had read Sartre's *The Imaginary*, that he and Sartre shared the same argument, and indeed, this had been the source of his confusion.

> I shall not repeat the arguments by which Sartre and I exposed the absurdity of Hume's view or of the other view that imagining is witnessing things existing or occurring inside a private chamber. What is

[55] See Ryle, Gilbert "Phenomenology Vs the Concept of Mind." In *Critical Essays*, edited by Julia Tanney. Oxon: Routledge, 2009. Pg 32.

more interesting, at least to me, is that after these insidious conceptual mis-constructions had been exposed, I was obliged to try and give the correct positive account, and in this conceptual search I got lost. I was, I think, on the right track in assimilating imaging, e.g. visualising, to the much more general notion of make-believe, about certain other varieties of which, like the notions of pretending and playing, I felt fairly clear. But when I found myself classifying visualizing as 'make-believe seeing' I felt conceptual embarrassments and these are always a sure sign that something has gone wrong.

This, I suggest, explains the gap which Ryle refers to in his Magee interview, and points to the fault line that underlines the puzzle in *The Concept of Mind*, the very puzzle that makes it such an interesting work. We shall now turn to the philosophical thesis of this paper in an attempt to untangle these confusions, and, in the process, untangle much of the analytic philosophy of mind that has followed.

Chapter Three : Towards A Methodology

3.1 The Difference between Linguistic Behaviourism and Logical Behaviourism.

3.1.1 David Chalmers

In contrast to all the different schools and subspecies of Psychological Behaviourism[56]I am using the term 'Linguistic Behaviourism' to refer to claims that describe or can demonstrate relationships between bits of linguistic behaviour, which, to belabour this classification a little, simply means making claims about the way that linguistic units or structures work and hold together as shown in the analysis of the way they behave in various circumstances and conclusions drawn from them concerning the nature of mind.

---

[56]See, for instance Russell, Bertrand. *The Analysis of Mind*. London: The Muirhead Library of Philosophy, 1951. For Russell's discussion of Watson, pg 52, or Thorndike's laws pg 53, or how Russell builds these into psychological definitions as part of his behavior cycles Pp 64 – 65. For Skinner's own formulation of a definition see Skinner, B. F. *Beyond Human Freedom and Dignity*. Middlesex: Penguin, 1976. Pg 104. Skinner analyzes behavior in terms of aversion and reinforcement which he builds into a definition of good or bad starting with survival contingencies. On pg 124 these natural contingencies of survival become conditioning in society. This conditioning is positive, pg 140, when behavior is followed by reinforcement either in rewards or praise. This completes his definition which he starts on page 48, with his discussion of dignity. Dignity is the illusion that arises from not knowing the true conditions of a person's conditioning while freedom is simply the illusion that allows a person to be conditioned by random chance and events rather than structured reinforcement for behavior. See also, for instance Sellars, Wilfrid. "Philosophy and the Scientific Image of Man." In *Science, Perception and Reality*, Pp 1 - 40. California: Ridgeview, 1991. Pg 24 – 30 for Sellar's Behaviorusitics distinctions. See Willem A. DeVires, Timm Triplett. *Knowledge, Mind and the Given*. Indianapolis: Hackett Publishing Company Inc, 2000. Pp 136-140 for the distinction between philosophical and methodological behaviorism. For a classification of Ryle as a 'Philosophical Behaviorist' see Stout, Rowland. "What You Know When You Know How Someone Behaves." *The Electronic Journal of Analytic Philosophy*, no. 7 (2002): http://ejap.louisiana.edu/archives.html.

This brief characterization I shall build into a description as the argument progresses. From this point on, what I have just said shall serve for the purposes of fixing what I refer to as Ryle's 'Linguistic Behaviourism'. This is one particular kind of argument Ryle uses. However the description of the analytic behaviour of language is one thing, and what he says follows about the mind from it is another while the justifications he gives for such reasoning is yet another thing again.

I should like to distinguish Linguistic Behaviourism from several species of Logical Behaviourism. By now there is such a wide variety of meanings for the term 'Logical Behaviourism' associated with 'Gilbert Ryle' that the simplest way forward is to show two species of interpretation and why the problems arising between them suggest that we need to start over and leave all the other readings by the wayside. To these ends the two most pertinent to the present project are those of Morris Weitz and David Chalmers.

Starting with the latter, Chalmers thinks that the philosophical concept of mind can be divided into two dominant concepts that arose from a division that began with Descartes and ended with Functionalism. These two concepts of the mind are the 'phenomenal' and the 'psychological'[57]. The introspective qualities of what thoughts are like Chalmers calls the 'phenomenal' and characterizes as 'feels[58]' while the psychological he gives a first approximation as what the mind 'does'[59].

---

[57]Chalmers, David J. *The Conscious Mind*. New York: Oxford, 1996.Pg 11

[58]Chalmers. *The Conscious Mind*. 1996 Pg 11 – 17. Pg 182 shows how these two processes parallel each other. This comes out in his treatment of judgments. For Chalmers judgments are beliefs with all of their phenomenal properties subtracted. This explains his earlier claim on pg 174, that judgments are purely psychological states. The difference is important for his central thesis. This difference comes out in the discrepancy between (2) and (3) of the conditions that make up the paradox of phenomenal judgment. Chalmers holds that whereas judgments about consciousness are logically superveniant on the physical, consciousness itself is not. This in turn explains the argument on pg 95 and the claim that phenomenal zombies are conceivable, since they would only employ judgements, and not beliefs, and in turn this explains how a phenomenal zombie might think he is conscious when he in fact is not, since the structure of his judgment is determined by his psychological state and not the content of his phenomenal experience. This parallel between phenomenal and psychological states later develops into the principle of structural coherence see page 219.

[59] This approximation is problematic. I take it he means 'does' in a functionalist in-put and out-put sense, although there are causal problems with configuring the Freudian position in terms of in-put and out-put analysis. There is a problem because some of the Freudian's drives are causally generic and in-built, and thus are not determined by in-puts and therefore can't be functionalist by the definition he gives. See the block quote from Chalmers, which I've reproduced below on Freud. For Freud himself on this matter seeFreud, Sigmund. "Three Essays on Sexual Theory." In *Psychology of*

Chalmers writes

> The phenomenal and the psychological
> aspects of mind have a long history of being
> conflated. Rene Descartes may have been
> partly responsible for this. With his
> notorious doctrine that the mind is
> transparent to itself, he came close to
> identifying the mental with the
> phenomenal. Descartes held that every
> event in the mind is a cogitation, or content
> of experience. To this class he assimilated
> volitions, intentions and every type of
> thought[60].

In David Chalmers' history this tradition is not seriously challenged
until Freud who, Chalmers argues, makes a move towards the
psychological by arguing that accessibility to consciousness is not
essential to explaining a state or its existence.

Chalmers writes

> It appears that Freud construed the notions
> *causally*. Desire, very roughly, was
> implicitly construed as the sort of state that
> brings about a certain kind of behaviour
> associated with the object of the desire.
> Belief was construed according to its causal
> role in a similar way. Of course Freud did
> not make these analyses explicit, but
> something along these lines clearly
> underlies his use of the notions. Explicitly,
> he recognized that accessibility to

Love, Pp 111-220. Victoria: Penguin, 2010. Pp 138-142 for the sexual drive in neurotics, and Freud,
Sigmund. *Civilization and Its Discontents*. Translated by David McLintock. London: Penguin, 2004.
Pp 17-25 for his discussion of the drives and the in-built causal efficacy of the pleasure seeking
principle.

[60]Chalmers. *The Conscious Mind*. 1996 pg 12

consciousness is not essential to a state's relevance in the explanation of behaviour, and that a conscious quality is not constitutive of something being a belief or a desire. These conclusions rely on a notion of mentality that is independent of phenomenal notions[61].

According to Chalmers the next stage separating the phenomenal from the psychological started with the behaviourist movement. The significance of the behaviourist movement, for Chalmers, was the rejection of the introspective tradition for an objective brand of psychological explanation. Chalmers argues that some behaviourists recognized consciousness but ignored it, and some denied it existed[62]. However Chalmers holds the overall significance of the behaviourist movement, when taken together with the Freudian, was the creation of a new Freudian-Behaviourist Orthodoxy which was ultimately established and preserved in the move towards Cognitive Science and Functionalism. Functionalism, Chalmers argues, is the converse of Descartes' argument that all psychological phenomena can be assimilated to the phenomenal. The reason why it is the converse, Chalmers holds, is that Functionalism, as defined by David Armstrong[63] and David Lewis[64] takes it that a mental state is defined by (a) the stimulation that produces it (b) the behaviour it produces and (c) the way it interacts with other states.

This history is important to David Chalmers' conceptualization of Ryle's Logical Behaviourism. Chalmers sees Ryle's Logical Behaviourism as a sort of precursor to Functionalism. I take it David Chalmers means 'reductive functionalism' in this earlier part of the book because later he goes on to announce that his overall project is the search for a non-reductive functionalist account of consciousness[65] and to these ends gives an account that focuses on the difference between

---

[61]Chalmers. *The Conscious Mind*. 1996 Pg 13

[62] Chalmers. *The Conscious Mind*. 1996 Pg 13

[63] As defined in Armstrong, David. *A Materialist Theory of Mind* London: Routledge and Kegan Paul, 1968.

[64] As defined in his paper Lewis, David. "An Argument for the Identity Theory." *Journal of Philosophy* 63, no. 1 (1966): 17-25..

[65] Chalmers. *The Conscious Mind*. 1996 Pg 229

beliefs and judgements[66], consciousness and awareness[67] and various bridging principles[68]. However we can see how an account of dispositions like Ryle gives might be important to the development of (reductive)[69] functionalism as conceived by Armstrong and Lewis, in particular defining a mental state by (a) the stimulation that produces it and (b) the behaviour it produces. Indeed Chalmers argues that between the historical rise of the Behaviourist-Freudian orthodoxy and (reductive) Functionalism is the period of Logical Behaviourism.

Chalmers writes

> In philosophy, the shift in emphasis from the phenomenal to the psychological was codified by Gilbert Ryle, who argued that all our mental concepts can be analysed in terms of certain kinds of associated behaviour, or in terms of dispositions to behave in certain ways. This view, Logical Behaviourism, is recognizably the precursor of much of what passes for orthodoxy in contemporary philosophy of psychology. In particular, it was the most explicit

---

[66] Chalmers. *The Conscious Mind.* 1996 As I noted he reveals that judgments are beliefs with all of their phenomenal content subtracted on pg 182. Prior to this on page 175 he lays the foundations for this move and reveals that the contents of first order judgements make up the contents of awareness.
[67] Chalmers. *The Conscious Mind.* 1996. The difference between consciousness and awareness is one of the central themes of the book and starts with the discussion of Ned Block's paper on pg 29. However, immediately preceding this he has a discussion of Armstrong's concepts of introspection and reportability. Pg 228 is where he finally reveals the principles for the basis of the distinction. The paper Chalmers draws from on page 29 of *The Conscious Mind* is Block, Ned. "On a Confusion About the Function of Consciousness." *Behavioral and Brain Sciences* 18, (1995): Pp 227-47.
[68] Chalmers. *The Conscious Mind.* 1996 NB The concept of bridging principles he, likewise, introduces early, but spells out on page 237 as part of the discussion about using coherence principles as epistemic levers. However on pg 225 he reveals that the structure of consciousness is mirrored in the structure of awareness. The discussion of bridging principles is probably most important, for our purposes, reflectively, on page 234 where he sets out the three fundamental projects that he thinks a principle of coherence will play in revealing the nature of the mind.
[69] Chalmers. *The Conscious Mind.* 1996 My parenthesis, he defines his project properly on page 229 but doesn't seem to clarify it any earlier. I take it this is what he means. There is a discussion on pp 104 -106 and a reply to Searle on pp 130-131. In the prior he argues that proponents of reductivism favor functionalism because it is the only tenable option, the latter is a rebuttal to Searle's position that consciousness must play a functional role which Chalmers thinks ignores the fact that consciousness is 'ontologically novel'. Neither capture the '(non)' of the (non)-reductive aspects of functionalism he develops later in the work.

codification of the link between mental concepts and the causation of behaviour[70].

This focus on 'associated behaviours' and 'dispositions to behave in certain ways' may be true in certain aspects from a historical point of view and certainly there are those who may have read Ryle this way. However there are ways in which this view might be inaccurate or problematic, two of which are central to this paper. Perhaps the least problematic to a reading of Ryle is the link between the causation of behaviour and mental concepts. I will deal at length with this elsewhere in the paper[71]. More problematic is what is missing from Chalmers' account. For Ryle was first and foremost an Ordinary Language Philosopher. His approach to the mind was firmly entrenched in the view that the mind reveals itself through the use of ordinary language. This is the second and more problematic element.

---

[70]Chalmers. *The Conscious Mind*. 1996 pg 14

[71] Ryle argues against conceptualizing the mind in causal terms. See Ryle, *Concept of Mind*, 1983 The Bogey of Mechanism Pp 74-80. See also 'Towards a Methodology' in this paper, and in particular the subsection 'Linguistic Behaviorism: Episodes and dispositions. What exemplifies the Linguistic Behaviorists claims' NB Ryle's argument against the Freudian psychologist is that a causal hypothesis must bottom out in an implicit understanding of language. See also 'Chapter Seven Motives, Moods and Reasons' the subsection titled '7.3 Dispositions, Phenomenology and Law-Like statements' in this paper for the way I untangle Ryle's position using Wolff's insight. Here we'll see that the pulling, or impelling 'power' in dispositions is actually a confusion between first and third person perspectives, and linguistic behavioral style arguments with phenomenological ones. This mix I will later call sympathy, and it occupies the indirect position between the two direct positions. See Chapter Seven: Motives, Moods and Reasons' and its subsection '7.3 Dispositions, Phenomenology and Law like Statements' in this paper. The two direct positions are, of course, the first personal direct position and the third personal direct position. See also footnote 219 for a discussion of the two direct positions in relationship to David Chalmers. See footnotes 457 and 458 for a discussion of how the first and third personal positions relate to the Chalmers and Daniel Dennett debates. Footnotes 362 and 200 contain the direct relationship between Sellars reading of Ryle's log keeper 'in forro interno' and the first and third personal position. See footnotes 377, 218 and 200 for the way this psychologistic thesis relates to Wilfrid Sellars' position in *Empiricism and the Philosophy of Mind*, and footnotes 200, 218, 229, 378, 377, 362 for the relationship between Ryle and Sellars' reading of Ryle. Footnote 193 explains the taxonomy of thinkers used in this paper and their relationship with Ryle and the critique of ordinary language style argumentation this paper builds up. Once the reader has reached the insight in '7.3 Dispositions, Phenomenology and Law-Like statements' they can return to this footnote and see how all of the parts link together. See also Ryle, *Concept of Mind*, 1983 Pg 52. Here he argues against the causal hypothesis on the basis of ordinary language. This is Ryle's implicit ordinary language position in the philosophy of mind. No doubt this is how he has been widely read and become a part of modern philosophical history from such readings, but, I argue, these are more or less a simulacra built up from Ryle's actual position.

In a discussion related to Augustine's "volitions" and the "Para-mechanical Theory of Mind" Ryle lays out a general criterion for mental states and mental acts:

> If ordinary men never report the occurrence of these acts, for all that. . . they should be encountered vastly more frequently than headaches, or feelings of boredom; if ordinary vocabulary has no non-academic names for them; if we do not know how to settle questions of their frequency, duration or strength, then it is fair to conclude their existence is not asserted on empirical grounds. . .[72]

Ryle's criterion for mental phenomena is that people must be able to talk about them. More than this, they must already be part of the currency in use in linguistic exchanges. If the ordinary language user can talk about such concepts then it seems prima facie that they cannot be inaccessible or undiscovered in the way that a psychological theory of mind, as defined by the Freudian-Behaviourist orthodoxy, might suggest. For Ryle legitimacy in a theory of mind demands that linguistic knowledge must already be known in some intrinsic sense to the competent ordinary language user. The orthodoxy must be one of ordinary language and not the unconscious or the strictly behavioural[73]. The exact nature of this 'intrinsic sense' is important to Ryle's position and I shall bring this out later in the paper[74].

Suffice it to say, at this stage of the paper, the problem with Chalmers' account of Logical Behaviourism, for our purposes, is that it leaves out precisely what is essential to Ryle's concept of mind. Ryle

[72] Ryle, *Concept of Mind,* 1983 Pg 64

[73] Ryle, *On Thinking,* 1979, See the introduction by K. Kolenda's , pp 1 – 17, and in particular pg 1, where he sketches out Ryle's basic move against a Cartesian or Behaviorist account. Kolenda writes "Ryle's basic move is well, even notoriously, known. He inveighed repeatedly against a twin mistake: to put the concept of mind into either a mechanistic (Behaviorist) or a ghostly (Cartesian) framework. . . Ryle undertook the task of reminding us of what we *pace* Behaviorist or Cartesian distortions, are perfectly familiar with."

[74] See 'Linguistic Behaviorism: Episodes and dispositions. What exemplifies the Linguistic Behaviorists Claims' in this paper.

was not just rejecting the Cartesian tradition of phenomenal consciousness. Rather he was rejecting the concept that the academic and scientific had anything more to teach us about the mind that we didn't already know or could not be gleaned from competent ordinary language use and its analysis.

3.1.2 Morris Weitz

Morris Weitz's concept of Logical Behaviourism [75] is a much closer reading of Ryle's position in *The Concept of Mind* than David Chalmers because he takes into account Ryle's position on ordinary language. Weitz bases his taxonomy of Ryle's analysis of ordinary language on propositional structures or model statements. Weitz provides the semantics to differentiate three such structures. While this is a closer reading it is still not entirely accurate.

Ryle himself bases his analysis in *The Concept of Mind* not on three model sentences, but on extra-sentential relations of words that arise from sets of behaviours of language at the level of nouns, verbs, epithets, adverbs and adjectives. The reason they are extra-sentential is that they contain appeals to properties, relations and characteristics that defy analysis at the level of whole model sentences. This is because Ryle's analysis contains structures that can pick out configurations that can differentiate between sub-groupings which can not be limited to a model propositional form, but which crossover and inter-relate whole bodies of propositions or models. Moreover Ryle does not abstract from

---

[75]Weitz, Morris. "Professor Ryle's "Logical Behaviourism"." *The Journal of Philosophy* Vol 48, no. 9 (1951): Pp 297 - 301. To be fair, Ryle does describe his project with a certain ambiguity in the preface, that could lend itself to Weitz's analysis. However, a closer reading of Ryle's actual arguments shows this position to be quite naïve. This reading is supported and clarified by Ryle's position in his paper on *Use, Usage and Meaning*. See Gilbert Ryle, J. N. Findlay. "Use, Usage and Meaning." *Proceedings of the Aristotelian Society; Supplementary Volumes* 38, (1961): Pp 228-229. Ryle differentiates between 'language' and 'speech', where language is 'having words' and speech is 'saying things with them'. Confusing the two results in equating the use of the sentence with its meaning. The use of the sentence depends on inter-sentential relationships with other words, whereas looking for the meaning, Ryle thinks, results in treating sentences 'as if (they) could be solecisms'. This accords with Dummett's concept of an implicit language theorist which I think that Ryle is. For this reason I have rejected the dominant tradition of a "Logical Behaviourist" reading that interprets Ryle's account of dispositions as using model propositional sentences, Weitz being the earliest concrete example I could find, and possibly even the origin of the error. The distinction Ryle makes in *Use, Usage and Meaning* influences my reading of him, and why later, as we will see I put him in the implicit language position in relation to meaning with Michael Dummett.

these relationships between words, to words and to the world to invoke truth conditions in the way a propositional model might require. To make this clearer, it is best if we turn now to what Weitz and Ryle actually said.

Weitz writes on the first model sentence

> There are, first of all, the categorical, those sentences which describe episodes, like "Jones looked for his dog," or "Jones solved the puzzle." These are simple narratives utilizing the many tasks and achievement verbs at the command of ordinary speech[76].

On the second type Weitz writes

> Secondly, there are sentences whose logical behaviour Ryle calls 'hypothetical' or 'dispositional'. Among them are sentences like "Jones is vain." "Jones is a careful driver," and "Jones knows French". None of these is a categorical, in spite of its surface similarity to "Jones sees a dog.[77]"

Weitz writes on the third type

> The third logical species of ordinary mind-sentences which Ryle's logical behaviourism discloses is one that he calls either 'mongrel categorical' or 'semi-dispositional'. . . So far as ordinary mind-sentences are concerned these mongrel-categoricals are embodied especially in sentences containing "heed" concepts; "noticing", "taking care", "concentrating on", "knowing what one is doing", and the like. Consider for example the difference between "Jones is a careful

---

[76]Weitz, *Professor Ryle's Logical Behaviourism*, 1951 Pg 296
[77]Weitz, *Professor Ryle's Logical Behaviourism*, 1951 Pg 297

driver" and "Jones is driving carefully". The first is completely dispositional; when ordinary people utter it, they mean that if Jones were to drive, under certain specified conditions, then he would obey traffic laws, be on the alert for other drivers and pedestrians, etc. But the second says more, and is spoken only if Jones is driving in such a manner[78].

If we take the second type of sentence in Weitz's taxonomy, the dispositional sentence "Jones is a careful driver," for example, he lacks the linguistic resources, on the propositional sentence-based structure of that taxonomy to distinguish between the specific types of dispositions which are a distinctive feature of Ryle's analysis of ordinary language. Take 'careful' from the example above.

'Carefulness' refers to a disposition of a certain type that belongs alongside a set of semantic distinctions which Ryle makes on the basis of the linguistic behavioural traits of adjectives and adverbs that bottom out in the behaviour of two distinct groupings of verbs which, Ryle argues reveals something about the nature of mind. "Careful" depends on the set Ryle calls 'capacity verbs' which are different from 'tendency verbs'. Ryle would argue that 'careful' and 'carefully' could never be used for the set of linguistic behaviours related to the tendency verbs; neither in an adverbial phrase nor as a dispositional description. Ryle would no doubt claim that Weitz is mistaken in his analysis and has fallen into the trap of assuming that there is a 'one-pattern intellectual process' for dispositions and thus that dispositions have a uniform exercise.

Ryle writes explicitly

Epistemologists, among *others*, often fall into the trap of expecting dispositions to have uniform exercises. For instance, when they recognise that the verbs 'know' and 'believe' are ordinarily used dispositionally, they assume that there must therefore exist

[78]Weitz, *Professor Ryle's Logical Behaviorism*, 1951 Pg 299 - 300

one-pattern intellectual processes in which
these cognitive dispositions are actualised[79].

Weitz's mistake, as I pointed out, is to assume there is a one pattern
intellectual exercise. Weitz, if not among the epistemologists, would no
doubt be among the *others*. Rather than limiting the role of natural
language analysis to a set of model sentences with a uniform exercise,
as Weitz's Logical Behaviourist's treatment of Ryle's dispositions does,
Ryle's own examination is explicitly interested in an investigation into
the specific sets of relationships and structures that arise from the
behaviour of natural language at a level that bottoms out, not just at
propositions and whole sentences, but in extra-sentential relations that
hold between groups and configurations of expressions that can be
defined by linguistic behaviours at the level of adjectives, verbs, nouns,
epithets and adverbs. Ryle is interested in sets of relationships that
hold between different words at the level of the dispositions themselves
and their exercise in linguistic discourse. Ryle is not looking for
uniform propositional models. In this case 'to know' and 'to believe' have
different sets of word relations and sub-sentential patterns that
characterize them. These different patterns arise because the verbs and
their cognates behave differently. That is what Ryle is interested in.

Ryle writes

> but even when it is seen that both (know
> and believe) are dispositional verbs, it has
> still to be seen that they are dispositional
> verbs of quite disparate types. 'Know' is a
> capacity verb, and a capacity verb of that
> special sort that is used for signifying that
> the person described can bring things off, or
> get things right. 'Believe', on the other
> hand, is a tendency verb and one which does
> not connote that anything is brought off or
> got right. 'Belief ' can be qualified by such
> adjectives as 'obstinate', 'wavering',
> 'unswerving', 'unconquerable', 'stupid',
> 'fanatical', 'whole-hearted', 'intermittent',

[79]Ryle, *Concept of Mind*, 1983 Pg 44 Italics, mine.

'passionate' and 'childlike', adjectives some
or all of which are also appropriate to such
nouns as 'trust', 'loyalty', 'bent', 'aversion',
'hope', 'habit', 'zeal' and 'addiction'. Beliefs,
like habits, can be inveterate, slipped into
and given up; like partisanships, devotions
and hopes they can be blind and obsessing;
like aversions and phobias they can be
unacknowledged; like fashions and tastes
they can be contagious; like loyalties and
animosities they can be induced by tricks[80].

As we can see from the above, Ryle divides verbs by the behaviour of
adjectives and the nouns that are qualified by them, and sorts them by
the specific adverbial structures these qualifications pose, into systems
of relations by the sets of linguistic behaviours that determine them. In
the broad sense, he creates two sets or 'families' of dispositions based on
a set of similarities between them, not at the level of whole
encapsulated sentences or specific propositional structures, but rather
at the level of relationships between the sub-sentential parts inside
sentences. This separation of dispositions is based on adjectives that
can describe the nouns derived from the behaviour of the verbs. On one
side of the configuration we have the 'capacity verbs' governing one
family of dispositions, on the other side we have the 'tendency verbs'
governing the other family. This difference between the two groups of
dispositions is further supported by the form of epithet applied to people
who simulate or fake one side or the other.

Both skills and methods can be simulated,
but we use abusive names like 'charlatan'
and 'quack' for the frauds who pretend to be
able to bring things off, while we use the
abusive word 'hypocrite' for the frauds who
affect motives and habits[81].

---

[80]Ryle, *Concept of Mind,* 1983 Pg 128
[81]Ryle, *Concept of Mind,* 1983 Pg 128

The ascription of Ryle's analyses to the level of the behaviour of propositions and sentential structures by Logical Behaviourists, of the sort proposed by Weitz, simply fails to take into account the full range of Ryle's arguments at the level of sub-sentential analyses for his descriptions. The trouble with a Logical Behaviourist interpretation of Ryle like Weitz offers is not only that it is too crude to capture these distinctions behind the configurations but in its crudeness, it ignores the behaviour of specific parts of natural language that Ryle is interested in. The problem with the Logical Behaviourism that Chalmers gave us was cruder still. It ignored the linguistic methodology altogether and reduced Ryle's position to a focus on dispositions to behave in certain ways and associated behaviours. Neither provides adequate grounds for an encounter with Ryle. As such, let us abandon the term 'Logical Behaviourism' and further define 'Linguistic Behaviourism' from the types of arguments and claims that occur in Ryle's philosophy.

3.2 Linguistic Behaviourism: Episodes and dispositions. What exemplifies the Linguistic Behaviourist's Claims?

So far we have looked at two attempts to define Ryle's philosophy of mind. Both attempts were too coarse grained to capture the nature of the types of claim that Ryle advances in *The Concept of Mind,* nor do they exactly seek to capture why Ryle thinks such claims are fundamental to exploration and understanding the nature of mind. An important question to ask given that the prior two attempts failed to capture the nature of these arguments is 'what exactly exemplifies Ryle's linguistic claims?'

Thus, to offer another example of a Linguistic Behavioural claim from *The Concept of Mind*, we might look at the way that Ryle bases one 'leg' of his distinction between episodes and dispositions, on the behaviour of specific sets of verbs.

He does this in three stages of language analysis.
If we start at the complex end, he argues

> (S)ome dispositional words are highly
> generic or determinable, while others are
> highly specific or determinate; the verbs
> with which we report the different exercises
> of generic tendencies, capacities and
> liabilities are apt to differ from the verbs
> with which we name the dispositions, while
> the episodic verbs corresponding to the
> highly specific dispositional verbs are apt to
> be the same[82].

The specific claim argued by Ryle about the disposition-episode distinction here is that the distinction bottoms out in descriptions of dispositional verbs in either generic and determinable, or specific and determinate distinguished in terms of linguistic behavioural traits.
Ryle writes

> Dispositional words like 'know', 'believe',
> 'aspire', 'clever' and 'humorous' are
> determinable dispositional words. They
> signify abilities, tendencies or pronenesses
> to do, not things of one unique kind, but
> things of lots of different kinds[83].

To clarify one might say that what Ryle is arguing here is that 'know', 'believe' and 'aspire' qualify themselves as a set of verbs governing determinable dispositions, that is, they are far more wide ranging in the dispositions that they can govern, than the far more limited and determinate set to which verbs like 'philosophize' or 'bike ride' belong. That is to say, on Ryle's account, verbs that govern by their limitation to a core domain of activity are determinate.

---

[82]Ryle, *Concept of Mind,* 1983 Pg 114
[83]Ryle, *Concept of Mind,* 1983 Pg 114

But why are they dispositions to begin with? How does Ryle establish the class of dispositional verbs themselves? Indeed, it is important to the task of establishing the nature of a philosophic claim, to ask what Ryle establishes the class of dispositional verbs themselves on? Ryle writes

> The verbs 'know', 'possess' and 'aspire' *do not behave* like the verbs 'run', 'wake up' or 'tingle'. We cannot say 'he knew so and so for two minutes, then stopped and started again after a breather', 'he gradually aspired to be a bishop', or 'he is now engaged in possessing a bicycle'[84].

One set of verbs, the determinable set, 'know', 'posses' and 'aspire' do not behave like another set of verbs, the group that contains 'run', 'wake up' or 'tingle. This second set, of course, are the episodic verbs that Weitz tried to capture in his first model sentence; the 'categorical narratives' and what Chalmers characterized as 'associated behaviors'.

To grasp this difference between episodes and dispositions, on the side of the episodes, we might say of someone, say one Reg, that if Reg is running down the road, or if Reg ran down the road, if Reg has started running down the road, that the behaviour of these verbs are governed by a set, or sets of instances in time, and thus belong to a specific episode embedded in a time frame with a content, namely, that of Reg running.

In contrast, Ryle's dispositional terms, the afore-given examples 'know, aspire and possess', are not episodic in the same way that the time we saw Reg running down the road being chased by Rod, refers to a single event, though the event in question contains lots of smaller events, i.e. now Reg is ducking Rod, now Reg is dodging Rod, now Reg is running back this way to escape Rod – or the way that 'wake-up' refers to an episode. One moment Reg is asleep. The next he has woken up. The bit between is the waking up. One could argue, it seems obvious that people do not go around in a state of 'waking up', and it would be a peculiar thing if they did, but they do go around, and as they go around,

[84]Ryle, *Concept of Mind,* 1983 Pg 112

they go around possessing skills, abilities, motives and inclinations. They carry their dispositions with them.

It is worth noting, that within the fabric of a language, a normally episodic term, may become a dispositional term, in the weak sense like 'that time Reg *knew* where Rod was hiding'[85]. However, Ryle's original point, to be fair, is that episodes and dispositions are different things because their verbs are governed by different sets of linguistic behaviours. One exists as a series or sequence of structured events, and parts of sequences, while the other, the dispositions exist outside these structures, though, of course, dispositional tokens or sets of tokens, may underlie the episodes. For instance, take someone who may be a smoker. In this case the tendency verb 'possesses' governs the dispositional term "the smoker" which of course involves a habit, or more accurately, in keeping with a modern viewpoint, an addiction. That is to say the smoker *possesses* an addiction. In some sense the addiction is with them while they are in the shower, while they are driving the car, while they are asleep in their room. In another sense it has a time stamp. The smoker started smoking while in high school. They quit in August of last year. The addiction, as a family of dispositions, governed by tendency verbs, can have a time stamp. The time stamp has within it many episodes or instances in which the person, while a smoker, lit up a cigarette, but it does not have an episodic framework. That is to say from High School to August of last year the person was a smoker. Ryle doesn't think a time stamp is an episode[86].

---

[85]For discussion on this objection see Palmer, A. "Thinking and Performance." In *Knowledge and Necessity: Royal Institute of Philosophy Lectures 1968/9*, edited by Godfrey Vesey, 3, Pp 107-118. London: MacMillian and Co, 1970. Pp 112-114 See also Ryle, *Concept of Mind*, 1983 Pg 258 NB This is an example of an 'occurrence' as a 'remember-when' mixed with a 'knowledge-how' configuration, since it would answer to 'how did he know?' but the method, skill or means is an instance of 'knowing-that' and 'as such' is not subject to law derivation as specified on pg 136-137 of Ryle, *Concept of Mind*, 1983. It is important to the overall project of this thesis to see why this is so. Law derivation, as I will show, is actually an illusion built on analogical structures that arise from confusions between first and third personal positions. To understand why this is not subject to law derivation see 'Chapter Seven : Motives, Moods and Reasons' and '7.3 Dispositions, Phenomenology and Law-like statements' in this paper. See also footnotes 71, 219, 457, 458 and 530, as well as footnotes 200, 218, 229, 378 to see how these analogical structures arise between first and third personal positions as confusions of indirect positions that carry-over assumptions from direct positons. This is perhaps best left until the later stages of the thesis, but the reader is invited to comeback to this point once they have grasped the insights on offer.

[86]Ryle, *Concept of Mind*, 1983 Pg 112 – 120. See 120. You can infer dispositions as capacities from skills and abilities from demonstrations of competence. If someone speaks Chinese, for instance, you

If we move on from dispositional verbs like 'posses, know and aspire' and have a look at episodic verbs, Ryle gives us two classes of these[87]. By examining sets of verbs like 'look', 'seek', 'build', 'compete', 'race', 'search', 'explore', we can quickly see that they don't in fact behave like another set of verbs which he also lists; 'see'[88], 'find', 'discover', catch', 'solve', 'score', and so on[89]. The former set of episodic verbs are task verbs and the later are achievement verbs.

Now, to resume the issue raised by Weitz and Chalmers, these types of distinctions we have looked at, being based solely on the structures connected to configurations of dispositional and episodic verbs, exemplify what I have described as Behavioural Linguistic claims because they are claims based on the way that specific components of language behave. The idea Ryle has here, which I am attempting to bring out, is that investigation into the behaviour of the linguistic terms will reveal the true nature of the mind's relation with the world. This next part is important because I explore what will become an important move later in the paper. That is, later in the paper I shall look into what the authority of the linguistic behavioral arguments rest upon.

In my analysis of Ryle so far you have two types of argument. One argument rests upon the analysis of linguistic behaviour. The other argument involves the authority that upholds linguistic conventions. When these two arguments are brought together they create a normative basis that Ryle uses to either forward his own claims about the mind, or to negate claims made by other philosophers.

---

can infer that they know-how to speak Chinese. Ryle also has a modal theory of 'can', see pp 120 – 124. Suffice it to say that this goes beyond the scope of this paper.

[87]Ryle, *Concept of Mind*, 1983 Pg 145, 146

[88] The distinction between 'see' and see, is important for Ryle's two accounts of the imagination. See Ryle, *Concept of Mind*, 1983. Pg 233.

[89]Ryle, *Concept of Mind*, 1983 Pg 125

3.3 Ryle's Two Most Fundamental Insights on the Mind and the Relationship between these Insights and Linguistic Behavioural Style Arguments.

Ryle thinks that Linguistic Behavioral arguments are good arguments because language reveals something fundamental about the mind's relationship with the world. This is based on two 'radical insights'[90] from which Ryle develops his idea of the normative value for Ordinary Language. I will introduce these here, but offer a more detailed analysis of the arguments supporting these 'radical insights' later in the thesis[91]. Following that I will explain how I intend to use Linguistic Behaviourist arguments, in order to draw out the phenomenological content in 'ordinary language' and illustrate them in the developing taxonomy of Ryle's arguments[92].

For the moment, let us return to my primary and introductory exposition and analysis of Ryle. Ryle's first argument or 'radical insight' according to my reading, is that the 'world is the place where minds happen'. By that I mean that Ryle's most fundamental argument is that the world is literally where we find the mind. This, of course, if we use Richard Menary's taxonomy of embodied and embedded forms of cognition[93] is the radical position both of a strongly embodied and strongly embedded mind.

Richard Menary writes

> A strongly embodied and embedded position would view cognition as, at least, sometimes involving more than neural activity but also bodily activity in the embedding environment.[94]

---

[90] I use 'radical insight' simply as a label, for ease of reference, to refer back to these.

[91] See 'Chapter Five : Ordinary Language Arguments' and the subsection titled '5.1 An analysis of Ordinary Language claims and their ability to affirm or negate on a reputed source'.

[92] See 'Chapter Five Ordinary Language Arguments' the subsection titled '5.2 Phenomenological arguments' in this paper.

[93] Menary, Richard "Dimensions of Mind." *Phenomenology and the Cognitive Sciences 561-578* 9, no. 4 (2010): Pp 561-578.

[94] Menary, *Dimensions of Mind*, 2010 Pg 562

Further, from Menary's taxonomy Ryle would occupy a position on the far right wing of Menary's Enactivism, and one middle to far right on the wing of Menary's formalization of the 'extended mind' hypothesis.

Menary writes

> Embodied mind *strong*; some of our mental and cognitive processes and states involve processes of the body acting in and on the environment. (Enactivism) . . . Embedded mind *strong:* mental and cognitive processes and states are integrated with states and processes found in the environment. (Extended mind)[95].

And indeed Ryle writes

> The statement 'the mind is its own place', as theorists might
> construe it, is not true, for the mind is not even a metaphorical 'place'. On the contrary, the chessboard, the platform, the scholar's desk, the judge's bench, the lorry-

---

[95]Menary, *Dimensions of Mind*, 2010 Pg 562 NB I am using this taxonomy, rather than Susan Hurley's "The Varieties of Externalism." In *The Extended Mind*, edited by Richard Menary, Pp 101 - 154. Massachusetts: The M.I.T. Press, 2010., or Menary's other taxonomy in Menary, Richard. "Cognitive Intergration and the Extended Mind." In *The Extended Mind*, edited by Richard Menary, Pp 227 - 243. Massachusetts: The M.I.T. Press, 2010., because I am deliberately avoiding the debate between Andy Clark, David Chalmers, Fred Adams and Ken Aizowa because it takes the paper too far afield at this early stage, however once we move from an Ordinary Language position like Ryle's back into what I shall call an 'Autophenomenological source' it reintroduces these problems about causality and the interaction that arises between the mind and the world. For the clash over Clark and Chalmers 'parity principle' and the Adams and Aizowa's 'mark of the cognitive' criticism involves causal issues that complicate the picture given by Ryle's refusal to mark the boundary between cognition, language or mind in the world. The place the mind happens, Ryle thinks, is in the world itself and hence his rejection of Cartesian and mechanistic causation paradigms and dualist problems in the relation between mind and world. In brief however the novelty of Ryle's solution is that it avoids dealing with causal problems directly, but instead concentrates on ordinary language discourse to talk about the mind while putting the mind in the world in talk about the world. In the extended mind debates Ryle's position, I argue, would be on the far right of Clark and Chalmers. He thinks the world is the place where the mind happens and we already know how to talk about it. See Fred Adams, Ken Aizowa *Defending the bounds of Cognition*, see pg 68 for the Mark of the Cognitive. See also Pp 70 - 72 for the argument on natural content and derived content. Ryle would also reject this sort of dichotomy.

driver's seat, the studio and the football field are among its places. These are where people work and play stupidly or intelligently[96].

In fact, Ryle's position in *The Concept of Mind* may be *too* strong for Menary's taxonomy, since Menary argues one need only demonstrate or argue that cognition "sometimes involve(es) more than neural activity but also (includes or involves) bodily activity in the embedding environment"[97]. Ryle goes beyond this. For Ryle the world is the place where the mind happens, both in bodily form, as in the examples he gives of 'knowledge how'[98] but also and importantly as the place of discourse.

Now to Ryle's second radical insight. He thinks people already know how to talk about the mind in the world. He thinks they know how to describe the mind in the world. They can talk about the mind's conduct in terms of whether someone does something intelligently or stupidly. These are the 'inheritance conditions' attached to people's conduct.

Ryle writes

> We possess already a wealth of information about minds,
> information which is neither derived from, nor upset by, the arguments of philosophers. The philosophical arguments which constitute this book are intended not to increase what we know about minds, but to rectify the logical geography of the knowledge which we already possess[99].

What ordinary language speakers do not have, Ryle argues, is a map or a drawing board of how those concepts fit together.

---

[96]Ryle, *Concept of Mind*, 1983. Pg 50
[97]Menary, *Dimensions of Mind*, 2010 Pg 562
[98]Ryle, *Concept of Mind*, 1983. Pp 125 - 129
[99]Ryle, *Concept of Mind*, 1983. 9 - 10

It is, however, one thing to know how to apply such concepts, quite another to know how to correlate them with one another and with concepts of other sorts. Many people can talk sense with concepts but cannot talk sense *about* them; they know by practice how to operate with concepts, anyhow inside familiar fields, but they cannot state the logical regulations governing their use. They are like people who know their way about their own parish, but cannot construct or read a map of it, much less a map of the region or continent in which their parish lies[100].

This is an interesting point. To see why consider Dummett's paper *What Do I Know When I Know a Language?*[101]For it would seem to follow from Dummett's argument that he should, prima facie, argue that what Ryle is attempting to do is impossible since it seems Ryle wants to make explicit what people know implicitly. It is that 'about' in the Ryle quotation that I put in italics which causes the problem.

The central argument of Michael Dummett's paper involves two types of knowledge, implicit and explicit. Assume that explicit knowledge entails a propositional body of statements. Assume also that explicit and implicit knowledge cannot be assimilated into the same mode of exercise in competency. This is because one form, the explicit, involves a propositional body whereas the other, Dummett argues, occurs in the exercise of the knowledge itself and can be found in types of knowledge where the expression is intrinsically difficult. Since the exercise of implicit knowledge can occur in activities in which the explicit exercise is difficult then it would seem that explicit knowledge must be separable from implicit knowledge. This can result in a pragmatic contradiction where the person knows how to do something but cannot tell you how they did it. If asked explicitly they might say 'I

[100]Ryle, *Concept of Mind,* 1983. Pp 10 – 11 NB Italics, mine.
[101]Dummett, Michael. "What Do I Know When I Know a Language?" In *The Seas of Language*, Pp 94 - 105. Oxford: Oxford University Press, 1993.

don't know' when clearly the performance demonstrates the implicit knowledge that they do.

Now a theory of meaning in language, according to Dummett, refers to an object that the two speakers share knowledge of, namely meaning. If the metalanguage means of expressing that meaning are the same as the body of expression, that is, explicit knowledge of the meaning is expressed in a body of statements by two speakers and so the metalanguage itself is a meaning about statements; we thus get either circularity, since each time you generate a new body of explicit statements you get a language body which assumes implicit knowledge to communicate or express it which calls on a fuller body of circular statements, or the possibility for a contradiction if you link the body of propositional statements, directly, to the implicit ability to articulate them[102]. The reason the contradiction arises is that explicit expression entails a body of explicit statements. But in the first order use of a language the mode of expressing a language is implicit. An explicit theory of meaning, once expressed, would be one that draws upon implicitly understood statements, and you end up with an explicit statement that is an implicit statement. If one argues Dummett's premise that implicit statements are not explicit statements the result is an explicit statement that is not an explicit statements; a flat contradiction. It is what it isn't or it isn't what it is.

Dummett writes concerning the circularity:

> Explicit knowledge is manifested by the ability to state the content of the knowledge. This is a sufficient condition for someone being said to have that knowledge only if it is assumed that he fully understands the statement that he is making; and, even if it is assumed that he

---

[102] While it is conceivable a theory of meaning could be expressed in some other medium, music for example, and not a propositional body of statements, it is a lot harder to argue for the feasibility of such a position. Nonetheless, there would still be an infinite regress, since the semantic musician, would then need to provide a theory of meaning for his symphonic bit of theorizing that would assume implicit knowledge that is prior to explicit expression of that implicit knowledge, and would require in turn further explicit statements about the implicit knowledge, in that explicit statement, and so on. My strategy focuses on the difference between Ryle's use of bits of ordinary language and his analysis of their behaviors and locating phenomenological 'wherewithal' for distinctions he makes where no concrete linguistic behavior is analyzed.

fully understands the statement that he is making; and, even if we make this assumption, his ability to say what he knows can be invoked as an adequate explanation of what it is for him to have that knowledge only when we can take his understanding of the statement of its content as unproblematic[103].

This is a problem because

In many philosophical contexts, we are entitled to do this: but when our task is precisely to explain in what, in general, an understanding of language consists, it is obviously circular. If we say that it consists in the knowledge of a theory of meaning for the language, we cannot then explain the possession of such knowledge in terms of an ability to state it, presupposing an understanding of the language in which the theory is stated[104].

Now while I think this is a strong logical point, the argument itself rests upon a number of assumptions. Firstly, Dummett's assumption of the mutual exclusivity of the implicit and explicit position could be attacked. One might argue for intermediate grades or transfer statements that can turn implicit knowledge into explicit knowledge without either needing to be explained nor utilizing further implicit assumptions that might implicate further explicit statements or require either more transfer statements or the infinite regress of transformational ones. Secondly, the necessity of identifying a theory of meaning with a propositional body of explicit statements, which in turn depend upon implicit standards of meaning in statements, and not say, an alternative medium can be attacked. In a moment, for instance, I shall present a Fodorian argument, that could, on logical grounds,

---

[103]Dummett, *What Do I Know When I Know a Language?*, 1993
[104] Dummett, *What Do I Know When I Know a Language?*, 1993

challenge the first and second assumption. The third assumption, of course, is protected by Dummett's argument about translation, that is, translation terminates in one's own language and here one needs a theory of meaning. This sort of argument could, in principle, be extended to cover another possible medium, like music[105] but one would still need something like intermediate grades or transfer statements to turn the implicit knowledge of meaning in a language into explicit knowledge without invoking the implicit knowledge in the process of making knowledge of meaning explicit and thus buying into the infinite regress.

So Dummett's argument, as I read it, presents a prima facie problem for Ryle's Linguistic Cartography. Ryle's map making efforts seem like they are in trouble since he would need to present his theory in a body of statements, and as such, would need to assume a theory of meaning in order to do so.

Ryle has already considered this point, in relation to what Ryle considers to be one of the central problems of psychology, but from a different angle.

Ryle says

> Indeed, supposing that one person could understand another's words or actions only in so far as he made causal inferences in accordance with psychological laws, the queer consequence would follow that if any psychologist had discovered these laws, he could never have conveyed his discoveries to his fellow men. For ex hypothesi they could not follow his exposition of them without

---

[105]See Simpson, David. "Language and Know-How." *Phenomenology and the Cognitive Sciences* 9, no. 5 (2010): 629–643. NB Now, that is not going to be my response to this argument. My argument is going to take a different approach than that. But first, it's worth pointing out that David Simpson takes a more naturalistic interpretation of the Dummett paper. He thinks the argument ultimately rests upon the point that Dummett makes, that almost no natural language speaker can cite off rules for syntax or grammar.

Simpson binds Dummett's argument, into a 'fragility hypothesis' in a pincer movement, which simply affirms 'having a language and the knowledge in what enables communication of a language, is itself, becoming fragile'. The Dummett argument makes up one side of this 'fragility hypothesis' while Davidson's argument that linguistic conventions cannot explain language sharing abilities, since, possessing language sharing abilities pre-supposes linguistic convention simply in order to get the sharing up and off the ground, forms the other side.

inferring in accordance with them from his
words to his thoughts[106].

For Ryle, meaning must be conveyed inside language[107]. Language is
the only means we have of conveying meaning. On this view if a new
theory of human behaviour emerged then the psychologist would only
be able to tell his colleagues and peers about it, using language itself.
This is the point of Ryle's move towards finding sets of relationships,
properties, internal relations and characteristics using the Linguistic
Behavioural strain of arguments.[108] Ryle thinks that analyses of the
mind must be done inside language. Moreover it must be ordinary
language. For Ryle, going beyond ordinary language into models of
scientific causation say, would be to represent the facts belonging to
human action and thought in the idioms and analogies of something
else. He thinks the history of philosophy has been filled with these sorts
of mistakes.

That is why Ryle says

To explode a myth is accordingly not to deny
the facts but to re-allocate them. And this is
what I am trying to do[109].

Because

---

[106]This was the problem with Chalmers and Freud I pointed out earlier. It is also the reason why I
avoided the Extended Mind debates and "The Historian argument" of Ryle. See Ryle, *Concept of
Mind,* 1983 Pg 55 – 57 The novelty of Ryle's solution is it seems to allow us to side step the causal
issues that plagued Descartes by concentrating on the way people talk about the mind and the
mind's relationships with people's activities through the types of things people do in the world. See
also G, Steiner. *After Babel.* Oxford: Oxford University Press, 1975. Pg 26 and Parkinson,
Translation Theory of Meaning, 1977: See Pp 14-16 for his discussion on Ryle's Historian argument.
[107] See Ryle, Findlay, *Use, Usage and Meaning,* 1961.For Ryle's distinction between 'language' and
'having words to say things with' and 'speech' which involves 'saying things with them'. Ryle thinks
that language and the meaning of words as found in a dictionary is a different thing to using words
and saying things with them. The failure to distinguish between them leads to a view of sentences
as 'solecisms', where the meaning of each word produces what Dummett would call an explicit
meaning. This is a mistake, Ryle thinks, and a costly one. I've focused on the arguments in *The
Concept of Mind,* for the purposes of this paper, but such as they are my reading is decidedly
influenced by Ryle's view in this paper over some of the peer review literature like that advanced by
Weitz.
[108] Ryle, *Concept of Mind,* 1983. Pg 52
[109]Ryle, *Concept of Mind,* 1983. Pg 10

> A myth is, of course, not a fairy story. It is
> the presentation of facts belonging to one
> category in the idioms appropriate to
> another. [110]

That is to say, Ryle, I take it, simply wants to explain what features of natural language are significant, re-order the idioms that have been appropriated by different myths and in doing so, explore how these relate to each other, and some of the ways meaning is expressed inside these features. This will give him the elements he needs to construct a geography for the concept of the mind from the ways people speak about the mind in the world. According to this view the nature of the mind is already known and it is implicit in the way people use language. Thus Ryle sees his role as needing to show us how the pieces fit together and what that amounts to in the final concept. An interesting question here is whether Ryle is guilty of psychologism since he is no longer dealing with an explicit theory of meaning. It is worth asking in light of the Dummett paper but also because Ryle explores semantic relationships and configurations which depend for their structure on the relationship and behaviour of sub-sentential parts and not whole propositions. Ryle, for instance, is interested in what the relationship between 'archery' and 'careful' might be and how that relates to the ways people infer and talk about skills as a subset of dispositions and the verbs 'to shoot' and 'to aim'.

The primacy of the sentence as both Dummett and Robert Brandom[111] point out, has dominated analytic philosophy for the past century. For Dummett it begins with Frege's argument that words have meaning only in the context of the sentences they appear or can appear in. The idea is that the word's meaning is dependent upon the contribution the word makes to the meaning of the sentence it either appears[112], or that it can appear in[113]. For Brandom, it begins with the

---

[110]Ryle, *Concept of Mind*, 1983 Pg 10

[111]In Robert Brandom's *Articulating Reasons*. Harvard: Harvard University Press, 2001.

[112] Dretske, Fred. *Explaining Behavior. Reasons in a World of Causes*. Massachusetts: MIT Press, 1988. NB see Pp 62 – 64 for Dretske's treatment of natural systems of representation and 70 – 77 for his treatment of Frege's Sense and Reference distinction.

[113] See Frege, Gottlob. "Illustrative Extracts from Frege's Review of Husserl's Philosophie Der Arithmetik." In *Translations from the Philosophical Writings of Gottlob Frege*, Pp 79 - 85. New York: The Philosophical Library, 1952. Frege, Gottlob. "On Concept and Object." In *Translations from the Philosophical Writings of Gottlob Frege*, Pp 42-56. New York: Philosophical Library Inc, 1952. And

'Apperception of the I' in Kant. The 'I think', Brandom argues that Kant thought and Brandom reaffirms 'is the minimum unit of thought possible'[114]. This point is important and we will return to it. For the moment it is important to note where Brandom and Frege overlap and its significance for the wider argument in the paper.

Dummett writes

> Philosophers before Frege assumed. . . that what a speaker knows is a kind of code. Concepts are coded into words and thoughts which are compounded out of concepts, into sentences, whose structure mirrors, by and large, the complexity of the thoughts. We need language, on this view, only because we happen to lack the faculty, that is, of the direct transmission of thoughts. Communication is, thus essentially like the use of a telephone: the speaker codes his thoughts in a transmissible medium, which is then decoded by the hearer[115].

Why is this important? Dummett writes

> The whole analytical school of philosophy is founded on the rejection of this conception, first clearly repudiated by Frege. The conception of language as a code requires

Frege Gottlob"On Sense and Reference." In *Translations from the Philosophical Writings of Gottlob Frege*, Pp 59 - 78. New York: The Philosophical Library Inc, 1952. See also Crane, *Aspects of Psychologism*, 2014. Crane has developed an alternative reading of Frege to the one that has dominated and shaped Twentieth Century Analytic Philosophy and that this paper draws on.
[114]Brandom, *Articulating Reasons*, 2001Pg 158. NB He distinguishes between actions and cognitive thoughts. "Judgements are fundamental, since they are the minimal unit one can take *responsibility* for on the cognitive side, just as actions are the corresponding unit of responsibility on the practical side."
[115]Dummett, *What Do I Know When I Know a Language?*, 1993

that we ascribe concepts and thoughts to people independently of their knowledge of language; and one strand of objection is that, for any but the simplest concepts, we cannot explain what it is to grasp them independently of the ability to express them in language[116].

So what does Dummett think Frege's major contribution was? Was it in fact the 'primacy of the sentence' or was it the rejection of this theory that language is some sort of code for putting thoughts into? If the latter, then does Dummett go over to 'the other side?' Does Dummett, in fact, argue himself into some form of advocating 'psychologism'?

Dummett in the same paper writes

I am not here concerned with the particular features of Frege's theory, but only with the general line of approach to the philosophy of language of which it was the earliest example. Frege's theory was the first instance of a conception that continues to dominate the philosophy of language, that of a theory of a specific language. Such a theory of meaning displays all that is involved in the investment of words and sentences of the language with the meanings they bear. The expression "a theory of meaning" may be used in quite a general way to apply to any theory which purports to do this for a particular language[117]

.

---

[116]Dummett, *What Do I Know When I Know a Language?*, 1993

[117] Dummett, *What Do I Know When I Know a Language?*, 1993

Now if you recall the pragmatic contradiction I pointed out, where the person might be able to do something but if asked, not be able to tell you explicitly how or that they might reply that they don't know, and that the basis of the pragmatic contradiction arose from 'implicit' and 'explicit' conditions for the use of language, then it can be seen that the force of Dummett's argument is essentially that meaning can only be explicit if it is implicit. The impact of his position should become apparent. Any explicit statement of language implies its implicit use. Thus something cannot be explicit without circularity leading back to implicit use. Dummett thinks that implicit understanding of language must be fundamental to its use. He reinforces it, here, in the paper, given that "it is obvious that the speakers do not in general have explicit knowledge of a theory of meaning for their language, if they did there would be no problem about how to construct a theory"[118]from which we might argue it is an obvious conclusion that explicit knowledge is not the default setting. This, in turn, upholds his second condition, which is simply a stating of the obvious; that the speaker cannot have explicit knowledge without a contradiction. The speaker needs to turn the implicit knowledge expressed in his explanation into explicit knowledge without invoking the implicit ability to communicate it. This seems on first glance, to be a move towards a Pre-Fregean psychologism because it reduces an explicit theory of meaning, such as Frege advocates to either circularity or a pragmatic contradiction.

However as Simpson rightly points out, Dummett avoids the charge of psychologism "because of his insistence (that) knowledge of implicit use of language must be manifested in (that very same) use of language"[119]. Dummett argues against an explicit theory of meaning as something that can be expressed outside of language, but he has not gone all the way back to a pre-Fregean psychologism. To be sure, he is arguing against Frege's theory of meaning as a model for explicit theories but he has not argued all the way to the point of advancing a position that views language as a code for thought. He thinks that meaning needs to be implicit inside language and that a philosopher's

[118]Dummett, *What Do I Know When I Know a Language?*, 1993
[119]Simpson, David. "Language and Know-How." *Phenomenology and the Cognitive Sciences* 9, no. 5 (2010): 629–643. Pg 638

business is to spell out the form and parts of language in which meaning is itself manifested.

Dummett is anti-explicit but he is also anti-psychologistic. This is because an anti-psychologistic theory of meaning need only argue that implicit knowledge of meaning is prior to or necessarily undermines an explicit theory of meaning. This is because arguing that either implicit knowledge of meaning is needed to understand an explicit theory of meaning or that implicit knowledge of meaning is fundamental, intuitive, non-reducible, primitive, or foundational for an explicit theory of meaning is not the same thing as arguing that non-linguistic knowledge is prior to any fundamental relation to implicit or explicit knowledge of meaning. If Dummett argues that meaning bottoms out in, arises from or is grounded in implicit understanding of language then he has not gone over to the psychologistic side. He still upholds an anti-psychologistic theory of meaning. He has not abdicated to psychologism. Psychologism argues that regardless of claims whether knowledge of linguistic meaning is more fundamentally theoretical and explicit, or more implicit and practical, non-linguistic knowledge is prior to knowledge of linguistic meaning. Dummett has not argued for non-linguistic knowledge, but rather, he has argued for implicit knowledge as essential to a theory of meaning.

Ryle has the same general thought as Dummett, or at least this is what Ryle's strain of Linguistic Behaviourist arguments try to do. They analyse bits of natural language inside the idiom of their usage. As such Ryle argues a similar position to Dummett. The Linguistic Behaviourist side of Ryle and in fact the central block of his most consistent and clear reading is governed by the argument that implicit knowledge of meaning is fundamental at making precise the meaning inside of language. I shall argue later, however, there is an argument in Ryle which can be made to advocate a turn towards psychologism in skills and the practice of knowledge-how. But firstly, I think the most obvious reading of Ryle blocks this off with another argument. That is, chiefly, Ryle thought he could capture what was most important about the two different types of knowledge at the linguistic behavioural level with the knowledge-how and knowledge-that configurations even though the skills, abilities and capacities themselves might go beyond the limit of linguistic, verbal or propositional forms of expression as modes of knowledge. This is important because this is not how some people have interpreted or taken up Ryle's point as we will see. There is a small school of contemporary followers who have adopted Ryle's point as the view that something must be expressible as a proposition in order to be knowledge and I think Ryle would disagree with this and in a moment I will introduce the argument he has against it.

Secondly I just think that Ryle did not think that far into the notion of a theory of meaning to see why a problem or tension like this might arise between the expression of a bit of knowledge in a propositional form and the practical application of it as a demonstration of competence, or why that demonstration of competence would be important to a position offered in a theory of meaning in language. Dummett's paper was published a good three decades after Ryle's *Concept of Mind,* with a great deal more reflection than seems achievable, given the philosophical turbulence still in the air in those early to mid-post war years. The debates around the form a theory of meaning should take emerge out of Quine's *Word and Object*[120] and were taken up by Davidson whom Dummett is debating, at least a decade on from the publication of *The Concept of Mind* and continued on well into the nineteen sixties and seventies.

Now Dummett writes

> It is part of the business of a philosopher of
> language to explain in what specific feature
> of this use a speaker's knowledge of each
> particular part of the theory of meaning is
> manifested[121].

If we refer Dummett's quote back to Ryle, we can see that this agrees with what Ryle is doing in his Linguistic Behavioural arguments. So while Ryle at least might be seemingly cleared of psychologism the overall significance of this total argument and of my thesis is not. In fact, I will provide a number of arguments that will culminate in the charge that a considered and disciplined return to Psychologism is exactly what we need. That argument will not make sense until I reach the end of this paper and I'm ready to make it.

---

[120]Quine, Willard Van Orman. *Word & Object.* Massachusetts: The M.I.T. Press, 1960. NB original publication date was 1960.
[121]Dummett, *What Do I Know When I Know a Language?,* 1993

In the meantime it is important for our immediate purpose to note that insofar as Ryle moves from Linguistic Behavioural claims and descriptions of the way language behaves to an argument in which the perfect domain for doing philosophy of mind is inside a purified and idealized domain of common language free from scientific and terminological contaminations, he is making a normative move. I will turn to some of the arguments he makes supporting this, and some philosophically interesting positions relevant to it later in this paper[122].

Let us return to the thread of exposition I started on Ryle's use of the Linguistic Behavioural argumentation method.

3.3.1 Capacity and Tendency configurations.

We can understand something of Ryle's Linguistic Behaviourist method if we examine how he divides dispositional verbs. Ryle divides the dispositional verbs by their behaviour, into two further groups. These two groups, of course, are the 'capacity' and 'tendency' dispositional verb sub-groups whose behaviour is individuated by similarities in the structures exposed by the behaviour of the adjectives as well as the grouping of their dispositions under the 'fraud' and 'charlatan' epithets. We explored these earlier. He supports this division of 'capacity' and 'tendency'[123] verbs in several different ways but the following will suffice to note for our purposes. The adverbial form 'carefully', as part of the 'take-care' genus that Weitz partially identified earlier, needs a capacity to identify its execution. For 'carefully' to apply to someone they need to actually be doing something. The 'carefully' refers to the action that is being done. It needs a verb.

The difference between capacities and tendencies is in the 'inheritance conditions[124]' one can attach to verbs and nouns which

---

[122] See section '3.4.1 Ordinary Language, the Manifest Image and The Philosophy of Mind', in section '3.4 Linguistic Behaviorism, Ordinary Language arguments and the normative value of Science' in this paper. There I distinguish between Ordinary Language arguments which employ an *'it makes sense to say'* claim, and an Ordinary Language statement about why one should trust the authority of an 'it makes sense to say' claim on the basis of the authority it is supposed to be upholding. Such distinctions are held in a purely technical sense for the purposes of this paper.
[123] NB I shall now adopt capitals to denote these categories when I discuss them.
[124] The term 'inherence property' I've adopted from Ryle's intellectualist legend to refer to whether something is done intelligently or stupidly, carefully or heedlessly, which Ryle at various times refers to with adverbs, 'mongrel categoricals', verbs, adjectives, adverbial phrases and so on. See Ryle, *Concept of Mind*, 1983. Pp 130 – 134 for uses and types of 'heed concepts', pg 32 for the distinctive claim that intelligence is a practice not an antecedent or an event. See also the discussion

reveal something about the set of dispositions someone has. These inheritance conditions usually take the form of adverbial phrases, adverbs and adjectives attached to skills, capacities and abilities. As such the 'inheritance conditions' of the adverbial phrases, adjectives and adverbs form a linguistic support for the division of dispositions between capacity and tendency verbs because capacities and tendencies have different linguistic inheritance behavioural properties.

In general Ryle bases the division of the capacities from the tendencies on the subclause characterized by a 'how' or 'that' structure already familiar to us from the knowledge-how/ knowledge-that configuration. Adverbial forms of the 'take-care' sort (or in Rylean "semi-episodic and dispositional terms of the improper or adverbial mongrel categorical" in this case 'carefully'), attach themselves to capacity verbs, which govern the use and application of skills, abilities, and proficiencies: 'know-how' on the 'how' side of the knowledge-how/knowledge-that distinction. In contrast the linguistic behaviour exhibited by descriptions of a tendency, (which is governed by the family of behaviours that Ryle allocates to inclinations, motives and beliefs) is defined by either a 'that' clause or a 'why' interrogative.

For example, it does not make sense to have either 'careful motives' or 'careful beliefs'. One can be careful *about* their beliefs, but this refers to the 'how', the skills, or the methods employed in forming them. The person might be careful to check into things, or check up on things. They may exercise their scepticism regularly as part of good practice, or they may be selective in what they bring to light to consider as a belief. Similarly one can be careful 'about' one's motives. Here the 'about' may refer to attempts at concealing one's motives, being critical and self-conscious[125], or simply 'being alive to'[126] the ways in which one makes decisions.

---

on pp 67 – 69 for the question of fault and adverbial structures in relation to capacities related to heed concepts, pp 47-48 for the semi-episodic, semi-dispositional mongrel categorical which Ryle lists as examples 'alert', 'careful', 'critical', 'ingenious'. See Ryle, *On Thinking*, 1979. Pp 17 – 31 for Ryle's discussion of adverbs and 'adverbial verbs' of thinking. I refer to all of these structures, for simplicity, as 'inheritance properties' in, associated with, or related to an act, except where I refer to a specific construction and its linguistic behaviours.

[125] Specifically the sense in which Ryle allows self awareness. See Ryle, *Concept of Mind*, 1983 Pg 149 for where Ryle makes the ordinary language philosophy of mind 'implicit claim' that 'the sorts of things I can know about myself are the sorts of things I can know about you.' What he means here is related to the public accessibility of language. Ryle has two theories of consciousness. One is an 'occult' stream, implied by the sorts of 'phenomenological style arguments he makes. The other is an overt stream that involves making 'special status reports', an 'internal narrator' and a sense of the

Motives, inclinations and beliefs, thus explained, come under a different family of dispositions. Motives, inclinations and beliefs belong to the 'why-that' side of the linguistic behaviours. These are located on the other side of the how/why-that configuration which Ryle uses to separate the 'tendency' and 'capacity' verbs.

Indeed, Ryle writes

> Roughly, 'believe' is of the same family as motive words, where 'know' is of the same family as skill words; so we ask how a person knows this, but only why a person believes that, as we ask how a person ties a

term 'being alive to what one is doing' to which Ryle ascribes a certain importance. I will go into some detail on this, but for the purposes of curiosity, and to show where this is headed, the overt stream is chiefly made up of types of 'talk' and the way people use the term conscious, so for instance Ryle, *Concept of Mind,* 1983 pg 150 he distinguishes self-conscious as associated with embarrassment, and associates this with types of 'guarded talk'. Another use is similar to Ned Block's 'phenomenal' and 'access' consciousness distinction in his paper, Block, *On a Confusion About the Function of Consciousness,* 1995, where a person might become conscious of a noise only after it has stopped, or similarly they might lose consciousness from the knees down. It is what Chalmers identifies as a type of functional state he refers to as awareness. The phenomenal aspects of an experience might be present but the awareness isn't. On page 150 of *The Concept of Mind,* 1983 , Ryle refers to consciousness indirectly by referring to 'unconscious' 'phobias' and 'desires'. But here there is a problem, since they are unconscious the person would not be able to give a 'special status report' since they would not be 'alive to what they are doing' and this would clash with the conditions he lays out against volitions. That is Ryle argues that the person can't make a report about volitions since he would not know what to say about them. Ryle argues that *it doesn't make sense to talk about* how many volitions it takes to get out of bed and there is no ordinary language sense in which one encounters talk about the term. Here Ryle must decide one side or the other, either he allows volitions, even though people can't make reports on them, and he dismisses his argument against them or he rejects unconscious desires in the Freudian sense. I should argue on the basis of page 99 where he argues that a man 'finds out he is tired because he yawns' and the log keeping role of retrospection on page 160 of *The Concept of Mind*, and the general thesis that his philosophical condition would collapse if he allowed Freud's unconscious states. The conflict between Ryle's own strictures about the report ability of status reports would mean that Ryle would drop 'unconscious' 'desires' and 'fears' from his account since there are no common terms for Freud's states that men use and they would thus suffer the same state as Ryle's 'volitions' if Ryle were being consistent on this point. See 'Chapter Four : Theitic Consciousness and the Interlude' and subsections '4.1.1 Ryle's use of the terms 'consciousness' and 'introspection'' and '4.1.2 The Species of Mindologue' in this paper for Ryle's log keeper roles and the way Ryle allows the use or the terms 'introspection' and 'consciousness'. See also footnote 229 for clarification of Ryle's retrospection and Wilfrid Sellars' 'in forro interno', footnote 200 in this paper for the relationship between Sellars' reading of Ryle position on thought as an 'in forro interno' log keeper and Ryle's own arguments, and footnote 362 which explains Sellar's construction of a Rylean language'. Footnote 218 may be of some interest to the reader in this regard as it explains and clarifies the relationship between the 'in forro interno' and the first and third personal positions.

[126] Another piece of Rylean. We will explore this terminology at length later in the paper. For the moment refer to the distinctions in the above footnotes.

clove-hitch, but why he wants to tie a clove-hitch or why he always ties granny-knots. Skills have methods, where habits and inclinations have sources[127].

The reason Ryle thinks there are two terms on this side of the knowledge-how/ knowledge-that divide is that how has a double function as both an interrogative and as a relative adverb of manner introducing a sub-clause. One can ask, 'How do you know that?' and state 'this is how you do it'. Similarly it makes sense to ask someone *why they believe that*, but Ryle thinks less so to ask '*how do you believe that*'. Knowledge, as Ryle envisages it, is almost entirely based on 'know how' as an extra-linguistic process about the 'mind's doing' in the world.

The normative basis for a piece of linguistic behaviourist analysis is also different to the analysis itself. The normative force of such analysis rests upon two assumptions, or 'radical insights' as I explained above, and which I'll lay out some grounds for in Ordinary Language arguments below[128]. But the concern is still germane since the world as the place where minds happen, and the way that the mind encounters the world is revealed, Ryle thinks, through adverbial terms like 'carefully' and more generally the 'take-care' stream as the how of things to do, and the 'why' of people who do them. These occur on a 'linguistic' level, and not a normative level such as a claim that one is prior to the other, or in what way they in turn relate to the world.

The claim that 'knowledge-how' is epistemologically prior to 'that' is a different sort of claim to an analysis of the sets of relations that the two terms have to dispositions, types of verbs, and the sorts of adverbials and adjectives that qualify them. Enquiring into the latter we are merely at the level of a linguistic analysis. That is, we are not attempting to establish the truth of any claim nor respond to any questions to the effect of 'how do you know?' on etymological or normative grounds. Nor are we attempting to address the primacy of Cartesian or Pragmatist claims about whether 'how' comes before 'that' in the edifice of knowledge.

---

[127]Ryle, *Concept of Mind,* 1983 Pg 129
[128] See 'Chapter Five : Ordinary Language Arguments' and its subsections.

I shall now clarify this with some philosophically interesting examples. It is to these I now turn.

3.3.2 How do you know? The Primacy of Concern

Consider the following as an example of the point I am trying to make about Linguistic Behaviourist kinds of arguments. Fodor in his introduction to *LOT2*[129] makes a distinction between Neo-Cartesian positions of the sort he says he shares with Chomsky and Descartes[130] and Pragmatism which is more widely spread, he thinks, than an epidemic[131]. Neo-Cartesians, Fodor maintains, claim that action is the externalization of thought[132]. Pragmatists, Fodor maintains, argue something much like the opposite, that is, thought is the internalization of action[133]. Thus Fodor's Neo-Cartesians argue 'knowing that' is prior to 'knowing how'[134]. This is in strict contrast to pragmatists who Fodor claims, should have it that 'knowing how' is prior to 'knowing that' in the 'strict order of '*intensional*' explanation'[135].

---

[129]Fodor, Jerry. *Lot2*. New York: Oxford University Press, 2008.
[130] These seem strangely opposed to the arguments he makes at the start of *The Modularity of Mind*. See Fodor, Jerry A. *The Modularity of Mind*. Massachusetts: M.I.T. Press, 1983. The Defining characteristic of Neo-Cartesianism which he is at odds at in *the Modularity of Mind* is the assumption that mental structures can be explained by the propositional structures of mental states. See Fodor, *The Modularity of Mind*, 1983, Pg 6. He uses Milers limit on seven memory objects to contrast faculty psychology with Cartesians of this sort. Fodor, *The Modularity of Mind*, 1983 Pg 8. The Cartesians, Fodor thinks, would think the seven objects of memory are more like propositions while faculty psychology would view the memory more like a box in the head with the seven objects inside. This in turn also clashes with his view about 'Cognitive Science' in Fodor, *LOT2*, 2008 Pg 20 – 21, where he argues the difference between Cognitive Science and Cognitive Psychology was that Cognitive Psychology was based on Hume's associations, while Cognitive Science started with Turings' Computations. The move is significant for Fodor's position, since computations require a language of thought. The 'LOT' thesis that there is a language of thought seems at odds with the Faculty Psychology thesis in the earlier work, particularly since the earlier work contains arguments against the Neo-Cartesian position. I'm not sure if there's a paper published that reconciles the two views and if there is I have been unable to find it.
[131]Fodor, *LOT2*, 2008. Pg 9
[132]Fodor, *LOT2*, 2008. Pg 11
[133]Fodor, *LOT2*, 2008. Pg12
[134]Fodor, *LOT2*, 2008. Pg14
[135]Fodor, *LOT2*, 2008. Pg10

By 'intensional' Fodor means that 'knowing how' and 'knowing that' are both individuated by their content and that 'how' and 'that' do not obscure each other. In this sense 'intensional' is directly contrasted with 'transparent'. Transparent pragmatism, according to the footnote on page 10[136], places the 'how' prior to the 'that', but unlike the intensional form, how and that can obscure each other. Since Pragmatism maintains that the 'how' is prior to the 'that', the 'transparent pragmatist' would claim that there are things one knows how to do which one knows nothing about. This, of course, echoes the argument that I mentioned before, when I was laying out some of the logical disjunctions in 'Dummett's argument' and one way one could read Ryle.

The Fodorian transparent pragmatist would know how to do something in a non-linguistic sense but not know anything propositional or that could be linguistically framed in appropriate terminology about what he was doing. He might practically know *how* to fire a gun, but he wouldn't know anything *about* the theoretical elements of firing a gun. He'd have ability but no propositional knowledge. Whether this is possible or not is an interesting question. This is like the implicit and explicit distinction that Dummett draws. Both Dummett and Ryle make allowances for practical knowledge that is non-linguistic in some sense. The extreme version of that would be a transparent pragmatist. But one need not go that far in order to argue for psychologism in the capacity side of the dispositions of Ryle's linguistic treatment of mind. All one would need argue is that non-linguistic knowledge is necessary for meaning or a theory of meaning in talk about such practical matters. The argument I'm making in this paper echoes this, but it does so in terms of mental phenomenon and the nature of mind and not sets of skills, capacities nor practical knowledge.

To see the point of this, recall that Ryle divides the dispositions into two groups. One group is made up of the 'tendency dispositions' and these encompass beliefs, motives and inclinations. The other half of the dispositions is made up of the 'capacity dispositions'. The capacity dispositions contain skills, abilities and methods. The argument I am at pains to make in this section, is that the application of practical knowledge need not be linguistic or propositional. It is my view that

---

[136]Fodor, *LOT2*, 2008. Pg10 NB see footnote

arguing that exercising what Ryle would call a 'capacity disposition' entails that one must necessarily be able to frame the knowledge possessed in that skill in a set of propositions that answer a 'how' query to be able to carry that skill off is building a strawman for an argument, either in the need for specific propositional structures or in linguistic framing that need to be either present in the execution or represented by the person who possess those skills or abilities in order for it to be counted as knowledge. I argue that someone can know how to do something without knowing how to put into words what they did or being able to form propositions about it. I base that argument on two factors. People who are excellent at certain tasks like race car driving or surfing or forms of martial arts often do not know how to explain what they are doing. Further they are often not thinking in words or constructing propositions as they perform these skills and abilities. Indeed Ryle argues that demonstrating the skills of a 'knowledge how' disposition does not entail any linguistic treatment of the mind. That is Ryle himself agrees with this position and argues that one does not necessarily think in words when one performs certain tasks, as we shall see. However it is necessary to go through this argument stage by stage, even if it is a strawman, because there are philosophers following on from Ryle who argue that something can only be knowledge if it can be syntactically embedded in a linguistic propositional form in response to a 'knowledge how' query.

So that while Ryle and Dummett both make some allowances for knowledge that is non-linguistic and one might take this and argue that this knowledge is necessary to have linguistic meaning about the dispositional skill set: that is one might argue that meaning in talk about skills, capacities and methods relies upon at least some competence or comprehension of those skills, arguing the converse is a fallacy. That is arguing that the skill depends on having the ability to talk about it, either forming or linguistically framing propositions is a strawman, and careful logical analysis will reveal it is a logical fallacy: the one we call 'affirming the consequent' or 'fallacy of the converse'. I think the stronger argument concerns the mind itself and attacking Ryle's position on the mind. Indeed I follow Frege's maxim here that it is philosophically better to attack your opponent's strongest point.

To be clear what I am attacking in this paper is the assumption that the mind bottoms out in an implicit understanding of meaning that can be found in ordinary language. This is not like a private language, but rather, this paper forwards the argument that there must be non-linguistic knowledge that precedes and is fundamental to establishing or entertaining meaning in, and meaningful talk about the mind. My ultimate argument is going to involve the emotions but I can only make that argument once I have drawn out and countered other possible avenues for Ryle's ordinary language arguments about the structure of the mind, starting with dispositions, moving through propensities and finishing with occurrences by which time we will have exploded Ryle's linguistic model for his concept of mind.

Starting with the dispositions Ryle divides these into capacities and tendencies. The tendencies as a set of dispositions distinct from the capacities, contain verbs for motives, beliefs and inclinations. The capacities as a set of dispositions governed by verbs distinct from the tendencies contain skills, abilities and methods. There are any number of linguistic behavioural arguments Ryle uses to separate them, including the epithets of charlatan and quack for people who fake capacities, and hypocrite for those who fake tendencies, as well as the difference in the way adverbs like 'carefully' and 'wavering' behave. However, before we can start attacking Ryle's concept of mind, we need to block off the capacity dispositions and show they are a strawman. As indicated we agree with Ryle that this is a strawman argument, but some philosophers insist on building them.

We would only be interested in the primacy of a 'how' statement as an argument against these types of strawmen if it could be established that demonstration of knowledge-how, and it must be all forms of knowledge how, and not simply linguistic knowledge about answering the question how to form propositions about answering the question how, and further which must be established by analysis on purely linguistic behavioural grounds[137]. Establishing 'knowledge-how' on purely linguistic behavioural grounds is to argue that knowledge-how is linguistic all the way down. That is, to make this argument and convince us that an anti-psychologistic position on the capacity

---

[137] The Implicit Ordinary Language thesis in the Philosophy of Mind. See O'Shea, James R. "'The 'Theory Theory' of Mind and the Aims of Sellars' Original Myth of Jones'." *Phenomenology and the Cognitive Sciences* 11, no. 2 (2012): Pp 175-204. This paper draws on O'Shea's concept of the 'Theory Theory' in developing its position. The 'Theory Theory' as O'Shea has it is simply the idea that our capacity for explaining human behavior depends upon our capacity to apply mental concepts which depends on an implicit theory.

dispositions in not a strawman, one would have to convince us that all forms of exercising or demonstrating knowledge-how are based on language structures and that the exercise of any set of skills, capacities, methods or techniques can not be done unless they are framed or embedded in either ordinary language talk or linguistic propositional structures for their execution. That is one must show that language is somehow necessary for performing or demonstrating any type of knowledge and having knowledge-how without linguistic encodification is impossible.

It is important to note that Ryle goes beyond Linguistic Behavioural grounds in an argument that I'll present in a moment. Ryle's position is complex in that he thinks that what is important about knowledge insofar as it is knowledge-how can be captured with a Linguistic Behavioural argument. One can always ask 'how do you do it?' But that the skills, capacities, intellectual work, problem solving and so on, in these actives themselves go beyond 'linguistic framing'. So for Ryle the *question* can be captured by a Linguistic Behavioural argument, along with something essential about the difference between a 'Knowledge-how/ Knowledge-that' linguistic configuration by a Linguistic Behavioural argument. However not everything about how one answers that question is restricted to, nor the demonstration of a 'knowledge-how' as a skill need be linguistically framed or behavioural.

By 'linguistic framing' I mean the view that something can only count as knowledge if by asking 'how' something was done the answer can be framed in a set of linguistically expressed propositions. I think a potential problem with the 'knowledge-how' Linguistic Behavioural argument is the assumption that some philosophers have made, that just because the question involves linguistic structures, the answer necessarily will as well. Someone doesn't have to respond to a 'how' with a set of propositions or a bit of linguistic utterance. They might demonstrate by simply showing the person how to do it. In such cases trying to explain in words, or using propositions alone may be difficult or impossible. In some cases the only way to respond is by showing the person how to do it.

J. Stanley and T. Williamson, for instance, hold the view that something must be able to be linguistically framed in order to be knowledge in their paper *Knowing How*[138] which Cheng-hung Tsai

[138] Stanley, J. Williamson, T. "Knowing How." *Journal of Philosophy* 98, no. 8 (2001): 411-444.

criticizes along very similar lines to the argument I am making[139] except I am using Ryle's own argument. Tsai points out that Stanley and Williamson reduce knowing how to a propositional form of knowledge with a mode of practical presentation[140]. Actually in the original paper Stanley and Williamson pick up this approach from Carl Ginet[141]. But Tsai's criticism is that by embedding the 'how' in a syntactic analyses in which it denotes the set of true propositions that answer the question Stanley and Williamson leave their analyses open to the criticism that a person knows how, only if they have a context for stating the propositional content since the mode of presentation by itself is not enough[142]. The person, if they know how to do something, must be able to express it in the form of a propositional content. This is a problem for forms of knowledge which may be expressed through competence and that competence may be expressible in some sort of explicit criteria but the competence grounding the "how" itself is inexpressible in linguistic forms. I think Tsai is right on this point. I think Ryle's argument that some forms of know-how like untangling string are done either without 'mind chatter' or without any sort of internal language expressions and linguistic framing. Ryle's argument cuts off Stanley and Williamson's position. A race car driver may waffle on about the lean and the acceleration and talk loose and fast about cornering and do something with his hands and mouth making strange noises, but what he's talking about will not make sense. It will not be expressible in a way meaningfully to the listener, nor be anything more than blather or drivel to someone who has no idea about driving. Similarly he may be so bad at explaining it or simply not be able to express his competence in a set of propositions that answer the 'how' question. He might sit there and shrug his shoulders when asked how he did something. The essence of the type of knowledge he is trying to

---

[139] Tsai, Cheng-Hung. "The Metaepistemology of Knowing-How." *Phenomenology and the Cognitive Sciences* 10, no. 4 (2011): Pp 541-556.

[140] Tsai, *The Metaepistemology of Knowing-How*, 2011 Pg 543

[141] Stanley, Williamson, *Knowing How*, 2001. Pg 415; Ginet, Carl. *Knowledge, Perception and Memory* Boston: D. Reidel, 1975.

[142] See Berkeley, Istvan. "Gilbert Ryle and the Chinese Sceptic: Do Epistermologists Need to Know How To?" *The Electronic Journal of Analytic Philosophy*, no. 7 (2002): http://ejap.louisiana.edu/archives.html. Stanley and Williamson are Istvan Berkley's 'Propositionalists'. See also Scheffler, Israel. *Conditions of Knowledge*. Chicago: Scott, Foresman and Company, 1965.Pg 96-101. Scheffler would argue that assimilating how to that in this manner constitutes the very foundation of what Ryle refers to as the "Intellectualist Legend".

express may simply resist being made linguistic and simply can not be placed in propositional structures. In either case the race car driver might be able to clearly demonstrate competence on the track but lack the ability to state the knowledge he demonstrates with competence in a syntactically embedded structure of propositions corresponding to the how question. He still posses a wealth of knowledge about race car driving, but none of it is linguistic. This can lead to our pragmatic contradiction where someone might be able to do something and do it with both competence and confidence but when asked how they shrug their shoulders and simply can't explain it. Simpson picks up the examples that Stanley and Williamson use and questions their authenticity as examples of knowledge how[143].

Ryle thinks the distinction between knowledge-how and knowledge-that can be captured on a purely Linguistic Behavioural level in terms of how you might ask the question and whether that reveals a skill set, a capacity, a tendency, inclination, or a motive, but that the exercise of the knowledge itself, on either side can exceed that distinction itself. For Ryle the question can be linguistically framed, but the answer need not be.

A very careful study of Ryle can reveal this psychologistic pragmatic strain which occurs in the chapter on the intellect and I argue, settles the above debate.

Ryle writes

> On the other hand we have to allow that a person is doing genuine work in situations where no expression at all are being used, whether words, code-symbols, diagrams, or pictures. Tracing out the intricacies of a tangled skein of wool, studying the position of the game on the chess board, and trying

---

[143] In Stanley, Williamson, *Knowing How,* 2001. The question is raised as to whether winning a fair lottery is an example of knowledge how, or is buying a ticket the knowledge 'how' and the winning something that happens. Similarly, for digestion, is it Hanna who digests her food, or is 'Hanna' used as short hand for 'Hanna's stomach'. These types questions, I feel, are indicative of a much larger problem, and one that I argue could be solved on the basis of a first-person science of consciousness as David Chalmers points out in "How Can We Construct a Science of Consciousness." *The Cognitive Neurosciences* III, no. M.I.T. Press (2004). Pp 42 – 43. As such these type of arguments support the third thesis of this paper, of the three I laid out in the introduction. That is, they are indicative of the need for a theory of conscious.

to place a piece of a jig-saw puzzle would
usually be allowed to be cogitation, even
though un-accompanied by any self-colloquy[144]

.

Ryle, in contrast to Williamson and Stanley, bases the primacy of the non-linguistic 'how' on a couple of arguments. The most powerful argument Ryle offers on this point is the one above; that people just do not think in words, silent colloquy or propositions for some forms of knowledge-how. Elsewhere, he argues, some forms of 'knowing how' resist becoming linguistic, for example, the difficulty in explaining or describing the tying of a specific type of knot which because of the nature of the 'how' involved, resists being made linguistic[145]. Another is his claim that efficient practice precedes the theory of it

Efficient practice precedes the theory of it;
methodologies presuppose the application of

[144] Ryle, *Concept of Mind,* 1983. Pg 266. It's possible that this argument has been overlooked because it occurs in Chapter IX, while the original division between knowing-that and knowing-how occurs in Chapter II, and the linguistic behavioral configuration that divides capacities and tendencies into families of 'know-how' and 'know-that' on the basis of linguistic behaviors that is most quoted occurs in Chapter V. This of course is the argument I mentioned earlier. This is the argument where Ryle explicitly steps out of linguistic-thought structures for certain types of knowledge-how and capacity verbs. What he is describing here with the wool and the jigsaw puzzle are very much like the causal-coupling loop distinctions that Andy Clark and David Chalmers introduce in the extended mind hypothesis. See Andy Clark, David J. Chalmers. "The Extended Mind." In *The Extended Mind*, edited by Richard Menary, 27 - 41. Massachusetts: The M.I.T. Press, 2010. But taking this line of interpretation brings out other inconsistencies with Ryle's position on the 'Bogey of Mechanism' and Cartesian Causal problems that lead to dualism, and would take the scope of the argument too far afield at an early stage. Focusing on the Mind-body problem would only cloud and divert the scope of this paper before the completion of the argument. Therefore I only mention them in passing to show where they fit in. Also, to be clear I disagree with Ryle's point about chess. I should argue, in company with Sellars, Wittgenstein and Saussure that chess has its own language of moves and strategies and this is what one is thinking in when one plays it. Nonetheless I agree with Ryle that the other two are pretty solid examples. Also see Ryle's last letter to Daniel Dennett in Ryle, Gilbert. "Ryle's Letter to Daniel Dennett." *The Electronic Journal of Analytic Philosophy*, no. 7 (2002): http://ejap.louisiana.edu/archives.html, for Ryle's position on 'representation' and the section titled '7.1 The Game of Giving and Asking for Reasons' in 'Chapter Seven : Motives, Moods and Reasons' in this paper, where I discuss what I think Ryle's ordinary language position would be in relation to some of the strictures and limitations imposed by Brandom's form of representation.
[145] Ryle, *Concept of Mind,* 1983. See Pg 42, 44. Note Ryle's early use of 'heed 'here clashes with his use of 'heed concepts' on Pg 130. This clash parallels the two senses in which he uses 'habits', either as 'good habits' one learns, for instance while learning an instrument, and habits that can create 'commotions' such as when a smoker is on parade without matches. This clash disappears if we allow for a distinction between 'habit' and 'addiction' as we might now think of these two.

the methods, of the critical investigation of
which they are the products[146].

This, like the one before it is a Rylean argument, but like the one before
it, it is *not* a Linguistic Behaviourist argument. The former is
interesting in that it exceeds the limits of language, but also defines
those limits by which it exceeds language. This thought offers the
tantalizing possibility of a form of 'knowledge how' that precedes
linguistic expression. This is important, because, such an argument
could put Ryle inside the tradition of pre-Fregeian psychologism. Such
an argument, can demonstrate, that while Ryle may use Linguistic
Behaviourist arguments, there are parts of Ryle that exceed Linguistic
Behaviourist grounds on the basis of a Pre-Fregeian psychologism.
Specifically, it would count as evidence towards the pre-Fregeian
position that language about some forms of know-how, capacities and
skills was merely some sort of code for the thought involved in the
competence and exercise of those skills. The significance, for this part of
the thesis, is merely, that we have a pair of Rylean arguments about
skills, capacities and methods that exceed the domain of linguistic
behaviourist arguments.

That is to say Ryle advances arguments against an anti-
psychologistic position on the capacity side of the dispositions. Ryle
rejects anti-psychologism in the exercise of practical skills. Ryle does
not think a Linguistic Behavioural argument can capture everything
there is to know about the exercise, demonstration or application of
knowledge-how. Ryle thinks that when we engage in practical tasks
we're not always thinking in words or using language. There is a
knowledge-how that occurs prior to and can occur without language. So
while Ryle is anti-psychologstic about the mind, that is, he is anti-
psychologistic about the tendency dispositions, the propensities and the
occurrences, he is not anti-psychologistic about the capacity
dispositions. He divides the dispositions into tendencies and capacities
and he thinks that while the question of the capacities may be drawn
out with a Linguistic Behavioural argument and contrasted with
knowledge-that which reveals motives and beliefs, the way someone
answers knowledge-how need not be linguistically framed because there
are some activities and tasks in which the mind is not involved in

---

[146] Ryle, *Concept of Mind.*Pg 31.

forming either linguistic propositions, ordinary language discourse or linguistic structural knowledge. For instance one of these Ryle identifies as unwinding a skein of wool.

If we refer this back to what Dummett said about Twentieth Century philosophy of mind and paraphrase what he said about what Analytic Philosophers reject and apply this here we can see that talk about certain skills, capacities, techniques and methods is like "the use of a telephone: the speaker codes his thoughts in a transmissible medium, which is then decoded by the hearer"[147]. The decoding depends upon the hearer's own non-linguistic knowledge about those skills. Since people are performing tasks and in some of those tasks they are not thinking in words, then talk about those tasks requires people to encode their thoughts into words and sometimes they can't. We know the tasks do not depend upon their use of words of language because someone can be quite good at something without being very good at explaining how he or she is good at it, at all. He or she may simply not be able to explain how they are good at something, may lack the ability to form and linguistically frame propositions and may not even be able to show you. The might develop 'stage fright'. But when they do the task, exercise the skill or propagate their technique, they might still be rather good at it. As Dummett says 'the conception of language as a code requires that we ascribe concepts and thoughts to people independently of their knowledge of language[148]" which is of course the view that Dummett and analytic philosophy reject.

If some forms of knowledge 'how' are linguistically inexpressible, say, formulating propositions on how to tie a set of shoe laces then it becomes immediately possible to argue for pre-linguistic forms of expressible knowledge-how. The knowledge expressible is not expressible in words or propositions like Williamson and Stanley argue it must be, but in the exercise of competence. One can be entirely competent in a skill or set of capacities or techniques and not be able to tell you how they did something. They might be entirely incompetent at forming and expressing propositions about something or explaining it. Some people might be able to explain how they did something only after

[147] Dummett, *What Do I Know When I Know a Language?*, 1993

[148] Dummett, *What Do I Know When I Know a Language?*, 1993

thinking about it and finding words that might explain it. They might need to compound the thoughts containing their 'knowledge how' into language, using language like a code to convey their practical experiences and knowledge. All this is to say that from Ryle's position one can argue that the non-linguistic 'knowledge how' and the 'how' that precedes linguistically expressible form; both may share a body of knowledge. This will be knowledge which is uncomplicated by the implicit stipulations of Dummett's conditions for meaning, since such knowledge isn't spoken or linguistically structured prior to being encoded. By introducing two further steps; one, that pre-linguistic knowledge-how can contain meaning, and that such meaning can be expressed explicitly in the Dummett sense, as the exercise of competence, one could argue for a logical gap in Dummet's theory, and explore the possibility of finding a way out. In this way one could on defensive grounds argue for a return to pre-Fregeian psychologism at least in terms of a small slice of the capacities, skills and abilities. From here one could make an argument that a cogent reading of Ryle could place him, inadvertently, in this camp. But this would not be a Linguistic Behaviourist argument. Its domain is explicitly outside language and would depend upon some other domain of argumentation. But to raise a victory flag at this stage would be premature since we have only unpacked a strawman.

The goal of this paper is not to argue for non-linguistic meaning preceding and fundamental to talk about skills, methods and capacities, the goal is rather to attack the use of ordinary language arguments in the philosophy of mind by undermining the assumption that language is fundamental to the nature and character of mind. The way that attack is made is by not simply undermining the assumption that meaning in language bottoms out in implicit understanding, but that some non-linguistic knowledge is necessary for linguistic meaning in talk about the mind and specifically for this task I have chosen the emotions and Ryle's complex and detailed linguistic analyses for them. To do this we must be able to recognize the limits of a linguistic behavioural argument for this is what Ryle's concept of mind rests upon.

Let me draw another example where the argumentation exceeds Linguistic Behaviourist grounds. Alfred Jules Ayer in *The Problem of Knowledge* attempts to answer Ryle on the question 'how do you know?'

Ayer begins with a distinction. Ayer prefaces his answer by dividing the knowledge-how side of the knowledge-how/knowledge-that configuration into two forms; 'how do I know?', and 'how is it known?' and then collapses the two into a typical 'verifiantionist' criterion before he gets himself into typical 'verificationist' trouble with statements about the past.

A. J. Ayer writes

> (T)o ask how a statement is known to be true is to ask what grounds there are for accepting it. The question is satisfactorily answered if the grounds themselves are solid and if they provide the statements with adequate support. But here a distinction must be drawn between asking what grounds there are for accepting a given statement and asking what grounds a particular person actually has for accepting it. For example, if I am asked how I know the earth is round, I may reply by giving scientific evidence. . . . But the question may also be interpreted as asking not so much how this is known, as how *I* know it: and if I construe it in this way my reply will take a different form. I may mention some source from which I derived the information, some book I may have read or some person who has instructed me[149].

He then reduces both claims; 'how do I know' and 'how is it known' to the same claim

> Since nothing is known unless somebody knows it, there is a ground for saying that

[149]Ayer, A. J. *The Problem of Knowledge*. Middlesex: Penguin, 1956. Pg 70

the first type of answer to the question 'How do you know?' reduces to the second. Having justified a claim to knowledge by testing the scientific or historical evidence one may then be asked how these supporting statements are themselves warranted[150].

He then answers that question on page 71 by stating that in nearly all cases, the factual statement is reducible to confirmation by possible, if not actual experience.

Nearly always, it will be a matter of claiming that I should have certain experiences if I took the proper steps. It may be held even that these two claims are equivalent, on the ground that every statement of fact is ultimately reducible to statements about possible, if not actual experiences[151].

This of course leads him into problems with statements about the past on page 155[152].

The events which it was within my power to observe when I began to write this paragraph have already disappeared into the past. So the interpretation of the statements is the analysis of which a description of these events figured will have to be revised; the description of these events will have to be replaced by a description of whatever present or future events are

[150]Ayer, *The Problem of Knowledge*, 1956. Pg 70
[151]Ayer, *The Problem of Knowledge*, 1956. Pg 71
[152]Ayer, *The Problem of Knowledge*, 1956. Pg 155

regarded as evidence for them; and as they fall into the past, the revised version for them, will have to be revised again[153].

This leaves him with the problem that only future experiences can count as evidence for statements containing facts about past events. There is an immediate gap here, which is, that an ordinary language speaker might argue that *it doesn't make sense to talk about* past experiences in terms of future evidence. An ordinary language speaker might say that talking about the past in terms of what might happen in the future is entirely nonsense. I hope that the normative gap that opens up is now visible. This is important so I am going to move through it slowly.

That sort of claim, that only future experiences can count for statements about the past, and the one that it is based on; that all factual accounts of the world are based on or break down into a possible or actual experience are different sorts of claims from the strictly Linguistic Behavioural analyses like we have in Ryle.

That is to say, Ayer's argument does not appeal to the Linguistic Behaviourist criteria since it goes beyond ordinary language analysis to posit a criterion of truth. But – and this is also important for the other direction of the distinction I am making – Ayer's position doesn't border on psychologism, like the distorted version of Ryle's 'how' and 'that' argument, I sketched out earlier[154]because Ayer is still talking about meaning in Dummett's post-Fregean sense.

In contrast to Ayer's criteria of possible meaning, the family of capacities, skills and methods that Ryle locates on the 'how' side of the 'how-that-why' axis, as a set of distinctions, is different to interrogatives governing the 'why' and the 'that' set of clauses, which demarcate the family of dispositions that Ryle, of course, picks out, together under the family he classes as motives, beliefs and inclinations.

The reason is, to summarize the point, the types of analysis made by an Ordinary Language argument are forwarded on the basis of an analysis into ordinary language structures not truth conditions. Truth conditions in Ayer's sense of possible experience are extra-

---

[153]Ayer, *The Problem of Knowledge,* 1956. Pg 155
[154]See earlier in this section.

linguistic factors beyond the linguistic range necessary to undertake a Linguistic Behaviourist analysis of a piece of natural language. This distinction is important for the way in which one can make an 'Ordinary Language claim'.

It is to this we now turn.

3.3.3 Knowledge-How and making an Ordinary Language Claim.

The marker of the distinction between Linguistic Behavioural claims and Ordinary Language arguments is in the normative phrase *'it makes sense to say'*. Where such a normative phrase enters into a line of thought-jargon, slang, exposition, description, argument, justification, or as support for a claim, it changes the Linguistic Behaviourist's description into an Ordinary Language argument. For it maps the description to a domain of competency in Ordinary Language use.

In the vocabulary of the old Semiotics the phrase *'it makes sense to say'* maps a langue to a parole, except that here that order is reversed. Ryle's Linguistic Behavioural Analysis is a 'parole' mapped by an ordinary language 'justification' to the langue[155]. In the language of the semioticians the parole was a given instance of language manifestation, say for instance a speech utterance, while the langue was the body housing the various rules, conventions, syntax and the body from which various other important distinctions could be made like the diachronic and synchronic axis[156]. The diachronic and

---

[155]See for instance Barthes, Roland. *Elements of Semiology*. Translated by Annette Lavers and Colin Smith. New York: Hill and Wang, 1984. Pp 54 – 57. Levi-Strauss, Claude. *Structural Anthropology*. Translated by Claire Jacobson, Brooke Grundfest Schoepf. 1963: Basic Books. United States of America. Pp 2-4, 21. 89, 291. Levi-Strauss, Claude. *Totemism*. Translated by Rodney Needham. London. Merlin Press. 1962. Pp 49 – 55. Jakobson, Roman. *The Science of Language*. London. George Allen & Unwin Ltd. 1970. Yule, George. *The Study of Language*. Cambridge University Press. Cambridge. 1985. Pp 222 – 225. Derrida, Jacques. *Writing and Difference*. Translated by Alan Bass. University of Chicago Press. Chicago. 1978. Pp 278 – 300. Hawkes, Terence. *Structuralism and Semiotics*. Suffolk: Routledge, 1988. Pp 19 – 26. NB I am using Semiotics because it gives us another angle to approach the normative force in Ryle's arguments. Since the distinctions between langue and parole rely upon the speaker's knowledge of the language, they allow us to target the normative force in Ryle's arguments with a closer approximation to what makes the arguments work than some of the analytic schools that have interpreted Ryle's arguments in a more literal sense.

[156]Hawkes, *Structuralism and Semiotics*, 1988. Pp 18 – 20, Barthes, *Elements of Semiology*, 1984. Pp 54 – 57. The Linguistic Behavioural argument would analyse what semioticians refer to as a 'parole' to make a claim about the mind. The Ordinary Language justification for a claim about the mind

synchronic axis is one set of distinctions that takes place in the langue. The diachronic refers to the axis of language which is concerned with its historical dimensions while the synchronic is concerned with its current structural properties. While the parole as a manifestation of a langue may embody synchronic structural elements that may be either paradigmatic or associative[157] these features themselves will be determined by the distinction belonging to the relationships inside the langue itself.

On this view the langue, as such, contains all of the social conventions in use by the ordinary language speaker.

Saussure writes

> (The language is) both a social product of the faculty of speech and a collection of necessary conventions that have been adopted by a social body to permit individuals to exercise that faculty[158].

Thus the distinction is between the body of knowledge of social conventions and uses in ordinary language and a given concrete instance of manifestation. Hawkes puts the distinction between the langue and the parole in this way

> The distinction between langue and parole is more or less that which pertains between the abstract language-system which in English we call simply 'language', and the individual utterances made by the speakers of the language in concrete everyday situations which we call 'speech'[159]. . . The nature of the langue lies beyond, and determines, the nature of each manifestation of parole, yet it has no concrete

would however draw on the body of knowledge the natural language speaker posses of what semioticians would refer to as the 'lingua'.

[157]Barthes, *Elements of Semiology*, 1984. Pg 58 - 62

[158]Saussure, Ferdinand De. *Course in General Linguistics*, Edited by McGaw Hill. New York, 1966. Pp 9

[159]Hawkes, *Structuralism and Semiotics*, 1988. Pg 20, these are analogous to bits of 'ordinary language' which Ryle analyzes.

existence of its own, except the piecemeal manifestations that speech affords. . . (The) *Parole*, it follows, is the small part of the iceberg that appears above the water. *Langue* is the larger mass that supports it, and is implied by it, both in speaker and hearer, but which never itself appears.[160]

Ryle's Linguistic Behavioural arguments are analogous to the semioticans' parole, because they are concerned with bits of linguistic behaviour, in a way analogous to how the semioticans might break up concrete manifestations of language in order to make a distinction. Ryle's Ordinary Language arguments are analogous to the body of the accepted conventions in the langue except that Ryle's sophistications include normative values and philosophical arguments to support the Ordinary Language body as a source of authority for upholding his own arguments. Ryle's sophistications include appeals to a natural language user's intuitions and parapets to defend against as well as engage and natural language siege engines to challenge the claims of other philosophers. The two; the Linguistic Behavioural claim and the normative authority he claims is in the Ordinary Language argument, come together, when Ryle uses an Ordinary Language argument as a base for a Linguistic Behavioural analysis to either of these purposes.

The way the normative move between a Linguistic Behavioural analysis and an Ordinary Language argument occurs is when the analysis calls upon a 'it makes sense to say' *pivot* to hold up a behavioural description of a bit of linguistic phenomenon. The normative claim enters into that analyses of the linguistic phenomenon and thus shifts the linguistic phenomenon of the argument. It enters into the description at the point of seeking a justification that relies on the users' knowledge of that domain to justify and not simply describe a claim.

This is different from what I will call an 'Ordinary Language statement'. An 'Ordinary Language statement' is a statement or an argument supporting a statement about the authority of Ordinary Language as an authoritative domain for making arguments. An *Ordinary Language argument* is identified by an *'it makes sense to say'*

---

[160]Hawkes, *Structuralism and Semiotics*, 1988. Pp 20 – 21.

or a *'it doesn't make sense to say'* claim, or their equivalent, which engages the natural language speakers' intuitions. An *Ordinary Language statement*, however, is a statement purporting or claiming that Ordinary Language is a good and authoritative source for making claims. The distinction is important for the clash between the normative value of Ordinary Language arguments, and other rival sources like phenomenology, which this paper is concerned with[161]or a claim like the normative value of science and whether it is continuous with ordinary language, like we'll look at in the next section with Wilfrid Sellars, or even claims made on the basis of some theory of logic, like I'll briefly touch on in the appendix[162]. An 'Ordinary Language statement' so technically defined is one that would support Ordinary Language arguments or the normative force they are supposed to appeal to as a good, or better source than another source like those I listed above. I'm making the distinction on technical grounds for the purposes of clarity in this paper.

This gradient of a fine line distinction between Ordinary Language arguments and Linguistic Behavioural claims is important to observe. The reason is that the difference between a *justification for* and a *description of* a linguistic distinction is directly related to any further claims that support arguments about whether the behaviour of the mind in the world is embodied in some linguistic phenomena and thus can be found from a descriptive analysis of language. Such claims are either, (i) subsequent to the philosophical concern over whether languages either vary in their use, or are similar in some philosophically relevant, or interesting way, or (ii) whether such an assumption about the mind's behaviour being embodied in the world is in the form of language, and whether this can be discovered by analysis,

---

[161] See the section titled 'Counterfactual Logic, Linguistic Behaviorism and Ordinary language arguments' in this paper.

[162] For David Lewis's argument on the rivalry between mistaken and invalid usage and the English Subjunctive conditional and his arguments based on counterfactual logic for the way truth behaves see the section titled 'Counterfactual Logic, Linguistic Behaviorism and Ordinary language arguments' in the appendix to this paper. I have included his example at length because rival sources can exert a pull on us and an analogical case can demonstrate this. It also shows us a comparative case where the normative force in language that Ryle is drawing on to make his arguments is challenged by a different rival source. In this case the clash between normative forces that arises is one that originates in the linguistic behaviour of ordinary language and the logical structure of a mathematical theory involving truth conditions.

prior to any such concern over the way languages in their behaviour vary, in philosophically relevant or interesting ways[163].

That is to say, if the solution to the mind-body problem can be found in the particular way using verbs of one type, and verbs of another type interact, then two questions arise. Firstly (a), supposing such a solution was to be found in English dispositional terms, governed by capacity and tendency verbs, then do English capacity and tendency verbs actually behave differently to the French and Maori equivalents of capacity and tendency verbs, in any philosophically interesting and relevant ways? The upshot, of course, is then, that any solution or therapy for the mind-body problem and its inter-related cluster of symptoms, that can be derived from the English verbs, (or similar linguistic phenomenon) can be translated, either directly, or indirectly into French or Maori, and whether it is justifiable in that language in the same, or at least analogous ways, by the ordinary language knowledge possessed by an ordinary language user. Say, for instance old mate Reg from up the road and the conventions used by his equivalent *bon ami* Pierre from Villers-Bretonneux, over in France.

This is why Ordinary Language arguments that justify their claims on the basis of an average language user's knowledge are of their own accord, philosophically interesting claims, since they point to pockets of knowledge, commonplace in that language which raise questions about whether it could make sense, outside of that language and in another. This is important for Sellars position and the link between ordinary language and science, for instance, and the claim the two are continuous, and the mind body problem in general. If Ryle is right a decontaminated domain of ordinary language should reveal the nature of mind, on implicit grounds. If Sellars is right there is a lot of social baggage in ordinary language and what supports its authority are tied into social and standard conditions and fact stating roles. Both of these would be ordinary language statements, or statements that support why we should think of ordinary language as a normative source.

---

[163] For instance, the sorts of concerns with deep grammar raised by Katz, Fodor. "The Structure of a Semantic Theory." *Language* 39, (1963): Pp 170-210. Or whether the sorts of deep structures that Lakoff and Johnson argue are in English metaphors are mirrored in other languages or differ in philosophically interesting ways. G. Lakoff, M. Johnson. *Metaphors We Live By*. Chicago: University of Chicago Press, 1980.

To short order the above concern in a few words by an analogous case, suppose we found the solution to unifying the strong nuclear force with the electromagnetic force, in a Chinese fable, that derived its validity from an appeal to ancient wisdom contained in Taoist scriptures. Could those Taoist scriptures be translated in to the language of modern scientific German? Replace the strong and weak nuclear and electromagnetic forces, with the mind-body problem, the Chinese fable, with an Ordinary Language argument, and the appeal to the I'Ching, with fish and chips shop English, and you have the above concern with appeals to the domain of Ordinary Language, as a court of appeal, for justification, and whether such knowledge, or 'local pub lore' would in fact, *be* justifiable in another language.

From that, we get (b) which concerns itself as to what exactly upholds the claim about the solution? Is it, merely, the English users knowledge of the English language or is it something about that language contained in the normative description, implied by the act of appeal by justification itself? The two are not quite the same thing. There is a problem with self congratulatory justifications, or justifications that justify Ordinary Language arguments, on the basis of the user's own competence, not how well the justification either upholds arguments, or props up assumptions necessary to posit that the behaviour of the mind in the world is embodied in some linguistic phenomena. Might not it be the language user's knowledge of the domain, patting itself on the back, by justifying that use? Think about a boy who can justify that boys know more than girls, by saying to a girl 'did you know boys know more than girls?' the girl replying 'no, I didn't. How do you know that?' and the boy responding 'see, I just proved it'. Might not it be, that the language user knows how to talk about the mind's behaviour embodied in these linguistic phenomena, in one tone of voice, as Ryle might say, and then justify that claim by his linguistic competence in another tone of voice?

This is like Dummet's argument, except here the issue is not an attempt at an explicit theory of meaning that pragmatically must bottom out in a contradiction or circularity on implicit usage, but the normative justification that upholds a bit of linguistic behavioural competence, the appeal to that competence, and the claim that adjoins that competence that the solution reveals something important about language and the way people use it, namely, in Ryle's case, the behaviour of the mind in the world; the mind and body problem.

Let me put it this way. If we say, for instance, that being linguistically competent is knowing how to use tendency and capacity verbs, then the 'back' is to make a Linguist Behaviourist's analyses of the behaviour of those verbs from competency in the use of those verbs, and to say that this is the right account, on the basis of a re-appeal to that competency, is the pat. The pat goes on the back. Might it not be the case in generic readings of Ryle that all one is doing is making an analysis from competence of the way we use words, then building a hypotheses about the mind in the world, from that competence and then appealing to that competence for justification that these are the right hypotheses? This is why any argument that doesn't challenge this kind of reasoning rests on an assumption which you might call "the three fold assumption of Ordinary Language arguments". This contains (i) Behavioural Linguistic claims, (ii) the justification or Ordinary Language argument, and (iii) the user's own domain of competence propping up all three by linking (i) and (ii) and if that's all that there is to Ryle, then this line of thought is philosophically boring, and fruitless, in the sense of the bloke who says he's right on the basis that he knows he is right, and knows he is right, because he says it.

How else might a language user *justify* their claims, towards *justifying* arguments by appeal to a common domain? Since this is Ordinary Language argumentation certain constraints must be observed. In particular, we must not allow contamination by highly theoretical language. This, the justification step of ordinary language arguments may not appeal to an inter-linguistic, or supra-linguistic domain of language, like logic, a Platonic realm of queer objects, in a linguistic version of J.L. Mackie's Platonic moral objects[164], or a master language, in the old European sense in which the vernacular Romance Languages were subsumed under an appeal to Latin when stepping out of their vernacular domains, or on finer points of Continental Grammar?

What options do we have left, without an appeal to the language itself, in a self congratulatory way?

How else might it be done?

---

[164] See Mackie, J. L. *Ethics, Inventing Right and Wrong*. Victoria: Penguin, 1986.Pg 40

How else might an Ordinary Language argument justify a Linguistic Analyses without an appeal to competence in the domain of the Ordinary Language user?

Can it?

Can an Ordinary Language argument justify itself and thus justify the Linguistic Analyses, without referring itself to its own justification the domain of Ordinary Language?

It seems it can't.

It has nothing to justify its own claims, except itself.

It seems we are at the edges of Ryle's map, where Cartographers pencil in sea monsters and mermaids, and if asked, how they know they are really there, assert it is because they've drawn them in, and if asked why they drew them in, answer because they know they are there. Indeed this paper will show that there are no sea monsters on the edge of the map, and will undermine the authority of Ordinary Language arguments. But there are stages that must be systematically laid out before that can occur.

3.4 Linguistic Behaviourism, Ordinary Language arguments and the normative values of Science

In the section immediately prior I mentioned that I would go into some of the reasons for supporting the normative value of Ordinary Language statements. Ordinary Language statements are claims that support Ordinary Language justification for the normative value in a piece of Linguistic Behavioural analyses. I do this for two reasons. Firstly they are an interesting type of argument. They contain an appeal to what the man in the street might know, and it is a philosophically interesting question to ask what that rests on. The second reason is, briefly, sometimes there is a confusion or an obscurity between what Ryle actually said, and the way people have read him. Weed is a very good example of this confusion.

Indeed, and in relation to the first reason, one of the most important debates to arise last century is that between Sellars and Ryle on what supports science and ordinary language as positions one might take in regards to world views. This is different to the prior section where we were looking at the relationship between ordinary language and Linguistic Behaviourism and how Ordinary Language arguments might justify a Linguistic Behavioural claim. This prior section is important because later in the thesis I will draw attention to the normative force behind a Linguistic Behavioural argument and compare it with another type of argument Ryle uses, which I have yet to fully reveal, but will in due course be developed as the phenomenological strain in Ryle. I will then argue that Ordinary Language arguments are a bad source of arguments because they allow contradictory claims to arise from rival sources of analyses.

In this next section of the paper we are looking at arguments that connect Linguistic Behaviourist arguments with two conceptions of Ordinary Language. One conception, Ryle's, sees science as discontinuous with ordinary language and a peninsular off shoot. The contrary view is Wilfrid Sellars who thinks that science is continuous with processes in ordinary language in the pre-scientific stage. Sellars presents a different sort of Ordinary Language argument, one that is continuous with science and part of a story between the 'Manifest Image' and the 'Scientific Image' and the way that imperceptible objects can become visible through introduction into discourse and a 'space of reasons'.

Ryle's view, as we shall see, is that ordinary language is a generalized field for doing philosophy of mind in, and should exclude science and other idiosyncratic and specialized disciplines, as they result in myths, which of course are the facts of one category presented in the idioms appropriate to another. Ryle thinks using science and scientific paradigms like the mechanical world view, muddles the clarity of ordinary language where people already know how to talk about the mind and its conduct in the world, and all we have to do as philosophers, is to draw out this implicit understanding and correlate or demonstrate how the terms fit together. Ryle doesn't want to deny the facts rather he wants to reallocate them. Ryle's analysis of Descartes is concerned about the mind and the way mechanical theories of motion have been applied to models of human causation[165] or Ryle's reading of Hume's 'causal theory of consciousness' are examples of this process[166].

---

[165]Specifically, Descartes, Rene. "Discourse on the Method of Rightly Conducting the Reason and

Sellars as we shall also see has a unique view of the value and semantic meaning in ordinary language discourse. He thinks it is the product of pre-scientific stages that flower into science and failure to understand this leads to failure to understand the basic terms and logic of the forms inside ordinary language usage. For Sellars understanding ordinary language is part of this process critical for the large scale picture of science. Ordinary Language gets its authority or normative value from its associations with science as part of this process. In the Sellarian picture Linguistic Behavioural arguments get their authority by the process of continuation between science and ordinary language, and this is what he does in *Empiricism and The Philosophy of Mind* by unlocking the logic behind the behaviour of "looks talk". This is also important because Sellars presents a viable alternative to Ryle, one that I will consider later in the paper when I introduce the autophenomenology thesis[167].

Seeking for Truth in the Sciences." In *The Philosophical Works of Descartes*, I. London: Cambridge University Press, 1979. Translated by Elizabeth S. Haldane, G. R. T. Ross. See Part V, Pg 109 – 117, where Descartes first realizes the problem with a mechanical world view and causation in human actions and motions. Ryle, *Concept of Mind,* 1983. Pp 14 – 25 Also see pg 79 for the particular passage which parallels Descartes problem in *The Discourse*. Ryle's passage reads "Men are not machines, not even ghost-ridden machines. They are men a tautology which is sometimes worth remembering. People often pose such questions as 'How does my mind get my hand to make the required movements?' and even "What makes my hand do what my mind tells it to do?' Questions of these patterns are properly asked of certain chain-processes. The question 'What makes the bullet fly out of the barrel?' is properly answered by 'The expansion of gases in the cartridge'; the question 'What makes the cartridge explode?' is answered by reference to the percussion of the detonator; and the question 'How does my squeezing the trigger make the pin strike the detonator ?' is answered by describing the mechanism of springs, levers and catches between the trigger and the pin. So when it is asked 'How does my mind get my finger to squeeze the trigger?' the form of the question presupposes that a further chain-process is involved, embodying still earlier tensions, releases and discharges, though this time 'mental' ones. But whatever is the act or operation adduced as the first step of this postulated chain-process, the performance of it has to be described in just the same way as in ordinary life we describe the squeezing of the trigger by the marksman."
[166]Ryle, *Concept of Mind,* 1983. Pp 232 – 237, 240.

3.4.1 Ordinary Language, the Manifest Image and the Philosophy of Mind.

Allow me to begin by pointing out something that seems to plague much of the contemporary Rylean literature[168] and we can swiftly move past it. This is often that what Ryle actually wrote, gets confused with how people have read him, and how people have read him and the impact that has had on the modern history of twentieth century philosophy leads to confusions about what Ryle's actual position was. We shall only spend a few words on it before we move into Sellars and Ryle's positions on ordinary language.

Weed writes

> Head-scratching is objectively observable. Incestuous desire is not; nor is universal doubt, apprehension of infinity, or Cartesian introspection. Philosophers like Carl Hemple and Gilbert Ryle shared the view that all genuine problems are scientific problems[169].

I just don't think that was Ryle's view. Ryle thought that philosophical muddles about the mind actually originated in science. His view was that the complex theoretical languages of sciences create confusions when we try to extend them to the mind. The way to solve this was to elevate the status of ordinary language and examine the behaviour of the language we use in our every day talk about the mind.

Ryle himself in *The Concept of Mind* writes

> Whenever a new science achieves its first big successes, its enthusiastic

---

[167] See 'The Midway Map' of this paper and 'Motives, Moods and Reasons' and its subsections.

[169]Weed, Laura. "Philosophy of Mind an Overview." *Philosophy Now* Nov/Dec, no. 87 (2011). Pg 6. Weed is, perhaps some might think, rather aptly named for this type of error.

acolytes always fancy that all questions
are now soluble by extension of its
methods of solving its questions. At one
time theorists imagined that the whole
world was nothing more than a complex
of geometrical figures, at another that
the whole world was describable and
explicable in the propositions of pure
arithmetic. Chemical, electrical,
Darwinian and Freudian cosmogonies
have also enjoyed their bright but brief
days. 'At long last', the zealots always
say, 'we can give, or at least indicate, a
solution of all difficulties and one which
is unquestionably a scientific solution' [170]

.

Indeed, 'At long last', the zealots always say, 'we can give, or at least
indicate, a solution of all difficulties and one which is unquestionably a
scientific solution' is what Ryle wrote which is contrary to Weed's
assertion. Indeed Ryle thinks that the extension of a scientific method
by zealots who go overboard is what causes the problem. Ryle's
argument here is that such fanaticism by scientific zealots
contaminates the domain of ordinary language with philosophical
muddles and category mistakes which he is at pains to clear up and
repair. That's one of the central theses, indeed I should very well argue,
the central thesis, running through *The Concept of Mind*. One example
of such a muddle is the mechanical theory of mind. The mechanical
theory of mind has two parts. Volitions form one side and thus make up
one of those parts.

Ryle writes

The physical sciences launched by
Copernicus, Galileo, Newton and Boyle
secured a longer and a stronger hold

[170]Ryle, *Concept of Mind,* 1983 Pg 74

upon the cosmogony- builders than did either their forerunners or their successors. People still tend to treat laws of Mechanics not merely as the ideal type of scientific laws, but as, in some sense, the ultimate laws of Nature. They tend to hope or fear that biological, psychological and sociological laws will one day be 'reduced' to mechanical laws though it is left unclear what sort of a transaction this 'reduction' would be[171].

The physical sciences, Ryle thinks, cause contaminations which result in 'myths' like the 'para-mechanical theory of mind'.

Ryle writes

Volitions have been postulated as special acts, or operations, 'in the mind', by means of which a mind gets its ideas translated into facts. I think of some state of affairs which I wish to come into existence in the physical world, but, as my thinking and wishing are unexecutive, they require the mediation of a further executive mental process. So I

---

[171]Ryle, *Concept of Mind,* 1984. Pg 74 This is the reason I avoided using Aizona and Adams paper. I should argue that the 'mark of the cognitive' is a special application of what's at the base of the para-mechanical theory of mind since it individuates on the basis of causal factors. See Adams, Fred. Aizowa, Ken. "Defending the Bounds of Cognition." In *The Extended Mind*, edited by Richard Menary, Pp 67-80. Massachusetts: The M.I.T. Press, 2010. Pg 40, and pg 70, since individuation for the 'mark of the cognitive' depends on intrinsic representations with non-derived content, then cognition on this view would be little more than the origin of causal processing and representation. Menary clarifies this point. See the "Introduction." In *The Extended Mind*, edited by Richard Menary, Pp 1 - 25. Massachusetts: The M.I.T. Press, 2010. Pp 18-19. This position that Adams and Aizowa put forward on individuating cognition on a causal basis clashes with Ryle's Anti-Cartesian position on the problem of causation which Ryle argues leads either to "The Two World Myth", or the "Bogey of Mechanism". Specifically, Adams and Aizowa's position would be subject to classification of that passage on pg 79, Ryle, *Concept of Mind,* 1984, which was pointed out a few footnotes ago, where Ryle claims men are not machines. This, I argue, would lead Ryle to the position of viewing Adams and Aizowa's position as just another development on the para-mechanical theory of mind, with its faults, since Adams and Aizowa seek to individuate cognition on the basis of causation.

perform a volition which somehow puts my muscles into action. Only when a bodily movement has issued from such a volition can I merit praise or blame for what my hand or tongue has done[172].

And indeed, I'll demonstrate in the next section, how Ryle goes on to argue against 'volitions' as one of my examples of an Ordinary Language negation that makes its move by denying another philosopher's arguments on the authority of an Ordinary Language claim. But first, it is probably worth pointing out that the para-mechanical theory of mind has a right wing and a left wing. On the left wing we have things like the passions and the causal theory of consciousness, both espoused by David Hume. On the right wing of the para-mechaincal theory of mind, Ryle thinks, we have theories like the volitions, originally espoused by the Stoics and Saint Augustine.

It is instructive to place Ryle's approach beside Sellars view of the relation between 'Science' and common sense. Wilfrid Sellars argues from the other end of the spectrum that science is the 'measure of all things[173]'. This is because Sellars thinks science is continuous with processes already at work in ordinary language communities it just comes to eventually replace them.

Sellars writes

My point is rather that what we call the scientific enterprise is the flowering of a dimension of discourse which already exists in what historians call the "prescientific stage," and that failure to understand this type of discourse "writ large" -- in science -- may lead, indeed has often led to a failure to appreciate its role in "ordinary usage," and, as a result, to a failure to understand the

---

[172]Ryle, *Concept of Mind*, 1984. Pg 62
[173]See Wilfrid Sellars' *Empiricism & the Philosophy of Mind*. United States: President & Fellows of Harvard University, 1997. Pg 83.

full logic of even the most fundamental, the "simpluist" empirical terms[174].

This Sellars thinks is because [175]

> scientific discourse is but a continuation of a dimension of discourse which has been present in human discourse from the very beginning, then one would expect there to be a sense in which the scientific picture of the world *replaces* the common-sense picture; a sense in which the scientific account of "what there is" *supersedes* the descriptive ontology of everyday life.

The position that Ryle introduces in *The Concept of Mind,* is not, of course, the position he maintains in all of his writing. In *On Thinking,* his philosophy is implicitly in the domain of Ordinary Language, and his analyses for the most part are confined to Linguistic Behavioural descriptions. In *Dilemmas,* for instance, Ryle argues the view that science is just another perspective of the world, a 'peninsular offshoot' as Sellars might describe it. In *Dilemmas,* Ryle thinks the 'world of science' is rather like 'the world of poultry' or 'the world of entertainment'.

---

[174]Sellars, *Empiricism & the Philosophy of Mind,* 1997. Pg 80.
[175] In Sellars, *Philosophy and the Scientific Image of Man,* 1991, Pg 24 the scientific image starts off as a methodological development of the manifest image, see pg 20. What characterizes the Scientific Image is its use of imperceptibles, pp 18-19. However, as Sellars later points out, in "The Language of Theories." In *Science, Perception and Reality,* Pp 106 - 126. California: Ridgeview, 1991. Pg 120, if our theory is a good one (kinetic theory) we are entitled to say that the entities (molecules) exist. There is a genuine rivalry. However, as Sellars points out at the end of *Empiricism & the Philosophy of Mind,* 1997, pg 113, to ask how impressions fit together with magnetic fields is mistaken. This is because impressions themselves are theoretical entities that we come to perceive, pg 111, 115. I will bring this move out more in my discussion through out the paper.

Ryle writes

> We know that a lot of people are interested
> in poultry and would not be surprised to
> find in existence a periodical called 'the
> Poultry world'. . . It is quite innocuous to
> speak of the of the physicists world, if we do
> so in the way we speak of the poultry
> keepers world or the entertainment world.
> We could, correctingly speak of the
> bacteriologists world and the marine
> biologists world[176].

He goes on to support this view of science as a collective noun unifying all of the family resemblance concepts associated with science with two analogies. The first analogy is that of a deck of cards. Ryle argues we may view, for instance, the house of hearts in the highly technical vocabularies of either Bridge or Poker, but this doesn't privilege Bridge, as a more truthful representation of, say, the Queen of Hearts, than that which would place it as a Poker schema in a Royal Flush[177]. The Queen of Hearts is the domain of Ordinary Language, while the Poker and Bridge interpretations are technical vocabularies, analogous to the special sciences and chicken farming.

The second is an analogy between an undergraduate and an accounts keeper. The accounts keeper has one set of facts that obtain about the books in the library, that is, how much they cost, how old they are, and where they came from. The undergraduate has another set, about what he's read in them and how useful they are to his course of study. Ryle's point is that neither view is reducible to the other. Certainly not 'that all genuine problems are scientific problems' as Weed would have it.

Likewise, Sellars has an additional account in *Philosophy and the Scientific Image of Man* with strains that, rather interestingly, argue the contrary to Ryle's position in *Dilemmas*. These strains emerge

---

[176]*Dilemmas*. New York: Press Syndicate of the University of Cambridge, 1987. Pg 72 - 73
[177]Ryle, *Dilemmas*, 1987. Pg 86 - 87

from two earlier papers *Truth and 'Correspondence'*, and *The Language of Theories* as part of a refutation of Carnap and a larger riddle that Sellars is trying to systematically resolve. This riddle is most explicitly expressed in *Philosophy and the Scientific Image of Man* where Sellars characterizes the everyday world as the 'Manifest Image'. The Manifest Image contains, most notably, the Empirical Image. The Empirical Image contains 'all of Mill's inductive canons', Hume's philosophy, and is the grounds from which develops Sellars' classification of Substantive Dualism[178]. The Empirical Image, as part of the Manifest Image, is the field in which science is represented as the truncated domain of persons[179]

Alongside the empirical image, making up the Manifest Image, is the original or primal category, in which all objects are treated as persons. The wind for instance is treated as possessing a personality. It is characterized by its cheekiness[180] while the totality of the world, form the primal category of the one is what Sellars categorizes as 'The Perennial philosophy'[181]. Sellars, from the position of this paper, and those other earlier ones I identified would maintain, that the Scientific Image, while it methodologically feeds on the Manifest Image[182] may very well be like the world of poultry, a peninsular off-shoot. But Sellars would also argue, countering Ryle of *the Dilemmas*, that when the Scientific Image is strong enough it becomes a genuine rival to the Manifest Image[183].

How Ryle responds to this depends on how you read him. A misread, I argue, would be simply to say that he would affirm the ordinary language world view as having a stronger authority, either because it is bound up in our everyday experiences or because rejecting it would result in devastating consequences to our world view like this argument out of Fodor

Fodor writes

---

[178]Sellars, *Philosophy and the Scientific Image of Man*, 1991. Pg 11, what Wilfrid Sellars calls substantive dualism develops from depersonalization of the a) empirical b) categorical, form of the Original Image.

[179]Sellars, *Philosophy and the Scientific Image of Man*, 1991. Pg 12 - 13

[180]Sellars, *Philosophy and the Scientific Image of Man*, 1991. Pg 15 - 18

[181]Sellars, *Philosophy and the Scientific Image of Man*, 1991. Pg 20

[182]Sellars, *Philosophy and the Scientific Image of Man*, 1991. Pg 20

[183] See O'Shea, James R. *Wilfrid Sellars* Key Contemporary Thinkers. Cambridge: Polity Press, 2007. Pp 41-47

> (I)f commonsense intentional psychology
> really were to collapse, that would be,
> beyond comparison, the greatest intellectual
> catastrophe in the history of our species. If
> we're that wrong about the mind, then
> that's the wrongest we've ever been about
> anything. The collapse of the supernatural,
> for example, didn't compare; theism never
> came close to being as intimately involved
> in our thought and our practice – especially
> our practice – as belief/desire explanation
> is. Nothing, except, perhaps, our
> commonsense physics – our intuitive
> commitment to a world of observer-
> independent explanation does. We'll be in
> deep, deep trouble if we have to give it up[184].

Fodor thinks that giving up our ordinary language common sense intentional psychology would be a disaster. His is an either/or picture based on the ethical consequences of abandoning common sense ordinary language psychology. Ryle simply thinks science muddles the picture by doing work where it shouldn't and by people trying to force the mind into scientific idioms. Putting Ryle in the same either/or category would be a mistake. Trying to force scientific paradigms to regulate mental structures is inappropriate, Ryle thinks, in the way

---

[184]Fodor, Jerry A. *Psychosemantics, the Problem of Meaning in the Philosophy of Mind.*
Massachusetts: M.I.T. Press, 1987. Pg xii. I think this is as dubious an argument as Fodor's reasons for making it. For instance, he needs a commonsense picture of the world in order to deliver the propositional attitudes on page 10, in order to individuate them and get around 'meaning holism' on page 59[184]. Indeed the arguments I'm going to present next chapter on Brandom's intentionalistic account, mutatis mutandis, I might apply to Fodor's propositional utterances attached to the attitudes[184]. But what's interesting is the implicit normative appeal of the argument Fodor makes. The normative appeal the argument makes is directed towards a world view is directly involved with our own personal experiences. This type of appeal appears strongest when it is grounded in either an intentionalistic view like Brandom, Peters and Fodor canvas, or in the form of a Rylean argument that claims an Ordinary Language basis for a distinction, but lacks the currency of a Linguistic Behavioural analysis to pin it on. Arguing for the consequences of a branch of knowledge or what certain knowledge might do, I should argue, is an argument one might make but its proper domain is in ethics.

that trying to force ballroom dancing paradigms onto woodwork is also inappropriate. It results in the facts of one field being reallocated in the idioms of another. What is appropriate is whether ordinary language talk about the mind can teach us about how the mind interacts with the world and what that rests upon is the question of whether or not implicit knowledge of meaning in ordinary language prior to and fundamental for any explicit theory of meaning.

Consider the disagreement like this; if Sellars is right then ordinary language is part of a process continuous with and moving towards the Scientific Image and not a confusion of the facts of one field with the idioms of another. If Ryle is right ordinary language comes first because it contains implicit knowledge of meaning that is fundamental to the nature of the mind, and precedes, and perhaps even eschews any explicit theory of meaning.

Now, of course, there is this other part of Ryle that sits outside his Linguistic Behaviourist arguments, which claims that non-linguistic knowledge is prior to knowledge of linguistic meaning, which is part of the knowledge-how of the knowledge-how knowledge-that configuration. The position I'm going to be arguing for is that the fundamental nature of the mind, the knowledge-that side, contains non-linguistic knowledge that is essential to a theory of meaning dealing with the nature of the mind. I'm going to base that argument on an insight, and an argument that arises from the classification of Ryle's arguments and a tension between them. It is to this classification that we now turn.

3.5 Linguistic Behaviourist Claims,the Occult Stream of Consciousness. Propensities, Occurrences and the Agitational Calculus.

For Ryle, expressions for occurrences, as different to propensities, contain an interesting sub-species of linguistic expression that I think can further our purposes towards good philosophical practice, by illuminating this difference between Behavioural Linguistic claims from those of Ordinary Language arguments. This subspecies of occurrences is designated by Ryle as 'feelings proper' and can be isolated by their unique linguistic structure. They will be important latter in the this paper[185]. But first, we need to establish grounds for the Behaviourist and Ordinary Language strains in Ryle, how to distinguish between them, and why they are an interesting species of argument. Then we will be able to discern the Phenomenological strain of arguments from the difference between Linguistic Behaviourism and Ordinary Language arguments that shall emerge. To do that, it is best if we observe the difference between occurrences and propensities very early on.

Properly speaking, emotions as Ryle describes them in his 'moodology' have four different types. Two of those types are collapsible into each other. A third is what happens when those two collide either with each other, or contrary versions of themselves, or a factual impediment, and a fourth type, which is separate to the other three.

Ryle writes

> (T)he word 'emotion' is used to designate at least three or four different kinds of things, which I shall call 'inclinations' (or 'motives'), 'moods', 'agitations' (or 'commotions') and 'feelings'. Inclinations and moods, including agitations, are not occurrences and do not therefore take place either publicly or privately. They are propensities, not acts or states. They are, however, propensities of different kinds, and their differences are important.[186]

So under the designation of *propensities*, we have (1) Inclinations or motives (2) moods (3) agitations or commotions. Different to (1), (2) and (3) is (4) occurrences. occurrences can take place *publicly* and *privately*. A subspecies of these occurrences will become important to the overall project of this paper. Let us keep them in mind.

Now, let us turn to (1) of the propensities; the inclinations and motives.

---

[185] See 'Chapter Eight: The Occurrences' and its various subsections in this paper.
[186]Ryle, *Concept of Mind,* 1983. Pg 81

Both (2) moods and (3) agitations derive from the inclinations and motives.
Ryle writes about agitations

> A keen walker walks because he wants to walk,
> but a perplexed man does not wrinkle his brows because he wants or means to wrinkle them, though the actor or hypocrite may wrinkle his brows because he wants or means to appear perplexed. The reason for these differences is simple. To be distracted is not like being thirsty in the presence of drinking-water; it is like being thirsty in the absence of water, or in the presence of foul water. *It is wanting to do something while not being able to do it, or wanting to do something and at the same time wanting not to do it.* It is the conjunction of an inclination to behave in a certain way with an inhibition upon behaving in that way. The agitated person cannot think what to do, or what to think[187].

Motives and inclinations are not agitations, but they are stackable into agitations, and thus form moods. Something similar happens with habits which are the semi-agitated forms of the commotion. Ryle writes

> Motives then are not agitations, not even mild agitations, nor are agitations motives. But agitations presuppose motives, or rather they presuppose behaviour trends of which motives are for us the most interesting sort. Conflicts of habits with habits, or habits with unkind facts, or habits with motives are also commotion-

---

[187]Ryle, *Concept of Mind,* 1983. Pg 94

conditions. An inveterate smoker on parade,
or without any matches, or in Lent, is in
this plight[188].

Agitations thus presuppose motives and inclinations, the same way
commotions presuppose habits and addictions. An interesting question
to ask about these Behavioural Linguistic claims is whether people
actually talk this way or if Ryle is putting the linguistic behaviourist's
words in the ordinary language speaker's mouth.

The idea, however, is simple enough. We can get out our
notebook, sit down, and calculate agitations. Combine any two contrary
inclinations, or motives, or one inclination, or one motive with one
factual impediment, and you'll get a specific type of mood, an agitation.
Moods, of course, are the genera to which agitations are the species.

But Ryle also cautions us that

Mood words are commonly classified as the
names of feelings. But if the word 'feeling' is
used with any strictness, this classification
is quite erroneous. To say that a person is
happy or discontented is not merely to say
that he has frequent or continuous tingles
or gnawings; indeed, it is not to say even
this, for we should not withdraw our
statement on hearing that the person had
had no such feelings, and we should not be
satisfied that he was happy or discontented
merely by his avowal that he had them
frequently and acutely. They might be
symptoms of indigestion or intoxication[189].

But what are these feelings, these 'tingles' and 'gnawings'. How might
'tingles' and 'gnawings' be mistakenly applied to moods, and why are
moods, as propensities, different to occurrences? Moreover, why might

---

[188]Ryle, *Concept of Mind,* 1983. Pg 94
[189]Ryle, *Concept of Mind,* 1983. Pg 97

someone who has 'feelings' in this second sense of 'gnawings' and 'tingles' and claims to be in a 'mood' deserve our dubious glare? What separates these mistaken moods from feelings? Why is the title 'feelings' so improper?

Ryle writes

> Feelings . . . are *occurrences*, but the place that mention of them should take in descriptions of human behavior is very different from that which the standard theories accord to it. Moods or frames of mind are, unlike motives, but like maladies and states of the weather, temporary conditions which in a certain way collect occurrences, but they are not themselves extra occurrences[190].

Again Ryle speaks of *occurrences*, from the propensities and occurrences distinction, which I pointed out at the start of this section.

So it seems, moods can also contain "collections of occurrences" but are themselves not occurrences even though feelings are of course occurrences. Well that would make sense, since moods, inclinations and motives are propensities. But wait a moment we still do not have an account of what feelings are. What exactly does he mean by feelings?

Ryle writes

> By 'feelings' I refer to the sorts of things which people often describe as thrills, twinges, pangs, throbs, wrenches, itches, prickings, chills, glows, loads, qualms, hankerings, curdlings, sinkings, tensions, gnawings and shocks. Ordinarily, when people report the occurrence of a feeling, they do so in a phrase like 'a throb of compassion', 'a shock of surprise' or 'a thrill of anticipation'[191].

---

[190]Ryle, *Concept of Mind*, 1983. Pg 81

Now I will introduce my own piece of jargon, because these little 'throbs of compassion' these 'shocks of surprise' and the 'thrill of anticipation' will be important to our account, and what, I shall argue, is haunting Ryle's elaborate grammatical machinery. The term I shall use is 'flash-bangs' for the veritable and genuine quality the term evinces of the feelings.

Flash-bangs are, if a definition is wanting, Neo-Rylean semi-Linguistic Behavioural terms, that are in Original Rylean[192] feelings proper as a sub genera of occurrences, with specific structures that employ either bits of onomatopoeia left laying around from other linguistic phenomena, or fragmented bits of adjective connected, either, to a noun of emotion, in the genitive case, or an emotional adjective functioning as one.

Now they are 'semi' Linguistic Behavioural terms in my budding Linguistic Behavioural definition because they have a number of bizarre properties. The question of just how bizarre these properties are it is part of the job of this paper to determine. Now, just so the form of the paper is clear, later[193] I'll attack this point from several fronts,

---

[191]Ryle, *Concept of Mind,* 1983. Pg 81

[192] Later I will be introducing Rylean sub-dialects. Neo-Rylean are the set of handles I use for these. Original Rylean is the Ryle of Ryle in *The Concept of Mind.*

[193]See '5.3 Exemplar: Ghostographic analysis of Ryle's philosophical practice' in 'Chapter Five : Ordinary language arguments' and 'Chapter Eight : The Occurrences' and its subsections. Arriving at the insight on offer at the end of this paper requires a number of steps and the first of these steps is understanding my mini-methodology for breaking Ryle's arguments up and sorting through the various strains of appeal in them. This will prepare the way for demonstrating how linguistic and phenomenological style arguments can clash and produce rival claims, that if taken together, would produce a contradiction and show why using a vague 'ordinary language' source for arguments is problematic. We arrive at this stage in the argument in the section titled '7.3 Dispositions, Phenomenology and Law-Like statements'. The insight on offer here involves a set of distinctions made up from direct, and indirect views, first and third personal perspectives as well as the auto-phenomenological and hetero-phenomenological distinctions that arise from the Chalmers-Dennett debates. See the footnotes in the section titled '8.2 On the Normative Appeal of an Argument' in ' Chapter Eight : The Occurrences' for this debate, centrally footnotes 457, 458. The auto-phenomenological and hetero-phenomenological distinction arises, specifically, from an ambiguity in the Chalmers-Dennett debate over language and consciousness, and whether to treat language as the first or third personal point of view for the purposes of experiments. It is important to keep track of how this relates back to the reason-explanation accounts of motives and beliefs in '7.1 Game of Asking for and Giving Reasons' and the developing thread on Brandom and Ayer. If you imagine the Brandom-Peters-Plato debate like a turn-pike, it re-routes reason-explanations as linguistic behaviors back into the chain of arguments, beginning in this early part of the thesis, on flash-bangs, and the call for a return to psychologism. But that insight will only make sense, once we have completed each of the stages. Understanding 'Ghostography' in Chapter Five is the first of those stages. There is a taxonomy woven into the core thinkers I am using that is worth noting. If we start

before I push this Linguistic Behaviourist distinction to its limit, to show, where it breaks down, and how we can locate the occult stream of phenomenology in Ryle, by locating a point where he uses an Ordinary Language argument to justify or ground an argument that he doesn't have a Behavioural Linguistic analyses for. I will do this by introducing some of Sartre's phenomenology, and a specific type of argument that Sartre uses where Sartre brings consciousness before itself to examine some of its features in order to reveal what is obscure or concealed in Ryle's general method. I'm going to argue that this type of argument is found by locating an Ordinary Language argument without a Linguistic Behaviourist description attached and examining the basis on which it rests its distinction or point. In most cases it will be a 'phenomenological' argument. This customized mini methodology I developed to draw out the phenomenological content, and thereby cross classify, and taxonimize the arguments Ryle uses in *the Concept of Mind*, I have called this method 'Ghostography' because it reveals where all the ghosts haunting the work are hiding. There is an insight that I am going to argue pushes us towards a Pre-Fregean Psychologism that arises from understanding this method, but first I need to show you what that method is. To do that we need to take a close look at how Ryle systematizes emotions.

Ryle writes

> Feelings, in any strict sense, are things that
> come and go or wax and wane in a few
> seconds; they stab or they grumble; we feel

---

with the explicit side, Brandom and Ayer represent Fregeian views of anti-psychologism that advocates explicit models of meaning on language in the paper. Ayer is Pre-Sellarian and Brandom is post-Sellarian. If we cross over to the implicit side, we see that Dummett is anti-psychologsitc and implicit about meaning in language and Ryle is one step beyond this, he is anti-psychologistic and implicit about language and the mind. Sellars himself is more complex. I argue he is psychologistic in *Empiricism and the Philosophy of Mind*, but he is pragmatically anti-psychologistic about thought in *Philosophy and the Scientific Image of Man* since he wants to treat thought like language. Peters, if he were consistent, would be anti-psychologistic and implicit about the mind, and Post-Sellarian about linguistic discourse, but there is much slippage in his argument through out the *Concept of Motivation*. Peters argues that motives are non-events, they are dispositions and we only come to know them through episodic events related to social discourse which reveals the reasons which are keyed into actions and revealed as actions, but the motives are not the actions themselves. Suffice it, when it comes to building my taxonomy he is the closest I could find to an implicit anti-psychologistic view about the mind. Where he differs on dispositions is that Sellars dispositions as fact stating roles are capacities and thus outside the scope of this paper, while Peters fact stating roles are tendencies and thus inside the paper. Footnotes 378 and 377 may be of some interest to the reader on this point.

them all over us or else in a particular part.
The victim may say that he keeps on having
tweaks, or that they come only at fairly long
intervals. No one would describe his
happiness or discontentment in any such
terms. He says that he feels happy or
discontented, but not that *he keeps on
feeling*, or that he steadily feels happy or
discontented.[194]

The important term there, to differentiate moods from our flash-bangs, is of course, *'he keeps on feeling'* which I've italicized. So if, using my term, we were to try and conjure up, or imagine a 'flash-bang mood' we might make better sense of this distinction that Ryle is using. Imagine, now, a man who, in the middle of a live philosophical debate, looks up as his opponent starts scribbling esoteric symbols on a board, which involve backwards Es, diamonds, hooks, and little boxes, and he thinks back to that symbolic logic class he skipped in his first year of university, and feels a 'lump' or a 'pang' of something. Then imagine that he keeps feeling that little 'lump' or 'pang' every few moments as the other philosopher continues to draw more and more symbols. One for each little symbol, indefinitely, with no little gap in between. That's what a 'feeling mood' would be like for Ryle. Similarly, happiness would be an ongoing sequence of little 'burstings' and puzzlement would be the ongoing hatching of little internal quizzicals. Such would be a 'flash-bang mood', or in Ryle's original terminology, a propensity of occurrences that arises between feelings as mistaken moods and feelings proper. Ryle doesn't think such a thing exists, at least not formally[195].

So far, we can see that flash-bangs have a semi-episodic structure to them, but they also have this other part to them, this *'feely'* bit, that is, we feel them like pin pricks, and this of course, is part of the problem with a mere Behavioural Linguistic description of them because, as Ryle says;

---

[194]Ryle, *Concept of Mind,* 1983. Pg 97
[195] Ryle doesn't always uphold this distinction outside *The Concept of Mind*. See Ryle, Gilbert.
"Feelings." In *Collected Essays 1929-1968*, edited by Julia Tanney, II, 284-299. Oxon: Routledge,
2009. Specifically pg 286 of that edition, 'acutely' or 'intensely' in use with 'unfit', 'tired', 'worried' or
'cross' are examples in that paper of what I would argue is a 'flash-bang' mood.

It is an important linguistic fact that these names for specific feelings, such as 'itch', 'qualm' and 'pang' are also used as names of specific bodily sensations. If someone says that he has just felt a twinge, it is proper to ask whether it was a twinge of remorse or of rheumatism, though the word 'twinge' is not necessarily being used in quite the same sense in the alternative contexts[196].

And it is here, that we move *from a Linguistic Behaviorists claim to an Ordinary Language argument*, because we are no longer describing linguistic behaviour in the claim but calling upon the user's knowledge of common language.

Let us, in fact, consider the difference between 'a glow of pride' and a 'glow of warmth'. How are the two forms in fact different? Are they different? What is the difference between feeling 'warmth' and feeling that sudden glow of 'pride'? How would we tell the difference? One might, justifiably say that the difference is in the use of the expressions and although we can't actually tell the difference from the way the language behaves because the two structures may in all other respects behave the same way, we might still fall back on a common domain of language use. That is to say Ryle might claim that it depends on the common knowledge possessed by the average user of that language, and say 'well he knows the difference'.

But suppose this isn't enough, our philosopher insists, 'I know he knows the difference, but in your original project, you pointed out that he already knows how to use these concepts, but he doesn't know how to correlate them, and if your project is to be merit worthy, you also promised to give us a map that would show us how to correlate these differences. And the map must include this difference. If a map it is, it must be able to map it.

'Now' such a philosopher might declare 'where's the map?'

---

[196]Ryle, *Concept of Mind,* 1983. Pg 81

And indeed this is a problem for Ryle because the very way out of this problem between glows of pride and glows of warmth, he's cut himself off from.

One might answer 'well pride feels this way, and warmth feels that way' and he might, accompany this difference by standing our intrepid philosopher next to the fire, and then showing the philosopher a picture of a woman holding a baby. But is this enough, or does he need the stubborn philosopher to participate in the distinction in some way? Does he need the philosopher to 'feel' these differences?

And what exactly is it that one 'feels'?

If we all know what a glow of pride or a glow of warmth is and we can relate to them, then what exactly is it that is going on in that act of relation? Do we have to stop, take a non-sensory look inside of ourselves, and find that difference? Is this some sort of act of 'introspection' perhaps? It is here that we enter into the third domain of Ryle's arguments, the Occult Strain of phenomenology running through Ryle's Ordinary Language arguments.

Chapter Four : Theitic Consciousness and the interlude

4.1 Introspection

As David Armstrong points out one of the most puzzling aspects of *The Concept of Mind* is Ryle's position on introspection and consciousness. Armstrong writes

> As a physicalist I originally thought, when young, that Gilbert Ryle's *Concept of Mind,* read as a sophisticated behaviourism, might do the trick for the mind. I was always troubled, though, by the apparent denial of introspection. Ayer's clever remark that a behaviourist must pretend to be anaesthetized struck home.[197]

Indeed G. E. Myers in his taxonomy of Rylean Behaviourism and Rylean Behaviourists, followed by David Chalmers who we looked at at the start of this paper, tend to think the significance of a 'Behaviourist[198]' move like Ryle makes is to disengage from introspection and introspective states[199]. Likewise Sellars thinks that a Rylean language can solve the problem of introspection with a theory of public meaning[200]

---

[197]Armstrong, D. M. *Sketch for a Systematic Metaphysics*. Oxford: Oxford University Press, 2010. Pg 105

[198] Not in these sense I use it, in the more general sense like G. E. Myers defines it. See the footnote immediately below.

[199]Myers, G. E. "Motives and Wants." *Mind* Vol. 73, no. 290 (1964): Pp. 173-185. Pg 173

[200]Sellars, *Empiricism & the Philosophy of Mind,* 1997 see pp 87 – 88 for Sellars account of 'private reports' which Sellars argues a Rylean language can not posses. I agree with Sellars, Ryle, from a Linguistic Behaviourist perspective cannot argue for 'private episodes' in the sense of the phenomenological distinction in the Reader/Witness argument, or the Remember-How/Remember-When distinction although 'Original Ryle' does. Seen from this angle, my strategy is to focus on the inconsistency between Sellars reading and the arguments that actually occur in Original Ryle to bring out these phenomenological elements and exploit them as resources for an argument for a return to a Pre-Fregeian Psychologism.

and a back story on how observation languages become report languages which can thus provide the grounds for a theory about impressions as theoretical entities. But all of these views seem to clash with some of the types of argument that Ryle makes.

I think that getting this element in Ryle that has confused or puzzled many philosophers, as Armstrong admits, untangled is important because it makes Ryle's position clear and many of his insights accessible for future work. What I want to do, briefly, is draw out exactly what Ryle means by 'introspection' and 'consciousness' in as quickly, shrewdly and accurately a way as possible because it is a bit of a logical puzzle and it has bothered many philosophers for several decades now.

On the one hand Ryle argues against introspective acts of consciousness [201] but on the other hand he has this account of the mind's eye[202] which, as I brought out in the introduction, seems to require both introspection and consciousness. On the one hand, Ryle argues against introspection but allows certain uses of the term 'introspective' and on the other hand he has acts that utilize introspective scrutiny in many of his arguments. Understanding what is wrong with Ryle's attack on introspection involves understanding the nature of the attack, which has two stages, and then contrasting this with some of his other arguments that seem to require these very introspective acts he eschews.

For the first stage of his attack on introspection Ryle argues that introspection is a theoretical and technical term introduced by art.

> 'Introspection' is a term of art and one for
> which little use is found in the self-
> descriptions of untheoretical people.[203].

'Introspective' is used as an adjective for the type of person who pays more heed than usual to problems regarding 'his own character, abilities, deficiencies and oddities'[204] and fits in with one of the uses he

---

[201]Ryle, *Concept of Mind,* 1983. Pg 149
[202]Ryle, *Concept of Mind,* 1983. Pg 257, Pp 9-11
[203]Ryle, *Concept of Mind,* 1983. Pg 156
[204]Ryle, *Concept of Mind,* 1983. Pg156

allows for the term 'self conscious' which we will explore in a moment[205] but here it is being used in a dispositional sense as a personality trait much like an inclination or a tendency.

The second stage of Ryle's strategy is to attack what I term 'introspective scrutiny'.

> (I)ntrospection is described as being unlike sense observation in important respects. Things looked at, or listened to, are public objects, in principle observable by any suitably placed observer, whereas only the owner of a mental state or process is supposed to be able introspectively to scrutinise it.[206]

The difference here is that

> Sense perception, again, involves the functioning of bodily organs, such as the eyes, the ears, or the tongue, whereas introspection involves the functioning of no bodily organ[207].

On the other hand Ryle uses arguments that, themselves, rest their appeal upon acts involving introspective scrutiny. For instance Ryle's argument that 'seeing' and seeing involves perception and imagining depends on an introspective distinction and ability to differentiate between sense perception and visualization[208]. A distinction so fine, Ryle thinks, that Hume was unable to discern it.

Ryle writes

---

[205]See 'the Species of Mindologue' in the next subsection.
[206]Ryle, *Concept of Mind,* 1983. Pg 157
[207]Ryle, *Concept of Mind,* 1983. Pg 157
[208]Ryle, *Concept of Mind,* 1983. Pg 257 NB. This is the account of the "mind's eye" which I also mentioned in the introduction.

To see is one thing; to picture or visualise is another. A person can see things, only when his eyes are open, and when his surroundings are illuminated; but he can have pictures in his mind's eye, when his eyes are shut and when the world is dark. Similarly, he can hear music only in situations in which other people could also hear it; but a tune can run in his head, when his neighbour can hear no music at all[209]
.

In particular he needs these two senses of see and 'see' as well as hear and 'hear' to differentiate between two uses of the terms 'vivid' and 'lively' for which, he thinks, Hume was mistaken.

Ryle writes

Hume's attempt to distinguish between ideas and impressions by saying that the latter tend to be more lively than the former was one of two bad mistakes. Suppose, first, that 'lively' means 'vivid'. A person may picture vividly, but he cannot see vividly. One 'idea'
may be more vivid than another 'idea', but impressions cannot be described as vivid at all, just as one doll can be more lifelike than another, but a baby cannot be lifelike or unlifelike. To say that the difference between babies and dolls is that babies are more lifelike than dolls is an obvious absurdity. . . . Alternatively, if Hume was using 'vivid' to mean not 'lifelike' but 'intense', 'acute' or 'strong', then he was

[209]Ryle, *Concept of Mind*, 1983. Pp 236 – 237. NB I am aware I have introduced these sections before, but they are here reproduced for a different analyses, and convenience of the reader so they don't have to flip back to old sections.

mistaken in the other direction; since, while sensations can be compared with other sensations as relatively intense, acute or strong, they cannot be so compared with images[210].

Moreover Ryle writes

> When I fancy I am hearing a very loud noise, I am not really hearing either a loud or a faint noise; I am not having a mild auditory sensation, as I am not having an auditory sensation at all, though I am fancying that I am having an intense one. An imagined shriek is not ear-splitting, nor yet is it a soothing murmur, and an imagined shriek is neither louder nor fainter than a heard murmur. It neither drowns it nor is drowned by it.

For Ryle sensations are not like images, and imaged or visualized things are not like things one sees. An imagined or fancied sound is not like a heard shriek. To see and 'see' and hear and 'hear' are different things. It seems like he needs something like a theory of introspection of exactly the kind he eschews to differentiate between these aspects and ascriptions. If he tries to base the difference between perceptual seeing and visualizational 'seeing' on the fact that one can do the latter while his eyes are closed he is going to run into problems with (a) the non-sensory element of visualization, that is, because no sense organ is being used while the eyes are closed and (b) the difference between the two types of 'seeing' while the eyes are open. Likewise hearing and 'hearing' requires one to take stock of introspective qualities between the two acts to fully appreciate. Surely one can visualize while one's eyes are open or 'hear' a tune which someone else can not. What it seems Ryle needs is exactly what he eschews. That is Ryle needs a theory of 'introspection' as a process that 'involves the functioning of no bodily organ' to uphold this difference.

---

[210]Ryle, *Concept of Mind,* 1983. Pp 236 - 237

Again he runs into problems if he tries to pinpoint the difference between observation and sensation. His argument depends on the claim that observation unlike sensation involves a publically observable object that can be witnessed or observed.

> It is true that the cobbler cannot witness the tweaks that I feel when the shoe pinches. But it is false that I witness them. The reason why my tweaks cannot be witnessed by him is not that some Iron Curtain prevents them from being witnessed by anyone save myself, but that they are not the sorts of things of which it makes sense to say that they are witnessed or unwitnessed at all, even by me. I feel or have the tweaks, but I do not discover or peer at them; they are not things that I find out about by watching them, listening to them, or savouring them. In the sense in which a person may be said to have had a robin under observation, it would be nonsense to say that he has had a twinge under observation. There may be one or several witnesses of a road-accident; there cannot be several witnesses, or even one witness, of a qualm[211].

Moreover, Ryle derives specific criteria for the objects of observations that sensations, he argues, cannot posses

> The properties which we ascertain by observation, or not without observation, to characterise the common objects of anyone's observation cannot be significantly ascribed to, or denied of, sensations. Sensations do

[211]Ryle, *Concept of Mind,* 1983. Pg 198 - 199

not   have   sizes,   shapes,   positions, temperatures, colours or smells[212].

In order to assert that sensations do not have sizes, shapes, positions, temperatures, smells, colours and so on, he needs to be able to introspect on them in some non-sensory way to compare what qualities these perceptions do have with those they do not. Likewise he needs introspection to introduce what sort of properties experiences have. This is important in a Neo-Kantian sense as we shall see shortly. But let us keep to the main point. That is, if Ryle eschews consciousness, as I'll demonstrate in a moment, and he denies introspection in the manner stated above, then, I argue, it seems he has little or no capacity to differentiate between perception and imagination in his see and 'see' claim, or to base the introspective qualities that allow him to draw a difference between sensation and observation.

Add to this his claims about ordinary language and his unique method of linguistic analysis, and you've got a puzzle. In this short interlude I'm going to pick up one of the threads of that puzzle, namely Ryle's arguments against consciousness and introspective acts and show you where some of the cracks from the fault line between Ryle's use of consciousness and introspective acts, and his tirade against them run through *The Concept of Mind* before formally introducing theitic arguments and some of Sartre's phenomenology.

---

[212]Ryle, *Concept of Mind,* 1983. Pg 198. See also Mandlebaum, Maurice. *Philosophy, Science and Sense Perception.* Baltimore: John Hopkins University Press, 1964. Pp 172-175. Alongside the above argument, that to determine whether sensations lack sizes, shapes, positions, temperatures, smells or colours, Ryle needs a theory of consciousness. There is an even finer grain distinction that can be drawn out of Mandlebaum that parallels Locke's primary and secondary colours and implicates human consciousness. Mandlebaum points out that some philosophers might argue that before humans, two objects might have distance between them, but it would be difficult for them to argue that they had colour. The determination of colour or even the lack of it, on this view, would need a human eye and by implication the consciousness to operate that eye for colour or the lack of it to exist. This is a contentious issue that depends on deeper metaphysics which is why I've delegated it to footnotes. But the claim it makes is prima facie strong enough to support the argument I've advanced in the central body of the paper. Suffice it that my central argument is simply that these distinctions and the determination of whether one has them or not depends upon acts of introspective scrutiny and consciousness, particularly since Ryle's line of argument here is to argue that we can imagine what it is like to have telescopes and microscopes for observations but we can't imagine what similar apparatuses might be for sensations like itches, aches and throbs. Later I shall define consciousness and introspection using Jean-Paul Sartre's phenomenology and the apperception of the 'I' and appearance of the 'me'. For the moment see the section above and the footnotes in the introduction to this paper.

The reason I introduce Sartre here, at this stage in the paper, is specifically to make the inconsistency in Ryle's claims and his arguments explicit in the next section of this move when I start talking about Ordinary Language arguments[213]. However, due to the poverty of Anglo-Sartre scholarship, I will introduce a very short literature review pointing out some of the key inconsistencies that turn up in most of the English speaking articles before directly engaging Sartre's thought towards the above ends. I have tried to make this as brief as possible, identifying what I take to be the most common misconceptions, and collecting them together in either their root cause, if I could trace it, or in the most common expression. Doing this here has the attractive benefit of freeing Sartre up from the body of these mistakes for the duration of the thesis.

4.1.1 Ryle's use of the terms 'consciousness' and 'introspection'.

Firstly, Ryle's position in *The Concept of Mind* admits consciousness in the following trivial senses. He draws these from limited cases where people use the term in everyday conversations. The first instance being when someone walks into a room and they notice that something is different. Ryle's focus isn't on the act but rather on the sense that people talk about it. I shall argue that this is a mistake and part of the overall problem with a linguistic approach and is essentially what creates the illusion that I am systematically attacking, because it focuses on how people use words not on how they think. There is a gap that this paper will exploit but to see that gap we first have to develop an important insight and systematically draw out the full range of its implication.

Ryle writes

> People often speak in this way; they say, 'I was conscious that the furniture had been rearranged', or, 'I was conscious that he was less friendly than usual'. In such contexts the word 'conscious' is used instead of words

---

[213] These are classified by an appeal to the ordinary language user's notions by what it makes sense to say.

like 'found out', 'realised' and 'discovered' to
indicate a certain noteworthy nebulousness
and consequent inarticulateness of the
apprehension[214].

He also allows it in the sense that someone may be self conscious. This,
also, is trivially important for a species of 'mindologue' he uses called
guarded talk, which I'll go into, briefly in the next section of the paper
where I explain Ryle's types of mindologue. This 'use' of conscious and
self conscious is relevant to the immediately prior section of the paper
and is directly relevant to the first part of Ryle's attack, Ryle's first part
of the attack on introspection. In this first part of Ryle's attack on
introspection he treats introspection as a tendency disposition through
his analysis of the adjective 'introspective'.
    Ryle writes

People often use 'conscious' and 'self-
conscious' in describing the embarrassment
exhibited by persons, especially youthful
persons, who are anxious about the opinions
held by others of their qualities of character
or intellect[215].

This self-consciousness has an extended use in the form of self
knowledge one develops about the sorts of character and person they
are.
    Ryle writes

'Self-conscious' is sometimes used in a more
general sense to indicate that someone has
reached the stage of paying heed to his own
qualities of character or intellect,
irrespective of whether or not he is
embarrassed about other people's
estimations of them. When a boy begins to
notice that he is fonder of arithmetic, or less

---

[214]Ryle, *Concept of Mind,* 1983. Pg 150
[215]Ryle, *Concept of Mind,* 1983. Pg 150

homesick, than are most of his
acquaintances he is beginning to be
self-conscious, in this enlarged sense[216].

Lastly he admits the use of consciousness in the sense of 'sensations'.

Quite different from the foregoing uses of
'conscious', 'self-conscious' and 'unconscious',
is the use in which a numbed or
anaesthetized person is said to have lost
consciousness from his feet up to his knees. .
. [217]

This makes up the ground of his 'sensations' side for the 'sensation' and
'observation' distinction. Observations can not be anaesthetized in this
sense while sensations can. Along with the above Ryle also
distinguishes between when a person notices or pays heed to an ache,
itch or pain and realizes he is aware of it. Besides these two Ryle has a
sort of log-keeper or internal mechanism which is responsible for a type
of status report that can report this and other events or phenomena[218].
Ryle writes

It is certainly true that when I do, feel or
witness something, I usually could and
frequently do pay swift retrospective heed to
what I have just done, felt or witnessed. I

[216]Ryle, *Concept of Mind,* 1983. Pg 150
[217]Ryle, *Concept of Mind,* 1983. Pg 151 See also Chalmers. *The Conscious Mind,* 1996. Pg 228: for
Chalmers discussion of Ned Block's paper and the distinction between phenomenal and access
consciousness which parallels this distinction.
[218] Ryle, *Concept of Mind,* 1983. Pg 153. See Sellars, *Empiricism & the Philosophy of Mind,* 1997.
Section XII Our Rylean Ancestors, Subsection 48. Sellars calls this log keeper form of narration 'in
forro interno'. It arises from the language of the behavourism of the "Messianic Behaviourist" Jones,
and the sophistication that develops when an observational language that describes behavour moves
into a report language and people begin applying third personal observations to first personal
reports. See also footnote 200 for clarification between Ryle and Sellars reading of Ryle along with
footnote 361 for clarification of the relationship between a Rylean language as Sellars perceives it
and 'in forro interno' account of the mind. Footnote 228 also clarifies the relationship between Ryle's
retrospection and Sellars 'in forro interno' account, and footnote 193 clarifies my position of Sellars
in relation to Psychologism, which is that in *Empiricism and the Philosophy of Mind* I think there is
room for a psychologistic reading of Sellars, however *Philosophy and the Scientific Image of Man* is
pragmatically anti-psychologistic.

keep, much of the time, some sort of log or
score of what occupies me, in such a way
that, if asked what I had just been hearing
or picturing or saying, I could usually give a
correct answer.[219]

This last distinction is in line with a specific argumentative move that
Ryle calls 'being alive to what one is doing'. The argument here is that
the person must be able to report their actions or thoughts in a verbal
manner. Ryle thinks that if a person can not or does not report the
presence of a mental phenomena like that of Augustinian 'volitions',
Humeian 'passions' or Freudian 'castration fears' as part of 'being alive
to what they are doing' or as a status report for an activity like reciting
little Miss Muffet backwards, then they are not applicable to an account
of the nature of mind. This is what underlies his argument for an
implicit criteria like we investigated earlier in this paper in the chapter
on Dummett. For Ryle the person needs to be able to communicate their
thoughts.

4.1.2 The Species of Mindologue

Ryle thinks that in thinking to ourselves, in general, we are talking to
ourselves and that this procedure follows the natural course of
acquiring a language. As such there are various species of 'mindologue'

---

[219]Ryle, *Concept of Mind,* 1983. Pg 152. Also see Chalmers. *The Conscious Mind,* 1996. Pg 28, where
Chalmers distinguishes phenomenal consciousness, consciousness, access consciousness and
reportability which parallel the above distinction in Ryle. Specifically what Ryle is describing
parallels what is defined by Chalmers as 'access consciousness'. 'Access consciousness' is a state in
which the content of consciousness is poised to be used as a premise in reasoning, rational control of
action and rational control of speech. My strategy is to focus on anological constructs that arise from
what I claim is a pre-linguistic position accessible from the direct first personal perspective, and
codified with the third person perspective when it enters into a language 'role'. This insight depends
on developing the argument from 'flash-bangs'. That argument won't make sense until the end of the
paper. Suffice to say the linguistic codification of 'access consciousness' implicates, from the position
argued by the end of this paper, a third personal perspective. I will point out how the pieces fit
together in the footnotes as we go along. See also Tim Bayne, David Chalmers. "What Is the Unity of
Consciousness." In *The Unity of Consciousness: Binding, Integration, Dissociation* edited by Chris
Frith Axel Cleeremans. Oxford Scholarship Online: March 2012 @
http://www.oxfordscholarship.com/view/10.1093/acprof:oso/9780198508571.001.0001/acprof-
9780198508571 downloaded 05/06/2012: Oxford, 2003. Section 3, for Bayne and Chalmers' discussion
of Access Unity and Phenomenal Unity.

which a person develops, but essentially Ryle argues there is no real difference between thinking to one's self and speaking aloud except that the former is a sophisticated development.

Ryle writes

> This trick of talking to oneself in silence is acquired neither quickly nor without effort; and it is a necessary condition of our acquiring it that we should have previously learned to talk intelligently aloud and have heard and understood other people doing so. Keeping our thoughts to ourselves is a sophisticated accomplishment. It was not until the Middle Ages that people learned to read without reading aloud. Similarly a boy has to learn to read aloud before he learns to read under his breath, and to prattle aloud before he prattles to himself. Yet many theorists have supposed that the silence in which most of us have learned to think is a defining property of thought. Plato said that in thinking the soul is talking to itself[220].

Now, being able to keep one's talk to one's self is different from guarded talk.

> When talk is guarded and often we do not know whether it is so or not, even in the avowals we make to ourselves sleuth-like qualities do have to be exercised. We now have to infer from what is said and done to what would have been said, if wariness had not been exercised, as well as to the motives of the wariness[221].

---

[220]Ryle, *Concept of Mind,* 1983. Pg 35

The reason why talk is guarded in introspective and self conscious people, Ryle thinks, is because

> Our knowledge of other people and of ourselves depends upon our noticing how they and we behave. But there is one tract of human behaviour on which we pre-eminently rely. When the person examined has learned to talk and when he talks in a language well known to us, we use part of his talk as the primary source of our information about him, that part, namely, which is spontaneous, frank and unprepared[222].

This species of frank and spontaneous talk is of course the 'unstudied variety'. Studied talk is associated with the self conscious and introspective.

On unstudied talk Ryle writes

> In unstudied chat we talk about whatever we are at the moment chiefly interested in. It is not a rival interest. We talk about the garden from the motive that prompts us to inspect and potter in the garden, namely interest in the garden. We chat about our dinner not because we are not interested in our dinner, but because we are[223]

This if contrasted, is different again to propounding
On propounding, Ryle writes

---

[221]Ryle, *Concept of Mind*, 1983. Pg 176
[222]Ryle, *Concept of Mind*, 1983. Pg 173
[223]Ryle, *Concept of Mind*, 1983. Pg 174

The physician, the judge, the preacher, the politician, the astronomer and the geometrician may give their counsels, verdicts, homilies, theories and formulae by word of mouth, but they are then talking not in the sense of 'chatting' but in the sense of 'pronouncing' or 'propounding'[224].

That is to say

First, it makes no important difference whether we think of the reasoner as arguing to himself or arguing aloud, pleading, perhaps, before an imagined court or pleading before a real court. The criteria by which his arguments are to be adjudged as cogent, clear, relevant and well organised are the same for silent as for declaimed or written ratiocinations. Silent argumentation has the practical advantages of being relatively speedy, socially undisturbing and secret; audible and written argumentation has the advantage of being less slap-dash, through being subjected to the criticisms of the audience and readers[225].

This in turn gives him the grounds for arguing that what makes something an act of the intellect is independent of what makes it public or private.

Ryle argues

What makes a verbal operation an exercise of intellect is independent of what makes it public or private. Arithmetic done with

---

[224]Ryle, *Concept of Mind,* 1983. Pg 174
[225]Ryle, *Concept of Mind,* 1983. Pg 46

pencil and paper may be more intelligent than mental arithmetic[226].

We've encountered these 'inheritance conditions' before. They are the adverbs of manner that we use to describe someone's conduct. Ryle's argument is that we already know how to account for these and that their criteria is implicit and understood in the ordinary language speakers' vocabulary. We know what it is like to talk about someone's behaviour as intelligent or silly, and their actions as successful or unsuccessful[227].

Ryle writes

> What distinguishes sensible from silly operations is not their parentage but their procedure, and this holds no less for intellectual than for practical performances. 'Intelligent' cannot be defined in terms of 'intellectual' or 'knowing how' in terms of 'knowing that' ;'thinking what I am doing' does not connote 'both thinking what to do and doing it'. When I do something intelligently, i.e. thinking what I am doing, I am doing one thing and not two. My performance has a special procedure or manner, not special antecedents[228].

Acting a bit like 'consciousness' Ryle has a 'log keeper' which I mentioned before. This 'log keeper' role, if questioned can give a verbal status report of what the person is doing. It also replaces reflection with retrospection[229].

---

[226]Ryle, *Concept of Mind,* 1983. Pg 35

[227] See Ryle's "Verbs of success", Ryle, *Concept of Mind,* 1983. Pp 145-146

[228]Ryle, *Concept of Mind,* 1983. Pg 32

[229] See Ryle, *Concept of Mind,* 1983. Pg 160 for Ryle's concept of retrospection, autobiographical information and the concept of a mental log keeper. This explains why Ryle makes the claim on Pg 99 of *The Concept of Mind,* that a man knows he is bored because he says he is bored, why Ryle dismisses volitions and it can go some of the way to pinpointing the clash between Sellars '*in foro interno*' reading of Ryle's concept of mind and elements of Ryle's *The Concept Of Mind* that do not fit Sellars reading. See Sellars, *Empiricism & the Philosophy of Mind,* 1997. Pg 91. For while there is a clash between Sellars claim that Ryle's 'concept of mind' does not have the resources to support

These similes of 'over-hearing', 'phosphorescence' or 'self-luminousness' suggest another distinction which needs to be made. It is certainly true that when I do, feel or witness something, I usually could and frequently do pay swift retrospective heed to what I have just done, felt or witnessed. I keep, much of the time, some sort of log or score of what occupies me, in such a way that, if asked what I had just been hearing or picturing or saying, I could usually give a correct answer[230].

This distinction Ryle calls 'being alive to what he is doing'.

Of course an agent can, from time to time, if he is prompted to do so, announce to himself or the world "Hallo, here I am whistling 'Home Sweet Home'. " His ability to do so is part of what is meant by saying that he is in that particular frame of mind that we call 'being alive to what he is doing'[231].

This allows Ryle to construct a concept of mind that depends on Ordinary Language argumentation with an account of mental phenomena that can focus on linguistic differences and behaviours in the ways words are used and also one that is centred by an implicit theory of meaning in language. He can offer an account of thought that

---

private episodes, and actual arguments that 'Original Ryle' uses, like the Witness/Reader argument, my treatment of Ryle's arguments, I hope it can be seen, is designed to bring out the inconsistency in Ryle along similar lines to Sellars' reading, and thus to deal with the central strength of Ryle's position as one advocating an 'implicit theory of meaning' in the philosophy of mind and then to critique it. See footnote 200 for an example of the sorts of private episodes that Sellars' 'in forro interno' reading of Ryle lacks the resources to deal with, while original Ryle makes phenomenological style arguments using introspective scrutiny of consciousness. There are arguments like the Reader/Witness argument and phenomenological elements of the 'Remember when/Remember how' distinction.
[230]Ryle, *Concept of Mind,* 1983. Pg 153
[231]Ryle, *Concept of Mind,* 1983. Pg 170

relies on different types of mental speech, or 'mindologue' and individuate mental phenomena by the terminology in use by ordinary language speakers. This is important because it allows him to argue against other philosophers on the basis of an ordinary language criterion and to forward his own arguments on the same ordinary language basis. However it is questionable how many of these linguistic behavioural accounts he can cover given the finite limits of his theory. I am going to focus on three specific examples that go beyond his ability to account for with a linguistic behavioural criterion. These will be (1) the reader / witness argument, (2) the species of occurrences designated as 'feelings-proper' which I pointed out and called 'flash-bangs' so as not to lose them and (3) the exercise of anticipating one's next thought. These will allow me, with the help of Sartre's phenomenology, to isolate a phenomenological strain in Ryle's method of argumentation. Differentiating this phenomenological strain from the linguistic behavioural strain is important because later I will use it to pin-point a contradiction that arises from ordinary language descriptions which Robert Wolff thinks act holistically and a type of first person phenomenology that is active in some of Ryle's arguments. The result will be to show a contradiction arises from rival claims if we take them both as ordinary language arguments. The rivalry between these two sources will set the scene for the remainder of this paper.

4.1.3 Ryle's Diachronic attack on consciousness.

Ryle's line of attack on consciousness is directed at diachronic stages[232] or elements that the concept of consciousness has acquired. Ryle thinks

[232]The term 'diachronic arguments' I've 'liberated' from Semiotics where 'diachronic' is used to refer to the properties and shifts in language over time. In Semiotics the term diachronic refers to a distinction that depends upon a polarization between 'diachronic' and 'synchronic'. This polarization of binary opposites is of course problematic as the Post-Structuralists point out. I use the term more loosely in a 'family resemblance' way to approximate similarities that can be used to describe arguments that depend upon chronological shifts in the meaning, usage and context of words over time. This species of argument is familiar enough though I won't go much beyond defining them in a general sense for this paper. One might call them 'etymological attacks" as they make arguments based on the etymological roots of words. One might call them 'philological arguments' as Philology was often used in this way. These are like Nietzsche's philological excursions into the origin of resentment, Nietzsche, Friedrich. *The Genealogy of Morals.* Translated by Horace Samuel. New York: Dover, 2003. See pp 19 – 21 for example. Lyotard's differend , Lyotard, Jean-Francois. *The*

that 'consciousness' is a myth that grew up and out of the Protestant revolutionand developed into a species of what he calls 'para-optics'.

Ryle writes

> When the epistemologists' concept of consciousness first became popular, it seems to have been in part a transformed application of the Protestant notion of conscience. The Protestants had to hold that a man could know the moral state of his soul and the wishes of God without the aid of confessors and scholars; they spoke therefore of the God-given 'light' of private conscience[233].

This Protestant version of 'do-it-yourself' moral conscience, according to Ryle, gets picked up after the Reformation during the Enlightenment where it gains another-worldly aspect with both dualist theories of mind and causal theories of consciousness.

Ryle writes

> When Galileo's and Descartes' representations of the mechanical world seemed to require that minds should be

Differend: Phrases in Dispute Translated by Georges Van Dan Abbeele: University of Minesota Press, 1989. The sorts of archeological and etymological surveys like we find in Foucault's role and treatment of contradictions in the history of discourse, Foucault, Michel. *The Archaeology of Knowledge*. Translated by A. M. Sheridan Smith. Oxon: Routledge, 2005. Pg 166-173. or Katz and Fodor's attack on Chomsky's generative grammar, Katz, Fodor. "The Structure of a Semantic Theory." *Language* 39, (1963): Pp 170-210., or for Pritchard's arguments on the origin of moral philosophy Prichard, H. A. "Does Moral Philosophy Rest on a Mistake?" *Mind* 21, no. 81 (1912): Pp 21-37 . These are arguments that focus on historical shifts in meaning and the use of language overtime, and in different social and historical contexts to make their point. They are an interesting species of argument, not unrelated to 'ordinary language' arguments, in a general sense. Foucault's historical work on the role of resolving contradictions within scholastic discourse, parallels the clash between the 'correct' ordinary language use of the English subjunctive conditional, and Lewis' logical grounds for rejecting that use on the basis of truth values and a theory of modal logic. The appeal in these types of argument is wide, from grammatical and syntactical shifts to etymological contexts for the relations between certain contexts, to historical social conditions for authority and the like. For the purposes of this paper, however, we will limit ourselves to Ryle's claim in the present context, and what Ryle's 'diachronic' attack boils down to as a claim about consciousness and language in the present. I.e. the aspects of language semioticians were trying to target with the 'synchronic' axis'.
[233]Ryle, *Concept of Mind,* 1983. Pg 153

salved from mechanism by being represented as constituting a duplicate world, the need was felt to explain how the contents of this ghostly world could be ascertained,
again without the help of schooling, but also without the help of sense perception. The metaphor of 'light' seemed peculiarly appropriate, since Galilean science dealt so largely with the optically discovered world. 'Consciousness' was imported to play in the mental world the part played by light in the mechanical world. In this metaphorical sense, the contents of the mental world were thought of as being self-luminous or refulgent[234].

In Ryle's history the theory gets picked up by John Locke, in whom it becomes refined into a 'reflective' model in which, Locke claims, consciousness can turn back on itself and examine, or rather, reflect on its own operations by means of introspective scrutiny.

Ryle writes

This model was employed again by Locke when he described the deliberate observational scrutiny which a mind can from time to time turn upon its current states and processes. He called this supposed inner perception 'reflexion'[235],

---

[234]Ryle, *Concept of Mind,* 1983. Pg 154

[235]Our 'introspection'. See McCosh, James. *Realistic Philosophy.* Vol. II. New York: Scribner, 1900. Pp 56 – 59. What Locke specifically meant, according to McCosh, who I agree with, was something as follows. Locke took the schoolmen's vocabulary of 'phantasm', 'notion', 'species' whereby the school men took 'phantasm' as the representation of a thing, 'notion' as an intellectual operation involved in apprehending the thing, and 'species' to refer to the visible appearance and objects classified, and he reduced them all to 'ideas'. Ideas, for Locke, are produced by sensations which later become Hume's impressions which fade into Hume's 'ideas', but for Locke, these are produced by the primary qualities of the object. Primary qualities become 'ideas' via an 'impulse' in the sense faculties. McCosh, Realistic Philosophy, 1900 Pg 58. Once the ideas are produced, reflection then sorts them out in to their proper place in categories via 'semblances'. 'Reflection', on this view, is the focus of the mind on the inner faculties. For the distinction between 'ectypal' and 'archetypal' ideas, see

borrowing the word 'reflexion' from the familiar optical phenomenon of the reflections of faces in mirrors. The mind can 'see' or 'look at" its own operations in the 'light' given off by themselves. The myth of consciousness is a piece of para-optics[236].

Consciousness, then, Ryle maintains, is a myth that began with the reformation, underwent several modifications, or reincarnations, and ends up in with a 'reflective doctrine' in John Locke. This, according to Ryle's account, in turn gives us the causal theory of consciousness, which David Hume inherits, in which sensations impress ideas on us and from which their conjunction creates the sentiment of belief[237].

However I just don't think this is the case. I think something like introspection goes on and that Ryle in many of his arguments unwittingly employs it. I pointed out one case where Ryle uses introspection in his arguments with his 'see' and see distinction. Ryle uses see and 'see' to attack the degrees of vividness in Hume between impressions and ideas. Ryle also uses introspection in his distinction between observation and sensation and as we will see a whole range of other arguments. I call these arguments the 'phenomenological strain' in Ryle.

Gotterman, Donald. "A Note on Locke's Theory of Self Knowledge." *Journal of The History of Philosophy* 12, no. 2 (1974): 239-242. Ectypal ideas contain reflections on the operations of the mind along with abstractable representations of substances. This later type is analogous with our introspection, and at an approximation to the way that Ryle is using the term if we look at what his attack amounts to in a 'synchronic' sense. See also Ryle, Gilbert. "John Locke on Human Understanding." In *Critical Essays*, edited by Julia Tanney, I, Pp 132-153. Oxon: Routledge, 2009.
[236] Ryle, *Concept of Mind,* 1983. Pg 153
[237] Something that is perhaps troubling, reading back into Ryle from Richard Rorty via Sellars, is the notion that we may be able to introspect and we may also have consciousness simply because we've developed those abilities out of developments in history. That is, indeed Sartre may very well be right, one might argue, when he pin points the implication of and uses for introspective arguments, but one might try to counter this and argue that phenomenology is only possible because it developed out of a Western historical context. Prior to that context one could not introspect. See Rorty, Richard. *The Mirror of Nature.* Princeton University Press: Princeton, 2009. Pg 218-220. My reply is that one would have to reflect to see if that were true, and the instant somebody did, they will have discovered theitic consciousness. See also "'The 'Theory Theory' of Mind and the Aims of Sellars' Original Myth of Jones'." *Phenomenology and the Cognitive Sciences* 11, no. 2 (2012): Pp 175-204.

In the next section, I'm going to argue on the basis of Sartre's phenomenology, that indeed, introspection and reflection are intellectual acts that one can perform, that many of Ryle's own arguments utilize these sorts of introspective acts of scrutiny, and various species of reflections, and that this will reveal a fault line running through Ryle that will allow us to build a taxonomy of his arguments. Ultimately this tension in Ryle between the Linguistic Behaviourist strain and the phenomenological strain of arguments will develop into the claim that these sources are actually rivals and they can forward different and contradictory claims about the mind and mental phenomena. The outcome of this rivalry will be to determine which of the two is a better source for exploring the nature of the mind and which of the two is a stronger source for arguments to which I will conclude that the phenomenological strain is the correct answer to both. To develop the groundwork to make that move we need to first get clear on what a theitic argument is and what are its sources of introspective scrutiny.

It is with this task in mind that we now turn to Sartre.

4.2 Sartre, Phenomenology and the Apperception of the I.

Robert Brandom argues that the Pre-Kantian tradition took it for granted that concepts could be grasped prior to judgments.

Brandom writes

> One of Kant's epoch making insights, confirmed and secured by Frege and Wittgenstein, is his recognition of the primacy of the proposition. The Pre-Kantian Tradition took it for granted that the proper order of semantic explanation begins with a doctrine of concepts or terms, divided into singular and general whose meaningfulness

can be grasped independently of and prior
to the meaningfulness of judgments[238].

Once a doctrine of concepts was established, Brandom argues, then a
doctrine of judgments merely provided the means for how these combine
together, and a doctrine of inferences regulates the way they inter-
relate[239].

Brandom argues

> Kant rejects this. One of his cardinal
> innovations is the claim that the
> fundamental unit of awareness or cognition,
> the minimum graspable is the judgment.
> Judgments are fundamental, since they are
> the minimum unit one can take
> responsibility for on the cognitive side[240].

Brandom argues, both in terms of Kantian scholarship and in terms of
the right philosophical position, that the 'I think' must accompany
representations because the judgment is the fundamental minimum
unit of thought. The reason it accompanies a representation, Brandom
maintains, is that it expresses the formal dimension for the
responsibility for judgments[241]. Here Sartre would part company with
him on both occasions.

Sartre does not think that Kant maintained the presence of an 'I'
as an inhabitant of the mind that must accompany all our thoughts.
Further Sartre doesn't think we do either. Sartre uses a distinction
between de jure and de facto problems to mark out a feature in Kant's
doctrine of the apperception of the I between the presence of the I as a
possibility and the presence of the I as a condition of our thoughts.

[238]Brandom, *Articulating Reasons*, 2001. Pg 159
[239]Brandom, *Articulating Reasons*, 2001. Pg 159
[240]Brandom, *Articulating Reasons*, 2001. Pg 160
[241]Brandom, *Articulating Reasons*, 2001. Pg 160

Kant's position, Sartre maintains, is that it must be a de jure precondition of thought that it is possible that the I accompany our thoughts. *De facto*, Sartre thinks, the I is not always there. The mistake, Sartre thinks, is to assume that the de jure necessity of the condition for the possibility of the I to accompany any thought has a de facto basis for the presence of the I in any thought. Sartre thinks it is a necessary condition that presence of an I can accompany our thoughts but that the I is not actually always there. Brandom thinks the I must always be there because judgements are a minimum unit of thought and an 'I' is necessary to affirm or deny them. That is, what Sartre thinks is a mistake is what Brandom has argued and what Brandom has argued is that the I must be present as indicative of the minimum unit of thought. For Brandom what is fundamental to the nature of thoughts is based on the fundamental nature of judgments, and that is for the I to accompany our thoughts as a reality, de facto, rather than as a de jure possibility.

Sartre writes

> We must have to agree with Kant when he says that 'it must *be possible* for the 'I think' to accompany all our representations'. But should we thereby conclude that an I inhabits, de facto, all of our states of consciousness and really performs the supreme synthesis of our experience? It seems that this would be to distort Kant's philosophy. The problem of critique is a de jure problem: thus Kant affirms nothing about the de facto existence of the 'I think'[242]

---

[242] Sartre, Jean-Paul. *Transcendence of the Ego*. Abington: Routledge, 2004. Pg 2. NB italics his

To be sure Sartre would allege that Brandom has taken Kant's apperception of the I as ever present in our thoughts and fallen into the same blunder as Brochard.

Sartre writes

> (Kant) seems, on the contrary, to have clearly seen that there were moments of consciousness without an I, since he says 'it must be possible'. The real issue is rather that of determining the conditions of the possibility of experience. One of these is that I must always be able to consider my perception or my thought as mine, that is all. But there is a dangerous tendency in contemporary philosophy – traces of which can be found in neo-Kantianism, emperico-criticism, or an intellectualism such as that of Brochard – which consists in turning the conditions of possibility determined by critique into a *reality*[243].

And indeed Sartre, would argue, Brandom has committed the same mistake as Brochard by arguing that the I is ever present in our thoughts. In fact, Brandom has argued that Kant maintains that thoughts cannot occur without an I present to them. Judgements are marked out for Brandom by the 'I think' which Brandom maintains is the minimum unit to have thought.

Since our interest in this thesis is in the clash that occurs in Ryle's Ordinary Language Philosophy between Linguistic Behaviourist style arguments and unacknowledged sources and practices that arise from other ordinary language style argumentation, and given that we are seeking to differentiate whether non-linguistic knowledge is prior to linguistic meaning, and the role this chapter plays in that philosophical

---

[243]Sartre, Transcendence of the Ego, 2004. Pg 2

enquiry, I now turn to scholarship rather than try to engage at length passages of Kant and the German casing system[244]. Here, of course, Otfried Hoffe would agree with Sartre about Kant on this point. He explicitly states "the I think must be capable of accompanying all of our representations"[245]. For Kant, Hoffe affirms, knowing consists in the conceptions of representations into a unity[246]. On Hoffe's reading, at the most basic level intuitions and concepts create a representation in what Kant, of course, calls a synthesis. The confusion, if we are to side with Hoffe's reading, would seem to arise from two processes; the 'Transcendental Apperception of the I', and 'Original synthesis'; and their relationship to the representation created through the process of a synthesis.

On Hoffe's reading[247] the process of the Transcendental Apperception of the I is a condition for the possibility of knowledge and must exist prior to experience. Hoffe's analysis of Kant has three levels. Original Synthesis is the unity of intuition and concepts at the first level, for instance, in the case of a body, the synthesis would be the

---

[244] The sections in question in The Critique of Pure Reason are:
Kant, Immanuel. *Critique of Pure Reason*. Translated by Vasilis Politis, J. M. D. Meiklejohn: Orion Publishing Group, 2004. For the Max Muller vocabulary see Kant, Immanuel. *The Critique of Pure Reason*. Translated by Marcus Weigelt. Max Muller. Victoria: Penguin, 2007.
Book I, The Analytic of Concepts, Section III. The Relation of the Understanding to objects in General and the possibility of Understanding A priori. Pp 130 – 136
Book I, The Analytic of Concepts. Section III . Summary of the Presentation of the Correctness of the Deduction of the Pure Concepts of the Understanding, and of its being the only possible deduction. Pp 137-139
Appendix to Book II, Chapter III, The Amphiboly of the Concepts of Reflection Arising from the Confusion of the Transcendental with the Empirical. Pp 216 - 220
Appendix to Book II, Chapter III and Remark on the Amphibology of the Concept of Reflection. 220-233
In particular, see *The Amphiboly of the Concepts of Reflection Arising from the Confusion of the Transcendental with the Empirical*. NB. For Kant not all judgments require examination, but all judgments require reflection. But here, Kant is referring to comparative judgments. Pure reflection on the mind in the sense John Locke held produce ectype ideas about the operations of the mind, I take it acts of pure reflection in the Locke sense, at this level, are analogous to Kant's 'transcendental judgments' concerning the operations of the mind by comparison of representations with one another. See pp 216 – 220. The terminology I am using comes specifically from the Meiklejohn translation. See Marcus Weigelt's translation for the Max Muller vocabulary. Kant, Immanuel. *The Critique of Pure Reason*. Translated by Marcus Weigelt. Max Muller. Victoria: Penguin, 2007.
[245] Hoffe, Otfried. *Immanuel Kant*. Translated by Marshall Farrier. New York, Albany: The State University of New York, 1994. Pg 99
[246] Hoffe, *Immanuel Kant*, 1994.Pg 76
[247] Hoffe, *Immanuel Kant*, 1994. Pp 74 – 84.

concept of weight. The second level is where a judgment is made such as 'the body is heavy'. The third level is the point at which the possibility of the I arises from the transcendental apperception of the I in the form of a judgment.

Sartre thinks we can inhabit the world at Hoffe's second level of judgements and perception. This, of course, will become Sartre's unreflective consciousness. Brandom, however, thinks we need to inhabit the world at the third level which is the point at which Sartre's de jure possibility for the 'I think' to accompany our thoughts would arise. But Brandom has inherited analytic insights into the way that propositions work. As well Brandom has nous on relations and the way properties stand between objects, and methodological sophistications that have developed out of specifically analytic readings of Kant, which he freely admits. Sartre, on the other hand, has come to his Kant via Bergson and Hegel's Logic[248] along with continental readings like Brochard.

In particular, I think Sartre has engaged what Wallace called Hegel's Doctrine of Being, and explicitly so. Moreover I think that a coherent reading of Sartre is impossible without engaging Hegel's Logic, but unfortunately this goes against conventional wisdom in the Anglo-Sartre scholarship. So now I am going to provide a short literary review before returning to Sartre's argument about the apperception of the I.

---

[248] For details on both of them see next chapter footnotes. Sartre refers to all three of Hegel's central logical works; the Encyclopedia of Logic, The Greater Logic and the 1808-1811 Gymnasium lectures.

4.2.1 Did Sartre Read Hegel's Logic and Why is this Important?

Pace Spade, I think that Sartre has been influenced by Hegel's reading of Kant, specifically Hegel's doctrines of Logic. In the following I'm going to make a point of a short literature review hereto correct a number of mistakes which I should argue are endemic to the Sartreian literature before returning to the thread of the argument concerning consciousness, introspection and reflection in Ryle's arguments. By the end of this section we will have the tools we need to pinpoint reflection and introspection in Ryle's arguments and unpack the phenomenological thread in *The Concept of Mind*. In particular, I'm going to spend a few paragraphs making a short case against Professor Spade, because a number of these seem to follow on from him, the most important of which is his contention that Sartre did not read Hegel's Logic(s)[249].

Professor Spade writes

> Sartre's own attitude toward Hegel is perhaps a little strange to modern readers. Oddly enough, Hegel was almost totally unknown in France until after World War I, when Kojeve and Jean Hyppolite began to introduce Hegel to French intellectuals[250]. And the main work they were interested in was Hegel's Phenomenology of Spirit, not the *Logic* and not Hegel's other writings.

---

[249] I have adopted Sartre's 'habit' of referring to *The Encyclopaedia of Logic* as 'The Logic' because it coincides with Wallace's. See Hegel, G. W. F. *The Logic of Hegel: Part I of the Encyclopedia of Philosophical Sciences*. Translated by William Wallace. London: Oxford, The Claredon Press, 1873. Hereafter I will refer to the Wallace translation first, followed by the Geraets, Harris and Sutching. For the Geraets, Harris and Sutching see Hegel, G. W. F. *The Encyclopaedia of Logic: Part 1 of the Encyclopedia of Philosophical Sciences with the Zusatze*. Translated by W. A. Suchting T. F. Geraets, H. S. Harris. Indianapolis: Hackett, 1991.

[250] This point about the phenomenology and Hyppolite is ambiguous at best. The Logic was already well known, popular and one of Hegel's only works already freely available in French at the time Hyppolite began to lecture. See Heckman, John. "Introduction." In *Genesis and Structure of Hegel's Phenomenology of Spirit*, Pp xv - xli. Evanston: North-Western University Press, 1974. Pp xxiii-xxiv.

Kojeve and Hyppolite's interpretations nowadays are regarded as pretty unorthodox. Nevertheless this is what Sartre knew. So if you know something about Hegel on his own don't expect it to conform necessarily with what Sartre says about him. [251].

In contrast I take it Sartre read Hegel's Logic, and the reason is firstly, because he says that he did[252].

Such is certainly the point of Hegel. It is in the *Logic* in fact that he studies the relations of Being and Non-Being.[253]

Secondly, I believe him, because the part of Hegel's Logic[254] that Sartre is referring to there is Being and Non-being[255] the stage in which Hegel is considering a halfway point between Being and Nothing. Here Hegel

[251]Spade, Paul-Vincent. *Jean-Paul Sartre's Being and Nothingness Class Lecture Notes Fall 1995*. http://pvspade.com/Sartre/sartre.html, 1996. NB Italics his, Professor Vincent Paul Spade, Lectures, Pg 11 -12

[252] Jean-Paul Sartre. *Being and Nothingness*. Translated by Hazel E. Barnes. Abington: Routledge, 2003. Pg 36

[253]Sartre, *Being and Nothingness*, 2003. Pg 38, Italics are his.

[254] See Sartre, *Being and Nothingness*, 2003. Part I, Section I, Sub-section III. *The Dialectic Concept of Nothingness*. Pp 36-40. As noted above I have adopted Sartre habit of referring to *The Encyclopedia of Logic* as 'The Logic'. He refers to all three in this section and throughout the rest of the book. The 'Greater Logic': the 1808-11 Gymnasium lectures and *The Encyclopedia of Logic*. See pg 37 and Barnes footnotes for her translation key. I have limited the focus of my discussion merely to *The Encyclopedia of Logic* for clarity. Where I use the term 'Hegel's Logic' or 'The Logic' I mean by it the German, to the French, as the one that corresponds to that translated by Wallace called "Hegel's Logic" and later Translated by Garaets et al, as *The Encyclopedia of Logic*.' For simplicity's sake I refer to the Wallace translation in the exposition and give the Geraets in the footnotes. Though of course Sartre is using the French translation, of '*The Encyclopedia of Logic*'. See Heckman's introduction to Hyppolite's lectures. Hyppolite, Jean. *Genesis and Structure of Hegel's Phenomenology of Spirit*. Translated by John Heckman Samuel Cherniak Northwestern University Studies in Phenomenology & Existential Philosophy, Edited by James M. Edie. Evanston: Northwestern University Press, 1976. Pp xxiv and xxiii.

[255]Hegel, *The Encyclopaedia of Logic*, Wallace, 1873.Pp 142 - 146 with a discussion of Zeno's Paradox on page 145; Hegel, *The Encyclopaedia of Logic*, Geraets, Suchting, Harris, 1991. Pp 140-142

looks at various philosophical problems like Zeno's Paradox alongside other philosophical arguments that arise from Plato and the like[256] about the states between Being and Non-being. Becoming, for example is not what it is yet, but also, it is not nothing. It is the negation of nothing, because it is something, however, it is not yet what it is because it is still becoming.

Further, I do not think Sartre got his analysis merely from Kojeve or Hyppolite because he uses direct quotes of other sources which Hegel uses, and where Sartre uses them, he either says he got them from Hegel, or he uses them to refer back to Hegel's Logic. For example in *Being and Nothingness*, page 38[257], Sartre quotes Spinoza's Latin from Hegel's Logic for the first time "Omnis in determninatio est negatio. . .'[258] and repeats it systematically at points through out Being and Nothingness. The quote of Spinoza's Latin from Hegel you'll find on page 147 of the Wallace translation[259] suggesting, at least, that Sartre had Hegel's Logic before him at some stage to read it. Moreover, on pages 36 to 42 of *Being and Nothingness* we find that Sartre goes on to quite a detailed discussion of Hegel's Logic adding more evidence that he sat down and read it. This is important to the interpretation of the text, because it affects how you read *Being and Nothingness* and view Sartre's overall project.

Indeed the translator, Hazel Barnes, in a set of notes she published after her translation was published[260] talks about passages of Sartre's analysis and Sartre's model of consciousness in a way analogous to the state between Hegel's being and becoming. The specific stage in Hegel's Logic is Being and Non-Being when being manifests in

---

[256]Hegel, *The Encyclopaedia of Logic,* Wallace, 1873.Pp 140 – 160; Hegel, *The Encyclopaedia of Logic,* Geraets, Suchting, Harris, 1991. Pp 138 - 137

[257] Sartre, *Being and Nothingness*, 2003. Pg 38

[258] Sartre, *Being and Nothingness*, 2003. Pg 38 NB The section where Sartre quotes Hegel, quoting Spinoza reads "What allows Hegel to make being pass into nothingness is that he has implicitly introduced negation into his very definition of being. This is self evident since any definition is negative, since Hegel has told us, making use of a statement of Spinoza's, that *omnis determinatio est negatio*. It does not matter what the determination or content is which would distinguish being from something else; whatever would give it a content would prevent it from maintaining itself in its purity. It is pure in determination and emptiness. *Nothing* can be apprehended in it." *Italics* his.

[259] Hegel, *The Encyclopaedia of Logic,* Wallace, 1873.Pg 147; Hegel, *The Encyclopaedia of Logic,* Geraets, Suchting, Harris, 1991. Pg 149

[260]Barnes, Hazel. "Sartre's Concept of the Self." *Review of Existential Psychology and Psychiatry* 17, no. 1 (1981): Pp 41-66.

becoming. 'Consciousness', Barnes writes in her notes, 'is an ego creating process'[261]. The past is represented in the ego by the appearance of the 'I' and 'me' complex[262]. Only when a person dies do they cease to become[263].

There are any number of parallels one could draw. For example Sartre's plenitude of reality[264] and the self identity of the past[265] seem at least reminiscent, if not directly engaging Hegel's discussion of the role of God as an ontological entity in the reality of the Eleatics though given from an atheist perspective[266]. Sartre's doctrine of the lack and the trinity of lacks, presents itself almost like an inverted Hegelian dialectic. So much so Barnes comments on it in her footnotes[267]. The

---

[261] Barnes, *Sartre's Concept of the Self*, 1981. NB see Sartre, *Being and Nothingness*, 2003. Pg 457 for the specific part I think she is referring to. Here Sartre reveals how all of the earlier stages co-inside with the 'upsurge' of the for-itself as an ego-making process. The for-itself goes after the in-itself, for-itself. Further explanation of this point would involve in-depth exposition on *Being and Nothingness* and here is not the place to do it.

[262] Barnes, *Sartre's Concept of the Self*, 1981. NB see Sartre, *Being and Nothingness*, 2003. Pg 241 the reason why Barnes is arguing that consciousness is like an ego creating process is because the past is where the ego appears as an in-itself of consciousness. This is the identity of the past where 'I am what I am' and traits like 'quick tempered', 'civil servant', 'dissatisfied' which form the 'cortege of the psyche' are static and fixed in time. This is the identity statement where a person is defined by their past traits, behaviors and actions. However, Sartre maintains that in the present 'I am not my past I was my past'. This is the process that collapses at and is the underlying reason for the transformation that occurs at the death of a person whereby the for-itself is transformed into an in-itself. The ego creating process ceases. We can no longer change what we were in life. A person becomes their past and an object for contemplation by other consciousnesses. To use a metaphor of Sartre's the for-itself is like the torso of a mermaid and the in-itself of the ego complex is like the tail of the mermaid. The in-itself of the past is where the I/me of the ego complex appears before the facticity of the contingency of the circumstances and objects in which consciousness first arose as an in-itself and, later, is revealed by reflection. This is the necessary condition of Sartre's earlier work with Kant and Brochard where Sartre argues the de jure necessity that *it must be possible* for an I to accompany our thoughts. The facticity is like the fossilized contingency and the ego is the pattern of a prior consciousness brought before a present one. This ego complex of the in-itself of facticty is in the tail of the mermaid. See pg 169 of *Being and Nothingness* for this metaphor. The unity of the for-itself occurs in an act of reflection as theitic consciousness to which the ego appears as an ensemble of qualities, virtues, lacks but which also reveal the world as a world of tasks with the upsurge of the for-itself and the trinity of lacks, creating 'hodological space' filled with utensils in instrumental complexes. This collection of individual character traits, which forms the ego and sits in the tail of the mermaid is the 'cortege of the psyche'. See pg 184 – 186. This is important because it structurally resembles Hegel's discussion of becoming, and Plato's arguments and his proof for immortality in the Phaedrus, which Hegel is no doubt drawing upon.

[263] Sartre, *Being and Nothingness*, 2003. Pg 138

[264] Sartre, *Being and Nothingness*, 2003. See Sartre's two questions pg 29 - 31

[265] Sartre, *Being and Nothingness*, 2003. Pg 139

[266] Sartre, *Being and Nothingness*, 2003. Pp 145 – 148; Hegel, *The Encyclopaedia of Logic*, Wallace, 1873.Pp 135-146; Hegel, *The Encyclopaedia of Logic*, Geraets, Suchting, Harris, 1991. Pp 137-136

[267] Sartre, *Being and Nothingness*, 2003. See footnote on pg 118 where Barnes considers translating

Origin of Transcendence, for Sartre, is when the 'in-itself[268]' first surges up as a synthetic unity of the lacking and the existent towards the lacked[269]. This is the synthesis of the whole, which is what eventually reveals the world as a world of tasks into which the for-itself enters as a relation with the world, creating instrumentality which appears before it at the unreflective level as an inescapable condition. Ultimately this inescapable instrumentality is what determines the infinite regress of utensils into the ensemble of instrumental complexes that makes up the triadic relation between the given[270], the co-efficient of adversity[271] and

---

Sartre's 'trinity of lacks' as a Hegelian dialectic, except that the for-itself is contingent upon the in-itself as facticity in the surge towards the lack, rather than as antithesis generated from opposition like she considers in her footnotes on Pg 110. If we view the trinity as an 'inverted dialectic', rather than try to see Sartre's relationship of the lacking to the lacked in the surge towards synthesis as merely a form of Hegelian negation, we could perhaps see the lacking surging towards a totality of synthesis of the lacked as an inversion of Hegel's opposition of the antithesis. Suffice it to argue there is more than ample evidence that Sartre is in some form of dialogue with Hegel's Logic, and this dialogue is essential and fundamental for understanding the work.

[268] NB I use in-itself and for-itself to refer to my reading of Sartre, and in-its-self and for-its-self, with the extra hyphen, to refer to Danto and his follower's overly simplified and faulty distinction.

[269]Sartre, *Being and Nothingness*, 2003. Pg 103 for the lack, see also pg 110 for the inverted dialectic of the 'trinity of lacks', and pg 113 for the series of arguments on transcendence.

[270]Sartre, *Being and Nothingness*, 2003. Pg 498 – 503. The 'given' is a part of a complex refutation of the behaviourist position, which is built on the 'Concept of the Other' and Sartre's analysis of the mind/body problem. The Concept of the Other is better represented in the literature but Sartre's analysis of the mind/body problem and the 'analogical man' significantly less so. The 'analogical man' is the product of the 'analogical reconstruction of the body'. The mind/body problem for Sartre, arises on this view, from a reconstruction of the body from a position approximating, roughly, the third person direct view, as defined by this paper. Movement of tendons and muscles are thus analogically reconstructed without consciousness from the third person point of view in the manner of pistons and using analogies of machines. The scientifico-medical fallacy, according to Sartre, arises from dissecting corpses and making analogies. For instance, a medical-scientific analysis of the eye might follow along the lines that the light strikes the cornea and travels through the aqueous humor, causing reactions and adjustments to the ciliary and orbital muscles, to the lens, along the visual axis, causing reactions in the optic disc and fovea, further stimulating the optic nerve and into the visual processing areas of the brain. Whereas an account of the body not built from analogical reconstruction would hold, rather than light striking the cornea and setting off a number of reactions, it would involve something along the lines of 'I looked towards the mountain'. The mind-body problem for Sartre is a confusion between third person analogical reconstruction and first person conscious experience. The lack of consciousness in the body from the third person position thus produces the problem of causation and the Cartesian myth that we looked at earlier, and which Ryle is attacking with his ordinary language solution. The 'given' for Sartre, (as opposed to Sellars' 'Given') is a complex refutation of behaviourism built along these lines. See Sartre, *Being and Nothingness*, 2003. Pg 348 for the analogical reconstruction, and pp 344 – 353 for Sartre's analysis of the mind-body problem. See Ryle, *Concept of Mind,* 1983. Pp 14 – 25 Also see pg 79 for the particular passage which parallels Descartes problem in *The Discourse,* which starts "Men are not machines" and Descartes, Rene. "Discourse on the Method of Rightly Conducting the Reason and Seeking for Truth in the Sciences." In *The Philosophical Works of Descartes*, I. London: Cambridge University Press, 1979. See Part V, Pg 109 – 117.

[271]Sartre, *Being and Nothingness*, 2003. Pg 504. The co-efficient of adversity arises because we are

the teleology of the ends organizing the primary and secondary possible in the last parts of the work[272]. Indeed it is the nihilation of a nihilation of the ego in the past as an in-itself that is not itself which creates the surge of the 'for-itself'[273] that makes reflection possible and able to move past the circuit of the self and out of hodological space.

Hodological space is the space of the everyday world. Sartre originally introduces this notion in his *Sketch for a Theory of the Emotions*. Here he borrows the term from Lewin[274]. For Sartre it is the space in which we go about and complete our tasks without reflection or engaging thought on the reflective plain. The reflective plain is the point where consciousness becomes aware of itself and its processes. On the non-reflective plain the self apprehends itself through qualities in the world like needs, desires, wants and so forth. This is what it discovers later when it reflects and becomes the reflecting consciousness for the reflective consciousness in the act of reflection.

Sartre writes

> What is important here is only to show that activity, as spontaneous, unreflecting consciousness, constitutes a certain existential stratum in the world, and that in order to act, there is no need to be conscious of oneself as acting – quite the contrary. In a word, unreflective conduct is not unconscious conduct. It is non-thetically

---

able to posit an end. This is the difficulty of a task constructed in our experience from the 'residuum' left over from brute existence when we transform the world into a world of tasks to fill the voids revealed as lacks in the upsurge of the for-itself. See Sartre, *Being and Nothingness*, 2003. Pg 490. The 'end' is a thematic totalisation of the 'tasks', 'likes', 'hates' to which being projects itself. The 'end' illuminates the world in light of the teleological instrumentality of the tasks produced by the lacks and the voids that arise in the circuit of the self. See 499, Sartre gives an example of this illumination as a light shinning on the 'dusty-road towards the room where the table is set in preparation for the meal'.

[272]Sartre, *Being and Nothingness*, 2003. Pg 493, the 'primary possible' is harnessed by the for-itself for dealing with secondary ends but never refers back to the secondary possible. The secondary possible is based on the stoic notion of an 'indifferent' see pg 491,

[273]Sartre, *Being and Nothingness*, 2003. Pg141

[274]Originally Lewin's term, see Sartre's *Sketch for a Theory of the Emotions*. Translated by Philip Mairet. Abington: Routledge, 2004. Pg 38

conscious of self; and its way of being conscious of self is to transcend and apprehend itself out in the world as a quality of things. In this way we can understand all those exigencies and those tensions of the world around us; in this way we can draw up a 'hodological' chart of our Umwelt, a chart that will vary in function with our actions and our needs. . . From this point of view, the world around us – that which the Germans call the Umwelt – the world of our desires, our needs and of our activities appears to be all furrowed with straight and narrow paths leading to such and such a determinate end, that is it has the appearance of a created object. Naturally here and there, and to some extent everywhere, there are pitfalls and traps. One might compare this world to one of those pin-tables where for a penny in the slot you can set the little balls rolling: there are pathways traced between hedges of pins, and holes pierced where the pathways cross one another. The ball is required to complete a predetermined course, making use of the required paths and without dropping into the holes. This world is difficult. The notion of difficulty here is not a reflexive notion which would imply a relation to oneself. It is out there, in the world, it is a quality of the world given to perception[275].

The notion of quality changes between his earlier works and *Being and Nothingness*. In the earlier works a quality is a reflection of the non-

---

[275]Sartre, *Sketch for a Theory of the Emotions*, 2004. Pg 38 - 39

thetic self in the world like hunger, need, desire, the difficulty of a task and so on. The non-thetic self reflected on becomes the thetic self, like recalling yourself being hungry at a BBQ. In *Being and Nothingness* a quality is separable from this reflective process. Later the reflection of the self in the qualities is replaced by the trinity of lacks and the voids which reveal the world as a world of tasks making up the pathways and passages of hodological space and the instrumental complexes of the world[276]. This is because Sartre offers an account and critique of Husserl's categories which he refers to as the 'attitudes of indifference[277]'.

In Hegel's Logic the negation of negation itself creates the conditions for the upsurge that produces a 'for-itself'[278]. This then provides Hegel with the grounds for a string of arguments that end in the creation of a quality and the quotation of Spinoza's in the Latin mentioned above. The determination that provides the grounds for a

---

[276] See Sartre's discussion of the Worumwillen on Sartre, *Being and Nothingness*, 2003. Pg 223 and his discussion of 'ontological solidarity for exploitation of the world' and 'stimmung" in Part III, *Being for Others, The Existence of Others*, subsection III pp 268 – 276. In particular his critique of Heidegger. I'm oversimplifying for brevity of exposition. The world, for Sartre, is not simply revealed as a world of tasks to the individual, the instrumental complexes of the world include 'worumwillen' the 'for whom' and 'stimmany' the relation of the self to others as co-worker. The nature of these relationships is too complex to do justice in a footnote.
[277] The two taken together can be seen as analogous to Wilfrid Sellars 'Empirical Image' in *Philosophy and the Scientific Image of Man*, 2001. Sartre thinks Husserl's categories are unnatural abstractions, they're the world devoid of the domestic characteristics that make up the mundane day to day existence. The character of this mundane existence consists in the difficulty of running for the bus, the irritation of spilling coffee on some documents, the annoyingness of the faulty elevator. Sartre thinks that science is the world stripped of these qualities, that is Sartre thinks that science is the world from the perspective of the 'attitudes of indifference', as pure exteriority, the relationship between one object and another, and in this sense his view of science is analogous to the Empirical half of the Manifest Image, i.e. all of Mills inductive canons and the world as purged of personality and the perennial philosophy of the one. See Sartre, *Being and Nothingness*, 2003. Pg 214 for the specific 'attitudes of indifference' which Sartre takes out of Husserl, and Pg 207 for Sartre's concept of space as a moving relation between objects stripped of both the instrumental complexes organizing them and hodological spaces structuring them within the circuit of the self. More importantly they do not posses the properties of the lacking to the lacked that allow the upsurge into being as temporal flight from the in-itself of facticity towards the possibilities that organize the projection of tasks onto the horizon of being. Husserl's objects of contemplation do not posses the essential properties of usefulness that characterizes the world as organized by the teleological structure of consciousness encountering tasks, which then create utensils in the projection of being towards the different types of possible, see Sartre, *Being and Nothingness*, pp 222 – 223, the world is revealed to the non-theitic self as a world of tasks. The tasks arise as demands or 'voids to be filled' which organize objects towards 'thematized' ends.
[278] Hegel, *The Encyclopaedia of Logic*, Wallace, 1873. Pp 151 – 152; Hegel, *The Encyclopaedia of Logic*, Geraets, Suchting, Harris, 1991. Pp 149 – 153.

negation is what allows the creation of quality. Likewise, Sartre argues that the internal negation of an external negation creates the grounds of the world, where-by an object can be brought before consciousness where its negation to consciousness as not consciousness produces a quality[279].

The exact mechanism for producing this quality in Sartre is particularly interesting. The for-itself creates a ground for the world by a negation; it is everything that is not the world. This is the first radical negation[280] that establishes the ground upon which an object appears. The for-itself then creates a 'this' by bringing forth an object, like say a dinner plate, from the background of the world. The 'this' of the dinner plate, to the world as background, appears as an external relation of the totality produced by the first radical negation, thus establishing the ground of the world by the for-itself to the world as a totality. However the 'this', the dinner plate once brought before the for-itself and considered apart from all external negations, the world, and any other 'thises'. The dinner plate as not the for-itself provides the conditions under which an internal negation can take place producing a quality[281]. That is as an internal negation, say the whiteness inherent in the translucency of good bone china, as not the for-itself, since consciousness is not 'white'. The 'not-whiteness of consciousness' as a centre of negations, to the 'whiteness' of the plate produces a 'quality' through a relation of the for-itself to the dinner plate as an internal negation. However, the plate to the background of the world is an external negation, and the background of the world to the for-itself is an

---

[279] The difference between the two conceptions of qualities is in the latter work *Being and Nothingness*. Here Sartre takes into account the object stripped of all instrumentality and Husserl's phenomenologically derived categories. Here he uses two different mechanisms. A 'trinity of lacks' which creates 'values' in the world and reveals objects in their instrumentality to tasks, and the purified mechanism of apprehending the objects stripped of their instrumentality which he thinks is a perversion. The former is what makes up hodological space, the latter is what makes up Husserl's categories of indifference. In *The Sketch for a Theory of The Emotions* he is merely looking at the role emotions play in acts of consciousness. What's important is to see how Sartre's dialectic of the lacks, what I am calling the 'trinity of lacks' which, without laying out an indepth thesis, is in some sort of dialogue with Hegel, and how this dialogue forms part of a critique of Husserl.
[280] Sartre, *Being and Nothingness*, 2003. Pg 192
[281] See his discussion of Cézanne and his refutation of Husserl's object pole in consciousness, Sartre, *Being and Nothingness*, 2003. Pg 209 -215. I'm simplifying his argument iin the body of the text of this paper for the purposes of exposition. The Husserlian stripped reality is an abstraction, not an underlying principle of hodological space. See also Sartre, *Being and Nothingness*, 2003. Pp 27-29, and his discussion of La Porte and the concrete and abstract distinction.

internal negation that produces an original ground for consciousness in the world. Quality is thus based on another internal negation, like the original radical negation that consciousness sets up as a background when it negates the world[282].

The mechanism of quality as an internal negation can naturally be seen to be in direct dialogue with Hegel's discussion of the finite and the production of quality by a negation of the finite by the infinite. This occurs also in the Doctrine of Being in Hegel's Encyclopaedia of Logic[283]. Sartre actually gives an example of this Hegelian-like mechanism early on with a sniper looking though a gun scope. First the sniper sees the world then he negates that to bring out the sentry he intends to shoot[284]
.

Just like in Hegel's Logic, once Sartre has produced quality he then follows the sequence of Hegel's process ontology and proceeds to produce quantity and presents this as his own negation of Husserl[285]. The means of making the move from quality to quantity, while structurally resembling and in dialogue with Hegel's Logic[286], I'd argue, Sartre has appropriated from Bergson[287]. Sartre following Bergson[288]

---

[282]Sartre, *Being and Nothingness*, 2003. Pg 210
[283]Hegel, *The Encyclopaedia of Logic*, Wallace, 1873.Pp 151 - 152
[284]Sartre, *Being and Nothingness*, 2003. Pg 32
[285]Sartre, *Being and Nothingness*, 2003. Pp 209-214
[286]Hegel, *The Encyclopaedia of Logic*, Wallace, 1873. Pp 151-155; Hegel, *The Encyclopedia of Logic*, Geraets, Suchting, Harris, 1991. Pp 149-157
[287]Sartre, *Being and Nothingness*, 2003. Pg 207, Sartre writes 'the continuous into the discontinuous defines space' by that, Sartre means a set of abstract principles that arise from the relationships between one 'this' and another 'this' to the relation between the two 'thises' which are not each other, as not space. This is the radical negation that sets up the 'grounds of the world' and the notion of the self to the 'this'. Space in its pure exteriority, for Sartre, is non-hodological space outside the circuit of the self. That is space unorganized by tasks would be another way of putting it. This space is a moving relation between the negation of objects in the field set up on the grounds of the world. Sartre thinks what lays at the base of this is the negation of the for-itself to the notion of space itself. The for-itself is not extended in space, rather space unifies objects that are not the for-itself, but are revealed against the foreground of consciousness as discontinuous, while suspended against the background of space which are continuous. This is the continuity which forms part of the transphenomena that Sartre identifies in his argument against Berkley which we looked at the start of this paper in *2.4 Sartre's Irreal* However even though the objects are unified in space as continuous the objects negate each other because they are not each other which makes them discontinuous. Thus, for Sartre, the movement of the continuous into the discontinuous defines space. This is an elaboration on *Hegel's Doctrine of Being* which Sartre, I argue, has picked up from Bergson's concept of number. In Bergson, see Bergson, Henri. *Time and Free Will, an Essay on the Immediate Data of Consciousness*. Translated by M.A. F. L. Pogson. New York: Dover, 2001. Pp 82 – 85, continuity and discontinuity in spatial terms allow for the development of the concept of number.

argues that space cannot be a being rather it is a relation that moves between beings[289]. Each being is differentiated by its external negation to the background of the world inside consciousness. Thus the desk and the stapler are both brought into consciousness as not the world. The world, as background, contains everything that is not consciousness, including the fine bone china dinner plate in the next room which we've seen and know about, and the dinner plates down the road we have never seen and do not know about[290]. But the relationship between the desk and the stapler is an external relation and is unlike the relationship between the for-itself and the world that sets up the original radical internal negation that provides the grounds for consciousness in the world. The desk is not the stapler. The stapler is not the desk. This is the moving relation of external negations that make up 'thises' of objects, against the background of the world. This moving relation of external negations provides the discrete grounds for objects in Sartre's concept of space. This, Sartre argues, is the origin of quantity. But this is only half the story because we don't normally see things this way. The 'inverted dialectic' or the 'trinity of lacks' is what makes up the other half. Let me explain.

The world is revealed, via the 'trinity of lacks' as a world of tasks. The tasks refer us to various utensils needed to complete them[291]. The lacks and the mechanism of perception via internal and external negations make up the instrumentality of the ensembles that inhabit hodological space'. In hodological space, I have to move the coffee mug to reach the stapler in order to staple some papers. The coffee mug to stapler is not merely a relation of measured distance, say, a grid of inches or centimetres squared, or sets of radii from some discrete point, or a series of transitive relations in logical space, but rather one of

---

Bergson holds that the objects must be continuous in one sense, to be counted together, but discontinuous in another sense, to be counted separately as individuated units. So for instance, Bergson would argue pebbles must be alike to be counted individually, but different enough in their discrete properties that they're not just the same continuous piece of rock. Thus, Bergson argues, the concept of number has origins in our experience and representation of spatial properties. This spatial representation of discrete units, Bergson holds, pp 105-107, get imposed on conscious duration and thus produces the phenomenal concept of individuated moments in time.

[288]Sartre, *Being and Nothingness*, 2003.Pp 207-208
[289]Sartre, *Being and Nothingness*, 2003. Pg 207
[290] In the manner of attributative and not a designational object, see Michael Devitt's "Thoughts and Their Ascription." *Midwest Studies in Philosophy* 9, no. 1 (1984): Pp 385-420.
[291]Sartre, *Being and Nothingness*, 2003.Pg 222

obstacles to completing my task. The internal and external negations stripped of these lacks produces Husserl's categories, i.e. unity, instrumentality, multiplicity, relation, within, without, all of which Sartre calls 'The attitudes of indifference'[292]. Sartre's critique is clear. Husserl's reality is not the reality of the everyday world, Husserl's space is not the space of hodological space, rather Husserl's reality is a perversion. The significance, of course, is that here Sartre again references Hegel's quotation of Spinoza's Latin in the Logic 'Omnis in determnninatio est negatio. . .[293]' Everything is determined by negation.

My argument is that any piece of scholarship that doesn't acknowledge Sartre's dialogue with Hegel's Logic(s) is a profound misreading.

4.2.2 Freedom, consciousness and Psychasthenia

Another criticism that seems fairly common in the Anglo-Sartre literature is that Sartre doesn't provide examples of extreme freedom or why people are concerned about it enough to adopt moral codes and values. Here is Gregory McCulloch's version

> The fact remains that few people have seemed to reach this point. Explicit subscription to Sartre's conception of freedom, is, if anything, even rarer than anguish. So there would still be a problem of explaining the apparent evasion. One could reply that the world still awaits the dawning; the full implications of the conclusion that the world in itself has no

---

[292]Sartre, *Being and Nothingness*, 2003.Pp 210-214
[293]Sartre, *Being and Nothingness*, 2003.Pg 202

> meaning has not filtered through yet. People just have not realized the consequences this has for our freedom and our values[294].

McCulloch seems to have missed the point of anguish. It is not simply what happens when we become aware that we no longer have sets of morals or procedures to guide us, and we are made aware of the sudden lack of restrains. It is also something that can only take place on a reflective level. The structure of our consciousness in the everyday world stops us from realizing an awareness of this freedom because it is generally unreflective. However movement from pre-accepted values into an area or situation without those accepted values, where we don't know what is expected of us, suddenly makes us aware of how free we are.

Sartre's point is that the structure of our relationship with the world prevents us from realizing our freedom and the anguish that the freedom generates at the unreflective level. We simply aren't confronted with it in our day to day affairs while engaged in the circuit of the self. It is only in moments when we are confronted with the choice whether to take a job or not, whether to commit an immoral act, whether to uphold a promise to ourselves that we realize the nature of our freedom and suffer anguish from it. Surely most people have experienced these moments in their life.

The hackneyed argument that Sartre claims the world has no meaning has a different but entirely related source. At the unreflective level the structure of consciousness prevents us from finding a certain type of meaning in the world. This is compounded by the fact that at the unreflective level one can also lose one's self in the world in a regress of utensils.

Sartre writes

---

[294]McCulloch, Gregory. *Using Sartre*. New York: Routledge, 1994. Pg 53

For human reality, being in the world
means radically to lose oneself in the world
through the very revelation that causes
there to be a world – that is, to be referred
to without respite, without even the
possibility of "a purpose for which" from
instrument to instrument with no recourse
save the reflective revolution. . . To be sure
these work clothes are for the worker. But
they are for the worker so he can fix the roof
without getting dirty. And why shouldn't he
get dirty? In order not to spend most of his
salary on clothes. This salary is the
minimum quantity of money that will
enable him to support himself; and he
'supports' himself so as to be able to apply
his capacities for work at repairing roofs[295].

At the unreflective level there is no room for posing questions of
meaning that will bring one face to face with the reflective conditions
for anguish. This, of course, is the point of Sartre's discussion of the
waiter, the grocer, the prison warden and so on[296].

Sartre writes

(B)eing is revealed to the for-its-self on the
ground of the world as an instrument-thing,
and the world rises as the undifferentiated
ground of the indicative complexities of
instrumentality. The ensemble of these
references is void of meaning, but in this
sense – the possibility of positing meaning
on this level does not exist. We work to live,
and we live to work[297].

[295]Sartre, *Being and Nothingness*, 2003. Pg 224
[296]Sartre, *Being and Nothingness*, 2003. Pp 67 - 68, 70, 81 etc,

Only when one reflects can the possibility for a different type of meaninglessness occur, and the possibility for inventing meaning, arises. This is why moral decisions can 'jolt' us on to the reflective level for the search for meaning simply because we don't know what to do. We no longer have sets of values to fall back on. We can not question the ensemble while we are in it, we can only question it as part of a self-discovery on the part of the for-itself, at the reflective level.

Sartre writes

> The question of the meaning of the totality "life-work" – "Why do I work? I who am living? Why live if it is in order to work?" This can only be posited on the reflective level since it implies a self discovery on the part of the for-itself[298].

This is the point where the for-itself moves beyond the circuit of the self in which it apprehends objects in hodological space. The objects in hodological space are organized into tasks by the voids that appear in the circuit of the selfness and which reveal the world through the trinity of lacks as the tasks to be completed[299].The question of meaning only opens up at the reflective level beyond the tasks of the world and the voids that appear in the circuit of the selfness that, in turn, reveal themselves as the lacking towards the lacked as transcendence. This is because reflective consciousness occurs in theitic consciousness which, of course, is consciousness that takes itself as its own object, but in doing so the prior or contemporary act of consciousness appearing before itself is still engaged in the world through the radical negation that causes the world to appear. Consciousness in the circuit of the self is engaged in the tasks of the world and can not engage these questions. It is only by reflecting on this consciousness either in a moment of crises or by meditating on a prior act of consciousness that the self can appear

---

[297]Sartre, *Being and Nothingness*, 2003. Pg 224

[298]Sartre, *Being and Nothingness*, 2003. Pg 224

[299] See Sartre, *Being and Nothingness*, 2003. Pp 222 -223 for a description of this process

to itself and the question of meaning, and in doing so brings the for-itself before the full realization of freedom as an act of self discovery. It is only in such a moment that the apprehension of anguish can be fully realized. To answer McCulloch, it is only in such moment that the full scope of freedom and anxiety arise. And I think if people are honest, most people would experience such moments on occasion in their lives.

4.2.3 The For-itself and the In-itself.

Another one of the common errors that riddles the Anglo-Sartre literature is that Sartre separates reality down into two separate and irreducible parts a bit like football teams. This division typically takes the form of a split between the For-itself and the In-itself with human consciousness playing on one side and objects on the other. Moreover the articles and books that make this mistake all seem to reference each other when they speak this way of the distinction in Sartre. The references seem to lead, through a warren, back to a work by an Arthur C. Danto of Columbia University[300]The number of works which reference or utilize Danto as a source are legion[301]. I should, by way of a general source introduce only a couple of interesting examples in the footnotes and focus on the central mistake in Danto's work and clearing it up in the main body.

This passage from Danto is I think where the mistake begins,

> In every case (Sartre) proposes that the analysis of the structures of consciousness yields two ontologically primitive types of being, conscious beings; these he terms etres-pour-soi, or beings-for-themselves and cannot exist as such without this awareness; and etres-en-soi, or beings

[300] Who also has the credit of being cited as a leading authority, and quoted on the blurb, on the back of the Routledege edition of *Being and Nothingness*. Sartre, *Being and Nothingness*, 2003. 'Back blurb'. This citation is significant as it seems to suggest this is the place for scholars to go when embarking on a study of Sartre and could be part of what is perpetuating confusions in the readings.

which exist in themselves, and which are objects for an alien consciousness, having no consciousness of their own[302].

Most recently Nigel Warburton writes

> The whole of Being and Nothingness rests upon a fundamental distinction between different forms of existence. Sartre draws attention to the difference between conscious and non-conscious being. The former he calls 'being-for-itself'; the latter, 'being in-itself'. Being for-itself is the kind of existence characteristically experienced by human beings, and most of Being and Nothingness is devoted to explaining its main features. Unfortunately Sartre provides no answer to the question of whether or not non-human animals can reasonably be categorized as examples of being for-itself. Being in-itself, in contrast, is the being of non-conscious things such as of a stone at a beach[303].

This is true enough of the characterization of the contrast in the first forty to sixty pages of *Being and Nothingness*[304], – Rather than refute

---

[302] Danto, Arthur C. *Sartre*. Glasgow: William Collins and Co Ltd 1975. Pg 50.

[303]Warburton, Nigel. *Philosophy: The Classics*. Oxon: Routledge, 2001. Pg 219. NB The Nigel Warburton source is particularly valuable. Most introductions or expository chapters on Sartre seem to recycle them. For instance, Stevenson, Leslie. "Sartre: Radical Freedom." In *Ten Theories of Human Nature*, Pp 181 - 200. New York: Oxford University Press, 2009. makes many of the same errors. Warburton seems to collect most of these in a single source. Again these texts are highly significant and could be what is perpetuating much of the misreading of Sartre I have been at pains to point out.

[304] The 'enough' used above is charitable. It could be argued that even this would be a misreading of what characterizes the Being-in -itself at the start of Sartre's process ontology from the perspective of what happens later to the ego creating process. On Pg 47, Sartre makes the distinction between a

these on a case by a case basis, I'll simply provide a sketch of the anatomy of a Satreian being to show what's wrong with the received conventional wisdom of a division between the in-itself and the for-itself as conscious and unconscious beings, and list some of the articles that make these basic mistakes in the footnotes[305].

To begin with at any one time the for-itself must contain an in-itself of facticity. This is because, Sartre asserts consciousness cannot

---

being-in-itself and a being-for-itself, and argues that the being-in-itself can not contain the nothingness of negation that defines his (Sartre's) mechanism for 'negative predication'. However, in order for theitic consciousness to arise, the for-itself must grasp a prior act of consciousness in the world, and this prior consciousness is a transcendent in-itself. This early distinction is essential to understanding Sartre's process ontology and fundamental for the means by which the 'I' of consciousness can arise. The distinction is the grounds Sartre later argues enables the conditions for the ego to arise within consciousness. See page 147. The ego is not a for-itself, but is in the in-itself of facticty and appears to consciousness with the grasping of a prior consciousness as an I or a me. Equating consciousness with the for-itself would flatly prevent the conditions that allow the ego to arise.

[305] Thomas Martin, for example, is an unfortunate case, in that he makes this mistake in his *Introduction to Sartrean Existentalism* on page 78 and again in his paper, *Sartre, Sadism and female Beauty ideals.* You can see clearly on page 91 that Thomas Martin seems to struggle with the distinction, and conflicting components of it, which leads to a general lack of clarity on it in his discussion of the body, on page 93, and again on page 98, and it has a clearly visible negative effect in his final analysis.

Indeed Thomas Martin lists Danto in his bibliography on page 102, and I should argue, that this is, perhaps the origin of the mistake, and the cause of is his confusion. Martin is an interesting case, in that, amongst the Sartrean literature, I think he is one of the ones who has actually read Sartre, but the over simplification made by Danto, and a lot of the poor quality peer review literature has, I should argue, led to a number of confusions in his papers. See Martin, Thomas. *Oppression and the Human Condition: An Introduction to Sartrean Existentialism.* Lanham: Rowman & Littlefield Publishers 2002. And Also Martin, Thomas "Sartre, Sadism and Female Beauty Ideals." *Australian Feminist Studies* 11, no. 24 (1996).

Willard D. Keim is another interesting case. See Keim, Willard D. *Ethics, Morality and International Affairs* Lanham: University Press Of America, 2000. In his analysis of Sartre at the start of his work on ethics Keim classifies Sartre's ontological position on realism, on the rather obvious basis of Danto's mistake, see pg 7, but then later, in the same chapter, on the basis of Wilfrid Desan's admirable analysis, I should argue, correctly identifies the structures inside the for-itself responsible for the upsurge, including the trinity of lacks. It's curious that he, himself, hasn't picked up on the tension between the two positions.

Indeed, Wilfrid Desan, on the basis of his essay on Sartre, *The Tragic Finale* is one of the few sources, from the papers I have read, that has actually picked up on the nature of the distinctions between Being in-itself and Being for-itself. Indeed, he does so through an admirably close examination of the structures inside of Sartre's *Being and Nothingness.* It seems a shame that this book wasn't picked up and had a wider influence, in place of that exerted by Danto's, since it is the far more accurate representation on any number of aspects and of even an earlier vintage. Indeed, being of an earlier vintage, it has perhaps, avoided the direct influence of Danto's work, and indeed, the influence of those who have contributed to both the peer review literature, and the general expository literature, via some influence or connection with Danto, or those who read him. See Desan, Wilfrid. *The Tragic Finale: An Essay on the Philosophy of Jean-Paul Sartre.* Cambridge: Harvard University Press, 1954.

arise in the world as its own foundation [306]. He introduces this idea rather early in *Being and Nothingness* and it ties directly into the differences between the imaginary and perception, and the Berkleyian argument occurring throughout *The Imaginary* and at the beginning of *Being and Nothingness*[307]. Moreover facticity resides within the experience of the world to be discovered and encountered from the world of unreflected consciousness, and it is from this encounter with the facticity of the world in which consciousness needs to arise as orientated towards an object[308]. This is the circuit of the self which we encounter in an unreflective state, and which I pointed out before. The circuit of the self occurs within the ensemble of instrumentality and the regress of utensils that Sartre thinks meaning cannot exist within. However, once consciousness arises in the world it has the power to reflect upon a previous consciousness. This previous consciousness was consciousness of an The object provides the grounds for the prior consciousness to orientate its fixation within the world and the object thus becomes the foundation for a theitic act. The reflection upon the prior consciousness is the point at which the 'I and me' complex of the ego appears. At the unreflective level there is no room for an I or a me. The world simply appears before consciousness with the 'foregrounded' object against the grounds of the rest of the world. In the circuit of the self immersed in the instrumental complexes of the world we don't see ourselves seeing a

---

[306]Sartre, *Being and Nothingness*, 2003. Pg 28

[307]Sartre, *Being and Nothingness*, 2003. Pp 6 – 8, Sartre's refutation of Berkley's 'esse est percepti' is based around the notion of 'resolution' I introduced in my historical thesis at the start of this paper. Perception has an element of 'surprise' or 'discovery' that makes it a passive activity and points towards a transphenomena of being.

[308]Sartre, *Being and Nothingness*, 2003. Pg 6 - 8 NB, to clarify the way the parts fit together: In Sartre's 'process ontology' the abstract properties of 'resolution', to use my term, which point towards the transphenomena of being, becomes the necessary grounds of contingency whereby consciousness arises in the world. Sartre, *Being and Nothingness*, 2003. Pg 107. This contingency when reflected upon becomes facticity which is apprehended at the theitic level. The theitic level, of course, is the level where the 'I' or the 'me' of a prior consciousness appears. Once the process is understood it can be seen how the in-itself of facticity contains the reflection of contingency from the pre-reflective cogito as presence to the world and simultaneously the in-itself of the lacked as the upsurge towards synthesis of the lacking, to the lacked, in what I mentioned in prior footnotes resembles an inverted Hegelian dialectic. See Sartre, *Being and Nothingness*, 2003. Pg 113, where Sartre reveals the latter, the trinity of lacks is what drives the process of revealing the world as tasks. This produces instrumentality from the decompression of being, which Sartre argues produces the possibles, and for-itself, as discontinuous from the grounds of its own contingency, as transcendence, and pivotal in the temporal flight from the present. This is a process that is critical to grasp for understanding the hierarchical structures of the primary and secondary possible in the last part of the work. Both sides need to be grasped at once, to see how leaving out one could be a distortion of human reality producing the 'categories of indifference' and the nature of Sartre's refutation of Husserl.

tree, we don't smell ourselves, smelling cologne. Sartre asserts that at this level an object of consciousness appears directly to consciousness. I shall toil over this argument shortly when we return to the apperception of the I. However the point to be made clear is at the reflective level the I of facticity appears when we reflect upon ourselves as being aware of an object in the past. This is the origin of the in-itself of facticity.

The trinity of lacks which Sartre first introduces on page 110[309] is, of course, the means by which the in-itself first surges up as the lacking, to the lacked, striving towards the synthetic totality of the completed as a surge towards the unlacking. Earlier I suggested that this had an 'inverted Hegelian dialectic' structure. Space limitations preclude me from going further into this aspect[310] suffice to say it is the synthesis of the lacked with the lacking in the surge towards the synthetic unity that produces the primary conditions that allow the for-itself to come into the world. It is important to separate the act of reflection from the mechanism that reveals tasks in the world, in order not to equate the for-itself with reflective consciousness. This equation is another common mistake in the literature[311]. The for-itself can occur

---

[309]Sartre, *Being and Nothingness*, 2003. Pg 110

[310] I should like to argue that he removes God from Hegel's Doctrine of Essences, and this inverts the dialectic. It's rather interesting once you realize what he's up to. The dearth of Anglo-Sartre scholarship prevents me from exploring these aspects of his work at any length, simply because I lack the depth of resources in any great aspect amongst the peer review community to tackle it.

[311] The common mistake, I take it, is that since Sartre argues that Descartes mistake was to assume that the 'I' and the 'think' of the cogito are on the same level in Sartre. See *Transcendence of the Ego*, 2004, pg 13, 14, Sartre, *Being and Nothingness*, 2003. Pg 6, that the equation of the for-itself with the appearance of the I, and the prior act of consciousness brought before it, as the in-itself, is what lays at the base of Sartre's criticism since Sartre argues that Descartes should have argued that 'I think therefore I was'. However the for-itself can occur within the circuit of the self. This is because there are two types of in-itself. There is the in-itself of the lacking, responsible for the upsurge that turns lacks in to tasks, and organizes the instrumental complexes, and there is the in-itself of the past which can appear before consciousness at the reflective level. The for-itself can move within the circuit of the self at a non-reflective level and also when the facticity produced by the contingency that, Sartre argues, is a necessary condition of experience and appears in a prior act of consciousness at the reflective level .I hope it can be seen why Danto's analysis is such a damaging one to anyone trying to grasp Sartre's process ontology. See also Wemin Mo, Wang Wei. "Cogito: From Descartes to Sartre." *Frontiers of Philosophy in China* 2, no. 2 (2007): Pp 247-264. See Pg 251 Mo captures an important aspect of the de jure distinction but would have, perhaps, benefited from reading *The Imaginary*. In particular the Cartesian argument. Sartre differentiates the cogito from thetic acts, but maintains knowledge in a Cartesian sense occurs at the reflective level. Otherwise a very good paper. See *The Imaginary* pp 174 – 175. Also *Being and Nothingness*, pg 6, where Sartre affirms his position on Descartes, and re-affirms that Descartes does not distinguish the reflective and pre-reflective level. In particular pg 178 of *Being and Nothingness* contains a clarification of this earlier

in the circuit of the self which only permits unreflective consciousness as 'selfness' in the world. The mistake is to equate the For-itself with theitic consciousness, possibly because theitic consciousness arises when a for-itself grasps a prior consciousness which contains an in-itself of facticity. But the for-itself can apprehend qualities as well in the upsurge towards being. Primarily it is at the unreflective level that qualities first appear as absences behind objects revealed as lacks to the lacking and voids in the circuit of the self. For Sartre, even though the mechanism he uses for the predication of objects is based on the nothingness of the negation of the predication, and is theoretically separable from the circuit of the self, Sartre thinks this is a perversion and a mistake that Husserl makes[312]. For Sartre human reality exists first as a lack and not something added after[313]. We don't apprehend the world in what Sartre refers to as the "attitudes of indifference" but rather we apprehend it in hodological space at the level of the instrumental complexes. This is important for the method in which anguish can arise from the realization of freedom in a moment of crises. Allow me to explain.

In Part II, Section 3, sub-section III of *Being and Nothingness*[314] Sartre explains the method by which lacks reveal tasks which occur at the unreflective level. The exact relationship between tasks and the for-itself is also revealed in this same section as that of instrumentality[315]. The for-itself occurs at the unreflective level in the present, but it is attached to the in-itself of facticity which, along with housing the structures for bad faith also houses the ego, which houses the I and the me which appear before consciousness *upon reflection*. The Circuit of Selfness is the world from the perspective of unreflective consciousness[316] and occurs within hodological space as the space of needs and desires.

---

position in *The Imaginary*, reflection is a type of knowledge and affirms the consciousness reflected upon. See also pg 182 for the distinction between pure and impure reflection to clarify this. A rare exception is Phyllis Sutton Morris who describes the pre-reflective level as pre-personal and differentiates pre-reflective and reflective acts of consciousness by the nature of their objects. The reflective level is thetic since it takes consciousness itself as its object. See Morris, Phyllis Sutton. "Sartre on Transcendence of the Ego." *Philosophy and Phenomenenological Research* 46, no. 2 (1985): Pp 179 - 198.

[312]Sartre, *Being and Nothingness*, 2003. Pg 209
[313]Sartre, *Being and Nothingness*, 2003. Pg 113
[314]Sartre, *Being and Nothingness*, 2003. Pg 222
[315]Sartre, *Being and Nothingness*, 2003. Pg 223
[316]Sartre, *Being and Nothingness*, 2003. Pg 127

The for-itself can occur in the Circuit of Selfness and when it does so its role is organizing the tasks which appear in hodological space. Indeed, Sartre maintains that the conditions that produce the for-itself are those by which tasks are revealed in the world. The reflective consciousness, however, can not enter the Circuit of Selfness. This is because at the point at which the reflective consciousness appears, unreflective thought disappears and you have apprehension of freedom and with it anguish. He gives examples of this like a solider reflecting on how he will hold up on the night before battle[317] or a gambler recalling a prior vow not to gamble when he comes in view of some card tables.[318]

Danto and his legion are wrong. The for-itself is simply not reducible to conscious things, while the view that the in-itself is detached unconscious matter and the view that cats and dogs are somewhere in between is erroneous at best. It's rather a little bit more tricky than that.

4.2.4 Phenomenological style arguments and acts of introspective scrutiny.

The mechanism I want to draw out of Sartre is the means by which an ego can appear as an object for the purposes of an argument. In Sartre the ego houses a prior or projected consciousness into the state of the present consciousness[319].For the purposes of explaining the following move and why it is significant to the overall thesis, I'm going to employ a brief figurative device. For the device to work I am going to imitate the subject and object casing of English grammar. I am calling

---

[317]Sartre, *Being and Nothingness*, 2003. Pg 53

[318] The difference between cases of anguish or fear if we take the case of the solider, is that if he were afraid of being shot or captured and tortured, he would be afraid of something outside of himself. However if he were filled with an anxiety of giving up valuable information while being tortured or that his courage should fail him under fire he is afraid of himself. The adoption of moral codes is the flight away from this fear of one's self. In the case of the gambler the anxiety occurs in the moment, at the tables when he is confronted with the choice of whether to gamble or not. See the distinction BN pg 53 with Kierkegaard and the distrust of the self on pp 61 - 64

[319]Sartre, *Being and Nothingness*, 2003. pg 127. The ego *is not* a for-itself but appears to consciousness as a transcendent in-itself. The ego houses the in-itself of past consciousness, these cast a shadow over consciousness which is the cortege of the psyche. See Pp 184-186

consciousness brought before itself an 'object consciousness' here and the consciousness that perceives the 'object consciousness' is the 'subject consciousnesses'.

The object consciousness is usually a memory of an event, as non-thetic, which when presented to the subject consciousness becomes thetic. It becomes 'thetic' in so far as the subject consciousness can reflect on the object consciousness apprehending the event or memory, and perceive a prior or projected consciousness as not the memory or the event itself. Likewise in a moment of crises, the object consciousness may be brought before the subject consciousness simultaneously such that thought becomes theitic and the subject is aware of itself.

To understand this moment of crises, perhaps it is best if we look at where Sartre seems to have first arrived at this theory. Sartre's theory of freedom as the realization of theitic consciousness 'in the moment' seems to have started out in Sartre's earlier works as a psychological theory to explain cases of Psychoasthenia. For an instance of these earlier works if we look to *Transcendence of the Ego* we find Sartre cites a case of Janet's where the woman became afraid each time her husband left for work. Her fear was that she would rush to the window and begin hailing people as clients like a prostitute might[320]. He describes this as a sort of 'vertigo of possibility' which is also the term that turns up in *The Imaginary*. In fact, he uses vertigo itself, as an example in *Being and Nothingness*, explaining the anguish of the sudden realization of one's freedom as the fear when one is near a ledge without a fence, that one might fling themselves over the side. The fear, likewise, arises in situations where we have to confront ourselves, or don't know what we are going to do, like the soldier the night before battle, or the man who has to give a public speech.

In Sartre's process ontology the prior consciousness brought before the present one is a deceptive in-itself which shares two relations with the present for-itself; intimacy[321] and the 'moiite'[322]. These, in turn,

---

[320] Sartre, Transcendence of the Ego, 2004 Pg 47
[321] See Sartre, *Being and Nothingness*, 2003. Pg 133. The 'intimacy' of the past deceives us when we apprehend it in the present. The in-itself of a past consciousness brought before the present consciousness has a homogeneity that often makes us take one for the other; the embarrassment of what happened yesterday, or the pain we felt from a break up two years ago can make us feel as though it were a part of the present consciousness thinking back to yesterday, or that the memory of

constitute the foundations for the mirage of the self that appears on the horizon of being. For the purposes of this paper and our analysis of Ryle we are less concerned with the anxiety of the moment like the gambler feels or Kierkegaard's fear of his own capacity to sin[323] and more interested in the structural aspects of the argument and the types of arguments that use theitic acts, that is, the bringing of consciousness before itself to reflect on some aspect of consciousness to illustrate a point.

The 'object' consciousness brought before the 'subject' consciousness has two sides to it. One side indicated by the 'I' in speech verbalizations refers to theitic acts that involved past actions. The other side usually indicated by 'me[324]' refers to the emotional contexts. This is like remembering but it is not just remembering, it is remembering with a sense of the self. It is not just remembering the orange kettle in the kitchen, but remembering yourself standing there and waiting for it to boil. It is remembering the sluggishness and frustration of waiting for it to boil and wondering why it is taking so long. In doing so we are bringing the self of a prior consciousness before the present one in order to illuminate something. When this is done in the context of an argument I call that a phenomenological style of argument.

The relationship between phenomenological style arguments and Linguistic Behavioural style ones in Ryle's Ordinary Language

---

the pain from two years ago is connected with the pain we still feel in the present when we think of the person. See also pg 142, and pp 244-247 of *Being and Nothingness*.

[322] The 'moiite' has a slightly different connection with a past consciousness or act of consciousness than intimacy. These two aspects are interesting for indexical reasons. Sartre raises the concept of the 'moiite' in *The Imaginary* in dialogue with Claparede and in discussion of Descartes' dream argument. The moiite is the self we identify with. Sartre gives the example of seeing a man running from a tiger in a dream, then suddenly finding ourselves as that man running from the tiger. Sartre. *The Imaginary*, 2004. Pg 170. The moiite is implicit in the homogeneity of intimacy with a prior act of consciousness, but it can also occur without the homogeneous aspect of intimacy, as a simple act of identification like in the dream of the man being chased by the tiger, or like the shopper in John Perry's article; Perry, John. "The Problem of the Essential Indexical " *Nous* 13, no. 1 (1979): 3-21.

[323] Sartre, *Being and Nothingness*, 2003. Pg 56. The moiite is analogous to the identification of Perry, by Perry, as himself, with the mystery shopper, but without the intimacy from the homogeneity of the prior consciousness since the memory of the shopper dropping the goods is missing from Perry's identification.

[324] Sartre, *Transcendence of the Ego*, 2004. NB Andrew Brown adopts the English 'I' for past reflections involving the self and actions and 'me' for past acts or stages of consciousness involving apprehending objects with emotional qualities. I'll maintain the distinction as a rough approximation of the phenomenological presence of consciousness before itself on either side for ease of reference and transparency with Brown's translation.

Philosophy allows us to build a taxonomy of Ryle's claims that will in turn allow me critique the normative Ordinary Language justification for these arguments by showing where a contradiction arises. The next stage will be to present these two styles of argument as genuine rivals and show which I think is the stronger of the two. This, in turn will lead to the grounds for arguing the three theses I laid out in the introduction and a push towards a Pre-Fregeian psychologism as so identified.

Identifying phenomenological style arguments from Sartre's phenomenology is critical for breaking up Ordinary Language assumptions and revealing the underlying types of arguments Ryle is using. For now though, let us return to the thread we started with Brandom and Sartre in regards to Kant's apperception of the I.

4.2.5 Theitic Consciousness and the Apperception of the I at the reflective level.

The simplest explanation for Sartree's theitic consciousness is that it is consciousness that takes itself as its own object. If consciousness is positional that is it has positional objects, then taking itself as its own object is the point at which it becomes theitic. Consciousness becomes theitic in reflection but it also becomes theitic in anguish when it grasps itself in the present. Consciousness can not grasp itself in the future because it has not existed yet[325] but it can draw upon the resources of the imaginary in positing an *irreal* future, and decide what it will do in a given instance or set of imagined conditions[326].

---

[325] NB New age claims for Quantum Jumping aside.

[326] Sartre. *The Imaginary*, 2004. Pg 187. In *The Imaginary* Sartre contrasts two types of futures, there is the future of the immediate present. This is like the future of playing Tennis, you anticipate where the ball will be in a few moments from its present trajectory. This future is attached with the present and has continuity. See *S.4 Sartre's Irreal* in this paper for a discussion of this continuity with relation to Sartre's argument against Berkley. See also footnote 287 for a discussion on Sartre's treatment of the continuous into the discontinuous. Then there is the irreal future which is detached from the present future. This lacks the 'resolution' of the present and the act of perception in the moment and thus lacks the qualities attached to the percept. This is like the future of next week, and hypothetical conditionals involving if-when, and visualizations of future events. If x happens, then I will do y, and when x happens I will do y. These 'irreal' futures when they enter the present future contain an additional affirmation in the immediate present, that is, as we approach them from out of

For the moment what I want to engage directly is Sartre's specific mechanism of consciousness and theitic consciousness in order to draw out and make explicit a pattern of inconsistency in Ryle's arguments. This will offer an insight into the way phenomenological data is stored in ordinary language and appealed to on normative grounds in an argument that I shall reveal at the end of the next chapter, but which, won't make sense until we get there.

When Sartre breaks with Continental Kantian Orthodoxy, he does so in order to argue that the unreflective level is the level at which we grasp things in the everyday world.

Sartre writes

> (T)here is no I on the unreflective level. When I run after a tram, when I look at the time, when I become absorbed in the contemplation of a portrait; there is no I. There is (only) a consciousness of the tram-needing-to-be-caught, etc[327].

At this unreflective level there is no room for an 'I' to appear. We do not see ourselves seeing a tree or smell ourselves sniffing a flower[328]. The I, Sartre thinks, were it to appear, would cut consciousness off from itself. This, I think, is the origin of the confusing assertion in the Analytic Anglo-Sartre[329] literature that argues that Sartre maintains

the detached irreal framework we 'image' and plan them in and they become immediate when we need to make a choice. This basic structure radically shifts in *Being and Nothingness,* gaining new levels of complexity. Sartre introduces, for instance 'the no-longer possible' and the 'future possible' as new levels of complexity.

[327]Sartre, Transcendence of the Ego, 2004. Pg 13

[328]Sartre, Transcendence of the Ego, 2004. Pg 8

[329]McCulloch, *Using Sartre*, 1994 Pg 6, 72-79. The problem with broad brushstroke generalizations on consciousness with Sartre is 'consciousness' has many internal levels to the ongoing 'process ontology' that Sartre identifies. Briefly, the for-itself contains reflection and the circuit of the self. There are sets of relations with the in-itself of lack, which produces the upsurge towards the for-itself in the facticity of the objects and circumstances of the hodological space found in the circuit of selfness which instrumentalizes the world for tasks, and the in-itself of prior consciousness which contains the de jure necessity of the possibility of the apperception of the I at the reflective level. The 'I' of course becomes realized when reflected on by the present consciousness in theitic states of

'consciousness is clear' or more confusingly 'consciousness is empty'[330] possibly because he uses the analogy of the opaque and clear in *Transcendence of the Ego*[331].

In contrast what Sartre actually argues in this particular analogy is that an I in the Husserlian sense that Husserl uses in the *Cartesian Mediations*, of an I unifying itself in time, would be *like an opaque blade*, in that it would cut consciousness off from itself. Ultimatley I think this analogy is misleading since the objects to which the properties of the irreal belong in *The Imaginary*, are themselves constituted by consciousness, and leads one to imagine rather something like how Ben Kenobi appears in *Return of the Jedi*, or the force fields and weapons Sue Richards uses in the *Fantastic Four* comics. All Sartre means is that there is no room for an I unifying time or experience at the unreflective level.

Sartre writes

> In fact, I am then plunged into the world of objects, it is they which constitute the unity of my consciousnesses, which present themselves with values, attractive and repulsive, but as for *me*, I have disappeared, I have annihilated myself. There is no place for me at this level, and this is not the result of some chance, some momentary failure of attention; it stems from the very structure of consciousness[332].

consciousness. Theitic states are states where an I appears as part of self consciousness. They are not always limited to reflective states. Attaining theitic consciousness by an act of apprehension of consciousness in the present by a present consciousness produces anxiety like the woman who was panicked by the thought of going to the window and hailing people like a prostitute each time her husband left for work, or the vertigo of an open ledge and the sudden realization that you could throw yourself over the side, or perhaps that sudden fear when you overtake someone on a country road and pass into the lane on the opposite side, and realize just how free you are to keep driving in that lane until you crash into someone coming the other way. Imaged consciousness itself isn't empty it can contain a real object aimed through intentionality as an imaged consciousness of someone who exists like Reg, with irreal properties, or it can contain an irreal object like a dragon or a woodland elf. Sartre lays down these fine line distinctions in *The Imaginary*.

[330]McCulloch, *Using Sartre*, 1994. Pg 73
[331]Sartre, Transcendence of the Ego, 2004. Pg 7

Let us thus return with this insight to the Brandom and Sartre debate over the apperception of the I which we started this interlude with, with the following philosophical grounds: Sartre would argue that when people say 'that's a car', they're not saying '*I think* that's a car', they're just saying, 'that's a car' which is all that is going on in consciousness. The 'I think', *de facto*, does not accompany consciousness. The I is not present in all of our thoughts, perceptions and emotions on the conscious level. We do not think as a car goes past 'oh I see a car' or 'I judge that is a car going past' but rather we might simply think to ourselves 'that is a car going past' or even just 'car going past'. We may not even think it in words, we may simply be aware of a car going past and perceive it doing so but not put it into mental words as Sellars would say 'in forro interno'. However Sartre argues, and here we might go along with him when he says the 'I' does exist as a necessary condition for the possibility for a thought and can be raised later by reflective consciousness as in 'I just saw a car go past'. That is, only in a reflective moment of consciousness like a moment of crises, self-analyses or reflection on a prior state of consciousness would we find ourselves at the point at which Sartre moves up to the next level of Hoffe's analysis of Kant's apperception and agree with Brandom. Pace Brandom not all past events, memories or thoughts have an I in them, even though, de jure, Sartre argues they have the potential for an I to become attached to them. This needs two further highly detailed clarifications in order to fence off possible objections or complications arising from certain obscurities.

Firstly I hold that it may be possible that there may be some thoughts where the possibility of an I at a present or future stage is not a necessary condition for the thought to exist. Secondly as Sartre argues, a person may remember an event, the way a tree or a land mark looked or the appearance of someone they know, for example, and not themselves be a presence in that recollection. Sartre does make that distinction in *The Imaginary* and *Transcendence of the Ego*.

In *Transcendence of the Ego* Sartre writes

---

[332]Sartre, *Transcendence of the Ego*, 2004. Pg 13

Now it is undeniable that the Cogito is personal. In the 'I think' there is an I which thinks. We here reach the I in its purity and it is indeed from the Cogito that an 'Egology' must begin. And so, the fact that can be taken as the starting point is this: each time that we grasp our thought, either by an immediate intuition, or by an intuition based on memory, we grasp an I which is the I of the thought that is being grasped, and which, furthermore, gives itself as transcending this thought and all other possible thoughts. If, for instance, I wish to remember a certain landscape I saw from the train, yesterday, it is possible to bring back the memory of that landscape as such, but I can also remember that I saw that landscape[333].

What Sartre is arguing here is that something distinctive happens in 'theitic' thought which takes itself as its own object, either in the present act of thinking or in reflecting upon a past event and one's self at the time of that event. It is during this reflection that the self can appear as an object of consciousness and various aspects of that self start to become apparent. The strange thing that happens of course is that the 'I' appears and is marked by either an 'I' or a 'me' linguistically where the prior consciousness of the in-itself appears before the present consciousness of the for-itself. We can recollect a landscape and this is simply a memory or we can recollect ourselves seeing a landscape and here thought becomes theitic. Here we recall our selves and how we felt, our impressions and the ambiances we perceived upon seeing a landscape.

[333]Sartre, *Transcendence of the Ego*, 2004. Pg 9

To bring this insight to a head when classifying these types of arguments that Ryle is using I will adopt this process of bringing a prior act of consciousness before the present one, as a source of appeal, and I will refer to arguments that do this as the phenomenological strain in Ryle. This insight will be important to the core argument of the paper which started with Dummett, Frege and Ryle, and the clash between psychologism and anti-psychologism so keep your eye on it.

4.3 Conclusion

In the above we considered Brandom and Sartre's position on the apperception of the I, did a little scholarship and phenomenology with Sartre, defined theitic acts, which are acts of consciousness that take as their subject reflection, or thoughts that posit a prior act of consciousness, which of course is marked by the appearance of the me, or I, at the phenomenological level, as a complex of the ego. We saw that Sartre thinks the apperception of the 'I' or 'me' is indicative of a past consciousness brought before a present one. Consciousness by itself is simply consciousness of objects, or as Sartre put it, we are 'plunged into a world of objects' at the non-theitic level. At the reflective level and at the point of anguish, Sartre argues, consciousness changes. It becomes theitic. That is to say it becomes aware of itself. It looks back on its prior self, from the present, for instance, and takes itself as its own object. These reflective acts, in which we study the character of consciousness by thinking back on prior acts of consciousness, and what they are about, the act of bringing consciousness before itself for a bit of introspective scrutiny, this is what interests us for the purposes of this paper. The reason being of course Ryle, inadvertently, has us doing the same or similar things in some of his arguments. So this view of consciousness before itself, as its own object of study, this brand of phenomenology Sartre develops, is useful for analysing some of the things that Ryle gets his readers to do in the process of making an argument. Here, we're interested, of course, specifically, in investigating the link between ordinary language and phenomenology of the reflective consciousness.

Prior to that we considered Dummett's point about meaning, and more specifically his arguments where he flirts with a Pre-Fregeian psychologism and stays just shy. Also I revealed the general thrust of this thesis would be to argue for a reconsideration of a Pre-Fregeian psychologism, in which language is merely a code for thoughts, and that I would do so in a number of steps, or insights, that couldn't be understood until we'd made each of them, but I nonetheless sketched them out.

The first of those steps, of course, involved developing a taxonomy of Ryle's arguments which we had already started to do with Linguistic Behaviourist claims. Here we saw the problem with, firstly, Chalmers claims that Ryle encoded the Freudian-Behaviourist orthodoxy and secondly Weitz's coarse grained Logical Behaviourist treatment. Both of which failed to capture various distinctions in extra-sentential relationships, and, among other sins; in Weitz's case he committed the sort of mistake Ryle was fervently averse to in treating all dispositions as the same fundamental type of operation. Chalmers disregarded the ordinary language foundations and justifications of Ryle's arguments. From Chalmers' deficiencies we were able to start a thread that high lighted Ryle's Ordinary Language species of argument. These are arguments that forward claims on the basis of a normative appeal to the reader or audience's knowledge of language conventions. These are characterized by *'it makes sense to say'* or *'it does not make sense to say'* claims. Weitz's deficiencies allowed us to start a thread based on analysing Ryle's Linguistic Behavioural arguments. These were arguments that use inter-sentential and sub-sentential relationships that allow us to analyse language at the level of verbs, nouns, epithets, pronouns, interrogatives and the sets of dispositions, propensities, occurrences and episodic structures that hold between them.

I, subsequently, argued that this Linguistic Behaviourist method of analysis, would eventually lead to a break up of Ordinary Language arguments as a normative platform to launch other types of arguments. Firstly because there are a number of distinctions Ryle makes which he just doesn't have the grammatical currency or linguistic behaviour to cash in on. These were the flash-bangs I asked you to keep an eye on. I also pointed to a gap in the linguistic behaviourist framework. The significance of this gap, as, I pointed out, would in turn, result in two individual, rival strains of analyses, each with its own normative force, and that the stronger of these two strains pushes us towards a theory of consciousness, and a move towards a reconsidered Pre-Fregeian psychologism that will start to take shape and emerge next chapter.

We then did some practical ground work towards these ends, and looked at what it actually means to be a Linguistic Behaviourist claim, and we also looked at some actual examples of Linguistic Behaviourist distinctions that Ryle uses. Here we redrew on our earlier work with Weitz, and adding a few more distinctions with 'flash-bangs' being among some of the most important.

More importantly we begun to look at how Linguistic Behaviourist claims cross over into the domain of Ordinary Language arguments by a normative quirk of a small expression 'it makes sense to say' or its negation 'it does not make sense to say' or alternatively 'people don't say' to ground a Linguistic Behavioural analysis in the domain of knowledge that an Ordinary Language user possess. These arguments perform a normative role that can either affirm by saying 'it makes sense to say' or negate by claiming 'it doesn't make sense to say' and their equivalents. These arguments are an interesting type of argument because their strength seems to be that they are not supported by correspondence to truth values or a criteria for true statements like Ayer's verificationist claims or J. Stanley and T. Williamson's semantic pairing of 'how' with all of the true propositions that answer the question. Rather, the sketch I suggested is much closer in likeness to the Semiotican's relationship between a 'langue' and a 'parole'. The linguistic analysis appears to draw upon an appeal to the user's knowledge of linguistic conventions and ordinary language intuitions in order to justify a behavioural distinction. This is particularly interesting because as arguments they do not correspond to various logical formulations for validity that we might use to asses them[334] since they don't incorporate truth values but instead rely on a 'court of appeal'.

---

[334]A standard formulation of validity like you'll find in any introductory text book on logic. For instance see Gensler, Harry G. *Introduction to Logic*. Oxon: Routledge, 2007. Pg 187, where he gives two rival definitions. "An argument is valid if it would be contradictory to have the premises all true and the conclusion false.' And its rival 'An argument is valid if the conjunction of the premises with the contradictory of its conclusion is inconsistent.

In this section I also ruled out a psychologistic attack on the side of the dispositions made up of the capacity configurations for two reasons. Firstly I think an anti-psychologistic account of the capacities is a strawman. There are any number of ways you could prove this including the extended-mind debates which I included in the footnotes. However it is necessary to go through these because there are people like Stanley and Williamson who build them as we saw earlier in this paper[335]. Secondly Ryle himself provides an argument against these. This leaves us with the tendencies' side of the dispositions from the tendencies and capacities configuration, and includes beliefs, motives and inclinations as well as propensities and occurrences. The occurrences contain moods which can be made up of contrary motives and/or inclinations, or either with a factual impediment or each other. Collapsing the linguistic behavioural anti-psychologistic foundations of beliefs and inclinations thus removes agitations as specific occurrences, but it still leaves feelings proper or 'flash bangs' as I labelled them and these will be important.

In the next section we will be looking at specific instances of how Ryle uses Ordinary Language arguments to affirm his own position or make arguments that negate the claims of other philosophers before moving into a principled taxonomy of these arguments. In particular I want to draw attention to purported Ordinary Language arguments which lack a Behavioural Linguistic analyses and which still claim some sort of distinction can be made by the natural language speaker of the language. I am going to argue that some of these arguments draw upon introspective scrutiny or phenomenological style arguments that posit theitic acts to make their point. Distinguishing these types of claims allows us to untangle a problem that will arise later in the thesis, which ends in contradictory claims about the nature of dispositions. This will allow me to suggest that what underlies these claims is not a unified ordinary language world view or a unified normative source, but rather, different types of arguments that draw on different types of normative force and that these rival forces can result in contradictory claims. This will set the scene for the last part of the paper which is where I argue for what I claim is the stronger source

---

[335] See ' 3.3.2 How Do you Know? The Primacy of Concern', in '3.5 Ryle's Two Most Fundamental Insights on the Mind and the Relationship between these Insights and Linguistic Behavioural Style Arguments' in 'Chapter Three : Towards a Methodology'.

with the result being a psychologistic position characterized by the three central theses I identified at the start of this paper.

Chapter Five : Ordinary Language Arguments.

5.1 An analysis of Ordinary Language claims and their ability to affirm or negate on a reputed source.

In this sub-section I will be exploring the ways Ordinary Language arguments can be used to affirm or deny a claim about the mind by considering some explicit examples. Specifically the examples I will be looking at are those where Ryle makes a direct appeal to the reader's knowledge of language and practice to support one of his arguments or lodge an objection against another philosopher.

For instance Ryle argues that

> the language of 'volitions' is the language of the para-mechanical theory of the mind. If a theorist speaks without qualms of 'volitions', or 'acts of will', no further evidence is needed to show that he swallows whole the dogma that a mind is a secondary field of special causes. It can be predicted that he will correspondingly speak of bodily actions as 'expressions' of mental processes. He is likely also to speak glibly of 'experiences', a plural noun commonly used to denote the postulated non-physical episodes which constitute the shadow-drama on the ghostly boards of the mental stage[336].

Now in advancing this argument against the right wing of the para-mechanical theory of mind[337] Ryle makes a direct appeal to the domain of common language for justification, against it. This is an example of an Ordinary Language argument that negates.

Here is the negation.

Ryle writes

> Despite the fact that theorists have, since the Stoics and Saint Augustine, recommended us to describe our conduct in this way, no one, save to endorse the theory, ever describes his own conduct, or that of his acquaintances, in the recommended idioms. No one ever says such things as that at 10 a.m. he was occupied in willing this or that, or that he performed five quick and easy volitions and two slow and difficult volitions between midday and lunch-time. An accused person may admit or deny that he did something, or that he did it on purpose, but he never admits or denies having willed. Nor do the judge and jury require to be satisfied by evidence, which in the nature of the case could never be adduced, that a volition preceded the

---

[336]Ryle, *Concept of Mind,* 1983. Pg 62

[337] The right wing of Ryle's the para-mechanical theory of mind, of course, being the passions see Ryle, *Concept of Mind,* 1983, pg 91. Also Adams, Aizawa, *Defending the Bounds of Cognition,* 2010 NB One of the reasons I avoided going into the Extended Mind hypothesis debate is that Adams and Aizowa's 'mark of the cognitive' on a reading of Ryle would be a special version of this sort of para-mechanical theory since it individuates the 'cognitive' by specific reference to its cause in terms of causal mechanisms. However, the dissolution of Ryle's Ordinary Language solution, I argue, re-introduces the problem of causation in theories of mind, and specifically, I'm going to argue at the end of this paper presents a new problem of causation in models that utilize theories about consciousness. See footnote 93 and 95. Read 95 only after carefully completing each stage of the paper. Redefining the bounds of cognition in terms of the irreducible direct first and third person positions, is of course, the project of a psychologism and the causal fixation model the paper finishes on.

> pulling of the trigger. Novelists describe the actions, remarks, gestures and grimaces, the daydreams, deliberations, qualms and embarrassments of their characters; but they never mention their volitions. They would not know what to say about them[338].

Notice here, the justification for dismissing volitions is based on several things: firstly that the ordinary language user has no knowledge of 'volitions' so he would not know what to say about them. The argument simply put is that ordinary people do not use 'volitions' in their vocabulary so it follows that they must not exist. That a person doesn't know how many volitions are in an act, or how many they may have performed that particular day counts for evidence for this argument. Ryle gets around the argument that people need to be conscious of their experience in some way as to be able to report whether they experience volitions with his 'log keeper' account of the mind and the special status reports made from 'knowing what one is about'. These are all appeals to the language user's knowledge, as contrasted with, say, evidence for quarks or atoms. Allow me to bring out something unique about Ordinary Language arguments by making an analogy on these points.

Let us accept as non controversial that people's bodies are made up of matter and their thoughts are influenced by brain chemistry[339]. One might extend Ryle's thinking here to claims about quarks, and dopamine levels, that is, they don't exist because the Ordinary Language speaker doesn't know about them, or use them in their ordinary language speaking. To put the cap on the bottle, the domain of justification for dismissing volitions is the ordinary language user's knowledge and use of the language. Ryle's argument quite simply is since ordinary people don't talk about volitions, ergo, they do not exist.

---

[338]Ryle, *Concept of Mind,* 1983. Pg 63

[339]Sousa, *Emotions,* 2013. See section 7. The Ontology of the Emotions. Physiological Processes. For arguments sake lets accept De Sousa's claim "physiological processes are conceded by all philosophers to be involved in clearly prototypical cases of emotion" even though Ryle, amongst others, are philosophers who argue to the contrary. Ryle avoids the problem of brain chemistry, emotions, mental states and consciousness because he focuses on language and how people use it so that he can analyse this use and make claims about the nature of mind. However the breakdown of Ryle's Ordinary Language arguments into heterophenomenology and autophenomenology claims reintroduces the problems of brain chemistry and consciousness and their causality. See the problem of the 'Roid Rager in the void at the end of this paper.

This is troubling because ordinary language speakers don't refer to their dopamine levels, ergo, - by following Ryle's reasoning here one would be lead to conclude – dopamine levels do not exist.

Now one could reply, that we do feel a difference in dopamine and oxytocin levels[340] and we have ordinary language names for these types of things, like happiness, trust, and so on, but here we have a problem, since the only two ways we can experience these are flash-bangs and moods. Moods need things like motives, one's motive or inclinations makes one happy when it is unimpeded, and there is no internal clash of motives and inclinations. Dispositional talk, it seems, can't account for ordinary references to brain chemistry since it is neither causal or episodic. The other option is flash-bangs, and here Ryle is going to have a problem, because, to be aware of one's emotions in the sense one can be aware of the effects of neurological agents that stimulate one's brain chemistry requires something that looks a lot like an act of introspection, consciousness, or both, and as we have seen he has flatly ruled these out.

Retrospection can only keep a log made up of various types of 'mindologue' which we can retrieve as 'special status reports' provided the person is 'alive to what they are doing.' If shifts in the mental garden of brain chemistry don't result in the fruit of special status reports given in ordinary language then Ryle doesn't think they exist[341].

The only course left open is to argue for a flash-bang without an act of introspection, that is, we can have flash-bangs without introspecting to know what they are. Since he eschews a theory of consciousness and he has ruled out introspection, it is left up to his Linguistic Behaviourism to tell us the difference between 'a glow of pride' and 'a glow of warmth' as a linguistic matter without falling back on the claim that non-linguistic knowledge is prior to knowledge of linguistic meaning and some form of psychologism. His original project, of course, was to make explicit the logic of the implicit knowledge of the ordinary language speaker by examining the bits of linguistic phenomena. He promised us a map.

---

[340] Most commonly produced by caffeine and anti-depressants

[341] There is a tension here since he thinks that some forms of work don't result in 'mind chatter' like untangling a skein of wool or solving a jig saw puzzle. We can solve this inconsistency by taking it that what Ryle means is some capacities, skills, methods and abilities in their exercise do not utilize 'mental chatter'.

Now there is a way out of this problem. That is, one might be able to use a bridging principle between ordinary language and neuroscientific data. That is, if one focuses on these terms and treats the sentences that these descriptions and thoughts occur in one might be able to map neuro-scientific data on to ordinary language useage and vocabulary. All one need do to get around the above problem of introspection and what such states might *feel* like is to treat the sentences and descriptions in use as analogous to thought. The difference between a glow of pride and a glow of warmth, on this view, is found in some device like an fMIR or EEG. This is exactly what Wilfrid Sellars argues in *Philosophy and the Scientific Image of Man* and I will be considering his position later on[342].

Indeed Willfrid Sellars critique of Ryle in *Empiricism and the Philosophy of Mind*, as well as, the interrelated thesis of *Philosophy and the Scientific Image of Man*, I argue, are central to what happens next in the philosophy of mind once you move away from a position like Ryle's. My ultimate argument, however, is not that Sellar's position is wrong, but that phenomenological sources of insight drawn from introspective analysis are simply a stronger type of claim for an argument than descriptions which depend on some type or variety of Linguistic Behavioural analysis. Auto-phenomenology gives us better grounds for individuating mental phenomena than hetero-phenomenological sources and in particular Linguistic Behavioural arguments. This argument rests upon three claims I will sketch out so you can see the rough form of the argument I am building.

Firstly I will undermine an Ordinary Language Neuroscientific position with the claim that non-linguistic sources are fundamental for establishing meaning in some mental phenomena. I will do this by arguing that auto-phenomenological arguments are stronger than hetero-phenomenological arguments because a) hetero-phenomenological arguments usually contain hidden or occult phenomenological assumptions and appeals to sympathy in their foundations; b) when a hetero-phenomenological argument makes a claim, either as a Linguistic Behaviourist, or Neurological claim, it is still open to an auto-phenomenological argument, either as a second or third source to individuate on that claim. If part x of the brain lights up

[342]See 'Chapter Seven : Motives, Moods and Reasons' in '7.1 The Game of Giving and Asking for Reasons'.

on the neurological 'doo-hickey' in two different cases, it is still up to an auto-phenomenological source, like the person that the doo-hickey is wired up to, to say whether he felt anger in both cases or disgust in the second, and this information is based on non-linguistic knowledge. This is where the anti-cognitive, anti-implicit, anti-reductive-functionalist psychologistic thesis of the paper comes in. The 'anger' or 'disgust' is a linguistic codification of something happening on a non-linguistic level and is necessary for claims about linguistic meaning.

Secondly, in support of the above, I will argue that some mental phenomena are simply irreducible to linguistic utterances and behaviours. This is the argument I started with the 'flash-bangs'. I will advance this later by arguing that motives which might be given as answers or linguistic behaviours responding to the 'why?' side of the 'belief, motive and inclination' configuration bottom out in non-linguistic knowledge. This is the anti-implicit thesis of the paper. I will then argue that if we examine reason-explanation style answers given as moves in the game of giving and asking for reasons, we can see these linguistic behaviours bottom out in non-linguistic knowledge necessary for playing the game and which can't be established or justified in the game itself. Moreover I am going to argue that expressability within a language or explicit linguistic formulation is not a necessary condition of forming or holding beliefs. Lastly, tendencies, as one group of dispositions, will be divested of their ability to count as mental phenomena in the first person, and thus lose much of their power once reduced to linguistic descriptions from the third person. I only need to uphold one of these claims to make the argument. I offer all three as overkill.

Thirdly, to support the anti-implicit, anti-cognitive, anti-reductive psychologistic position I drew out above, I will argue that linguistic codification is a different sort of thing to some forms of mental phenomena. Upholding this third claim results in the argument that we need a theory of consciousness since the driving force behind the claim that linguistic codification is a different sort of thing to some forms of mental phenomena, I argue, depends upon the claim that consciousness precedes linguistic codification. We will see an example of this with the 'remember how/ remember when' configuration in a moment, which we can think about in linguistic behavioural terms but also in non-linguistic terms of some skill or some memory. Later we will see the same thing between emotions and the way we talk about them, except in this latter case, we can not individuate on the basis of linguistic behavior alone. That is, we can not tell the difference between a 'glow of pride' and a 'glow of warmth' simply on the basis of linguistic behaviorism. This is important because with the 'remember how/ remember when' configuration we can. There is an insight on offer here and I am leading up to it.

But for the moment, let us return to Ordinary Language arguments and Ryle's implicit theory of meaning as we consider the next type of argument.

Ryle writes

> However, when a champion of the doctrine is himself asked how long ago he executed his last volition, or how many acts of will he executes in, say, reciting 'Little Miss Muffet' backwards, he is apt to confess to finding difficulties in giving the answer, though these difficulties should not, according to his own theory, exist[343].

Now this is a sophistication on the prior argument. Instead of depending on an Ordinary Language criteria like whether or not people use volitions Ryle is drawing on the resources of a 'special status report', one which is consistent with his position on thought possessing a log keeper role and the process he defines as 'being alive to what one is doing'. Ryle's line of thought here requires self-consciousness without introspection. To make this distinction he can draw upon the vocabularies of unstudied talk we've already discussed[344]. He does, of course, as I pointed out, distinguish some mental operators such as playing chess or untangling a skein of wool which don't use linguistic expressions, symbolic forms, codes and so on, but these are on the capacity and tendency side of the distinction and don't come under the propensity dispositions like motives, beliefs and tendencies or the

---

[343]Ryle, *Concept of Mind,* 1983. Pg 64
[344]Ryle, *Concept of Mind,* 1983. Pg 168 – 169, 172.   See the 'Introduction' and 'The Species of Mindologue' for Ryle's use of 'consciousness' and 'introspections'.

occurrences like the flash-bangs. This will be important later in the argument being advanced by the paper.

Finally let's look at one last case, before we move onto Phenomenological arguments. This is an important example because it shows how Phenomenological arguments can sometimes occur directly as the other side of the Linguistic Behavioural coin that the currency of Ordinary Language Arguments are issued in. Ryle isolates and develops another distinction between episodes and dispositions on the basis of the way we use the infinitive 'to remember' in ordinary language.

We'll look at the episodic event first, which when recalled behaves like an occurrence. Ryle writes

> Quite different from this[345] is the use of the verb 'to remember' in which a person is said to have remembered, or been recollecting, something at a particular moment, or is said to be now recalling, reviewing or dwelling on some episode of his own past. In this use, remembering is an occurrence; it is something which a person may try successfully, or in vain, to do; it occupies his attention for a time and he may do it with pleasure or distress and with ease or effort. The barrister presses the witness to recall things, where the teacher trains his pupils not to forget things[346].

Note that this is the episodic case, as in 'to remember' (when). . .This forms one side of the linguistic usage which is the episodic case of an event. The following passage from Ryle illustrates the 'to remember' (how)side of the configuration and shows how 'to remember how' can be used in the sense of having not forgotten a skill like in the instance Ryle compared with the way the teacher trains his pupils. This forms the other side of the distinction and connects the verb 'to remember' to the

---

[345] The use given in the following quotation on the following page.
[346] Ryle, *Concept of Mind*, 1983. Pg 259

capacity side of the configuration governing the distinction between capacity and tendency dispositions.

> By far the most important and the least discussed use of the verb is that use in which remembering something means having learned something and not forgotten it. This is the sense in which we speak of remembering the Greek alphabet, or the way from the gravel-pit to the bathing-place, or the proof of a theorem, or how to bicycle, or that the next meeting of the Board will be in the last week of July. To say that a person has not forgotten some-thing is not to say that he is now doing or undergoing anything, or even that he regularly or occasionally does or undergoes anything. It is to say that he can do certain things, such as go through the Greek alphabet, direct a stranger back from the bathing-place to the gravel-pit and correct someone[347].

Note the difference between the two uses. The former is episodic in the sense of events and the later is dispositional in the sense of skills and abilities. Note too that this distinction *can be* based on the linguistic behaviour. 'Remember-How' is linked to the Linguistic Behaviour of capacity verbs. This forms one side of the knowledge-how/knowledge-that distinction. We add to this a further distinction and we can ask ourselves; why don't people use the term for the other side of the dispositional table? Why don't people use it for the motives, inclinations and beliefs somebody has? This gives us a further linguistic behaviour to differentiate capacities and tendencies since motives, beliefs and inclinations are not the sorts of things somebody can ask a 'remember how' or a 'remember when' question about. But someone might ask 'do

---

[347]Ryle, *Concept of Mind,* 1983. Pg 258

you remember why you did that?' if searching for a motive or a belief that can explain some behaviour.

Now in a shallow sense someone can say that remembering-how is episodic in the sense that a person may remember the episode, when by implication they remember how to do something, because they are recalling the instance that they learned something, and indeed in a phenomenological sense which we will get to in a moment there is something to this. When someone remembers or recalls a capacity or skill for the first time they might even think of that math teacher, or perhaps, finding themselves locked out of the apartment once more, they might think back to that other time when they were locked out of the apartment and recalled that they had to climb up the balcony face, scale the terrace and leap to the drainage pipe, and so forth, and that's how they got in. But this isn't the sense in which, we must admit, that Ryle's distinction can be put and there are arguments which Ryle raises against this[348]. What I am interested in is offering an insight into the way that some of Ryle's arguments work because some of them involve two sides and this makes them interesting arguments.

First we must purge the argument of all phenomenology before we can hold the phenomenological pieces in one hand, the non phenomenological ones in the other and see the fault line between Linguistic Behaviourist style arguments and introspective phenomenological ones. It is my aim that through this treatment I can lead the reader to the place where he or she might glimpse the first ghost haunting Ryle's account of the mind in *The Concept of Mind* and his grammatical machinery. There is an insight that will show why Ryle's problematic patterns are interesting, and thus introduce a sort of 'skeleton key' that will unlock the problematical structures haunting Ryle and reveal the need for a theory of consciousness.

So, in the Linguistic Behaviourist sense, we can see that there are two types of behaviour for the infinitive 'to remember'. That is, if we were to lay out the uses, we would find, that the word behaves differently when you apply 'when' and 'how' after it. One set of verbs will have one set of inflections, most typically infinitives i.e. to tie a knot, to drive a car, to find the square root, to use a hammer, to say

---

[348]Ryle, *Concept of Mind,* 1983. Pg 259

Little Miss Muffet backwards and count the volitions. The other will use episodic verbs, like 'ran', 'threw', 'hid', 'buried', 'sang', 'volitioned', and so on after 'remember when' along with a proper noun, pro-noun, or other noun of some type.

Next, we can see that an appeal to Ordinary Language in the way 'that it makes sense to say' that 'to remember' refers to skills when used in the sense 'having not forgot' used by schoolmasters, master builders and logic instructors. It also 'makes sense to say' that one doesn't talk about remembering-how in questioning motives or remembering beliefs, although one might talk about remembering practices that go with beliefs, like eating fish on Good Friday, or abiding by the Laws of Leviticus, but that remembering motives in terms of 'how' isn't something that it 'makes sense' to talk about. We might ask do you remember 'why' when inquiring about beliefs and motives associated with these practices. It 'makes sense' to talk about one as being motivated, or having conflicted motivations, or keeping motives in mind, or conflicts between motives and beliefs, but not to talk about motives and beliefs in the way one talks about remembering how to do group theory algebra, or the time 'when' uncle Rod dropped the Christmas ham[349].

Now, for the first glimpse of a ghost: when one actually recalls an event or when one actually thinks about a skill. That is consider talking with and responding to questions such as 'do you remember what you did last Christmas?' and, 'think about driving a car'. 'Okay? Got it?' 'Now you're backing down the drive, now you're pulling out and want to turn left'. When one is thinking of these things one is performing a different function, or engaging with the question in a different way than simply, and straight forwardly instantiating the behaviour of the language, or making an appeal to the sorts of things it 'makes sense to say'. To be thinking about that ham on the table last Christmas, requires us to become immersed in the events and step into the experience evoked. Without these distinctions it is ambiguous what Ryle is doing in some of his arguments. We shall see then that a piece of

---

[349] It should be noted in the above I'm simply offering an expositional account of Ryle, showing the unique method and analysis of treatment I've developed for his arguments, not making a cogent reading. Ryle, in spite of himself mixes the linguistic and phenomenal components of his arguments. Here I am at pains to show their distinctions. Ryle, in spite of himself, mixes the linguistic and phenomenal components of his arguments. Here I am at pains to show their distinctions.

clever phenomenology may be hidden behind an Ordinary Language argument and require what looks very much like an 'act of introspection'. This, if we recall is problematic to his thesis about consciousness, Protestant Para-optics and introspection. Similarly, one or another particular analysis or distinction may in fact be something to do with partaking in an act of 'phosphorescent consciousness' and not simply a bit of linguistic behavioural analyses.

We will now examine these.

5.2 Phenomenological arguments.

Not all of Ryle's strain of 'phenomenological arguments' can be read as having a Linguistic Behaviourist side like the 'to remember' and the 'flash bang' sub-strains we've looked at. It is just easier to discriminate the phenomenological content by using the treatment I have developed and to refer to it under the moniker of 'Ghostography'. In using this 'Ghostography' we only have to look for Ordinary Language justifications without a corresponding Linguistic Behaviourist claim. Ryle usually makes Ordinary Language arguments and offers the Linguistic Behaviourist analysis in grammatical terms or bits of jargon referring to linguistic behaviours like 'mongrel-categoricals' or 'heed-concepts'. Sometimes he offers a distinction that *seems* intuitively correct and justified from an Ordinary Language point of view but he doesn't offer the Linguistic Behavioural claim to demonstrate it. If there is no Linguistic Behaviourist claim then we ask the question 'what is the Ordinary Language argument justifying?' I submit that if one looks a little deeper one will see in the examples we've already gone through have a ghostly finger pointing towards a phenomenological content.

There is an interesting question this raises: are all of Ryle's Ordinary Language arguments which lack a Linguistic Behaviourist description phenomenological? The short honest answer is that I don't know. Those I have checked for the purposes of the argument of this thesis all tend to be, but I have not exhaustively checked all of them in every possible interpretation or reading[350]. But I have probably come

---

[350] I made an index of all the arguments he uses in *The Concept of Mind*, and went through them looking for patterns, before breaking them down. There are other types of arguments with other

close. Prima facie if Ryle is using Ordinary Language arguments as justification there must be a content he is justifying whether it is in the form of a negation as with volitions or in the form of a distinction where he doesn't offer grammatical criteria for his arguments. If there seem to be no immediate Linguistic Behavioural forms for such a criteria this would suggest it is at least possible there is a phenomenological content present to these arguments. That is to say where you find an Ordinary Language argument without a linguistic behavioural analysis, you will usually find a phenomenological strand of argumentation in Ryle.

There are also a handful of *pure* phenomenological treatments in *The Concept of Mind*, which don't have Ordinary Language justifications or Linguistic Behaviourist claims attached to them. These seem to me to infringe directly on his prohibition against consciousness and introspection to make their argument, and undermine the general line of his diachronic claim that consciousness is a piece of para-optics that arose from the Reformation.

As these can not be hidden behind or be justified by Ordinary Language or Linguistic Behavioural forms of analysis they are 'naked' phenomenological appeals.

Take the following as an example of a pure naked phenomenological argument out of Ryle. He writes

> The reader of a report of a race can, subject to certain restrictions imposed by the text of the report, first picture the race in one way and then deliberately or involuntarily picture it in a different and perhaps conflicting way; but a witness of the race feels that, while he can call back further

---

types of rival criteria for assessment or normative value than Linguistic Behavioral analysis, Ordinary language claims and phenomenological insight in Ryle and elsewhere. See for instance the section on David Lewis in the appendix for a clash between Linguistic Behaviorism and Logic in 'Appendix A Counterfactual Logic, Linguistic Behaviorism and Ordinary language arguments'. There are also diachronic arguments like those discussed earlier in this paper in '4.1.3 Ryle's Diachronic Attack on Consciousness'. See footnote 232. There are also moral arguments that exert their own force. The list goes on.

views of the race, yet alternative views are
rigidly ruled out[351].

To make that distinction one has to actually put one self in the shoes of both cases. One has to think back to a day at the track or of a football game, from a specific vantage point in the stadium and then think about another case where instead of watching from the stadium one has read the report in the news paper and thus compare the two.

Similarly we can illuminate this phenomenological difference between the race reader and eye witness with an analogy between a book and a film. While chapters and subplots can be written from the perspective of different characters, one is free when reading the chapter to visualize the character and action from whatever angle, position, place or even from inside the story line as the character itself. Not so with the movie. Here one is limited to frames, perspectives, and angles, chosen by the director and filmed by the camera man. The movie is more like the limited perspective of the eye witness.

To turn these in to Linguistic Behavioural arguments without phenomenological participation is very difficult. One can talk about seeing in the world and 'seeing' in the mind's eye, as two different uses of 'seeing'. It is very difficult to see how one can make that distinction in purely Linguistic Behavioral terms without the phenomenology.

On first appearance, if we take a witness to an event and a reader of a report about that same event and put them in different chairs and asked them questions, there are things the witness could not tell us, if Ryle is right, and that the reader of the report could, since the reader's view, as he might imagine it, is not hampered by a man sitting in front of him with a funny hat, unless the reader perhaps wishes to imagine the man there. The witness, if he was unfortunate enough to be seated behind the man with the silly big hat, may have had his view hampered, or be liable to any one of the endless varieties of contingencies his witnessing the event made him subject to on the day. The reader would not be limited to one perspective, in reading the

---

[351]Ryle, *Concept of Mind,* 1983. Pg 262

report, but only by his powers of conjuring up the details of the race's report.

The book reader, like the report reader, similarly, might imagine what the other side of the main character's face is doing during the final battle scene, or he might picture it from front on, or from any one of the other characters, while the cinema goer is limited to seeing just this side, as shot by the film crew. The movie, as an analogous case, with all its frames, angles, and perspectives, is shot from one perspective, like the witness's, namely, that of the director's. So, prima facie, there would seem that there should be some sort of linguistic behavioural difference since one is limited to one perspective, that from which he or she witnessed it, while the other is not.

But let us imagine a canny reader, who, for his own reasons, needs to make a convincing 'report' for he wishes to convince our canny Judge that he really was at the race track on the day in question. He, foreknowing the distinction between a witness and a reader, sits down and visualizes his entire day from start to finish as if he lived it. Since he has relative freedom to imagine the races from any perspective this should not be a difficult task, since he has only to imagine the one perspective, and to stick with it.

Given this is the case I argue that there is no discernible difference between the linguistic behaviour of the cunning reader and the genuine witness at the level of a purely linguistic behavioural analysis. Furthermore there is no "ordinary language" appeal or justification for this distinction in Ryle, and in fact, it is hard to imagine what an Ordinary Language justification or an appeal might look like. We can only understand this bit of phenomenology by stepping into it and thinking about it. The point of the argument only arises as 'the sugar melts' as Sartre likes to quote Bergson[352] one must use a subject consciousness to grasp an object consciousness, in the manner of the device I described earlier[353] in order to examine it and perform the part of the argument contained in the 'phenomenological' description. One must make consciousness itself, in this case the imagining consciousness, subject to an act of consciousness to see the distinction.

---

[352]Sartre. *The Imaginary*, 2004. Pg 8
[353]See '4.2 Sartre, Phenomenology and the Apperception of the I' in 'Chapter Four : Theitic Consciousness and the Interlude' in this paper.

As such, the argument for the distinction between the reader and the witness at the race track is a purely phenomenological argument, one that requires the twin processes of introspection, and consciousness, as outlawed by Ryle, in order for the argument to make sense. That is, by the very inclusion of this argument, in Ryle's *Concept of Mind*, Ryle has argued against one of his own leading theses. He has undermined his thesis that the self luminous Cartesian and Protestant Para-Optic strand of philosophy of mind is an unmitigated mistake. Rather as will become evident by means of my methodological analysis, Ryle's argumentative practice, has in fact, inadvertently demonstrated the existence of the Cartesian Theatre which he wishes to prove doesn't exist in the very act of getting us to perform these thought experiments for the purposes of his argument. Ryle demonstrates that the introspective act itself required for this argument fulfils all of the criteria of introspection by being deliberately conscious, attending to the phenomenal contents of consciousness twice, as well as drawing attention to non-optical and non-sensory elements. And in fact, the only way to avoid this conclusion once Ryle has pointed out that the above act contains the very properties of consciousness he wishes to deny, is if Ryle then claimed not to be able to perform the introspective act.

Next let us consider the following two examples.

> Similarly, I can give you the fullest possible advice what to do, but I must omit one piece of counsel, since I cannot in the same breath advise you how to take that advice.

Next is the Pure Phenomenological argument about anticipating one's next thought.
Ryle writes

> (W)hile normally I am not at all surprised to find myself doing or thinking what I do, yet when I try most carefully to anticipate what I shall do or think, then the outcome is likely to falsify my expectation. My process

of pre-envisaging may divert the course of my ensuing behaviour in a direction and degree of which my prognosis cannot take account. One thing that I cannot prepare myself for is the next thought that I am going to think[354].

It is not immediately obvious, at least from verbs, nouns, adjectives, nor from any immediately obvious idioms, that one can not anticipate one's next thought. It is not until you try the exercise that you begin to see the problem. Likewise the problem of advice, is not prima facie obvious, otherwise we wouldn't go to people for advice. Yet, in a way similar to Sartre[355] one cannot give the advice, and then give advice on how to take that advice, and then give further advice on how to take the two former bits of advice, and so on. There is a regress. Likewise you have to try anticipating your next thought in order to make the trick work. Let us note, this is a phenomenological argument – at least so far as in the above example. It is a pure, naked, phenomenological argument because one cannot account for this phenomena in Behavioural Linguistic terms or find a way to justify it as an Ordinary Language argument with what it 'makes sense to say' unless one tries it out for one's self.

Perhaps it is worth noting that there is no reason why a phenomenological distinction or point can't enter into ordinary language use under the aegis of a metaphor or new linguistic behaviour involving the invention of new verbs that bring this process of thought out, or using inverted commas, or as a set of nouns that posit entities as a theoretical species of thought processes, that may, given time, become publically observable phenomena. But, of course, once we start allowing new phenomenological arguments, processes, experiences, exercises and techniques, the introduction of such 'new words', 'forms of words', or letting ourselves conjure up grammatical forms not easily found in everyday use then we have moved away from the domain of purely ordinary language in the sense Ryle thinks it best to do philosophy of mind in, and stepped into the realm of Psychological Nominalism, and

---

[354]Ryle, *Concept of Mind,* 1983. Pg 188
[355]Sartre, Jean Paul. *Existentialism Is a Humanism.* New Haven: Yale University Press, 2007. Pg 32, 45.

the shifting ontological sands of Wilson Cloud Chambers and Manifest Images that Wilfrid Sellars posits to describe it.

Now consider another example form *The Concept of Mind*

> A man is interested in Symbolic Logic. He regularly reads books and articles on the subject, discusses it, works out problems in it and neglects lectures on other subjects. According to the view which is here contested, he must therefore constantly experience impulses of a peculiar kind, namely feelings of interest in Symbolic Logic, and if his interest is very strong these feelings must be very acute and very frequent. He must therefore be able to tell us whether these feelings are sudden, like twinges, or lasting, like aches; whether they succeed one another several times a minute or only a few times an hour; and whether he feels them in the small of his back or in his forehead. But clearly his only reply to such specific questions would be that he catches himself experiencing no peculiar throbs or qualms while he is attending to his hobby. He may report a feeling of vexation, when his studies are interrupted, and the feeling of a load off his chest, when distractions are removed; but there are no peculiar feelings of interest in Symbolic Logic for him to report. While undisturbedly pursuing his hobby, he feels no perturbations at all[356].

This, of course, is a demonstration of the difference between flash-bangs on one side, and motivations and inclinations on the other. It is important to see that one can not actually make this argument without

---

[356]Ryle, *Concept of Mind,* 1983. Pp 84 - 85

the audience thinking it through for themselves. One needs to think the argument through before it makes sense.

However, notice two things that happen when you do think it through.

Firstly, while we can attach Ordinary Language argument to it in the form of '*It doesn't make sense to say that* a man was so agitated by his interest in math that he couldn't study it' or 'a man that was so patriotic that he couldn't go to war for his country' this is a different sort of argument from getting the audience to sympathize, or asking the audience outright, how many times in an hour or so of study does one feel the impulse to study logic. It asks you to "take a (non-optical) 'look' at what is passing in (your) mind" in the introspective sense.

Now notice the second interesting thing; there's no Behavioural Linguistic claim attached either here, to the "'introspective' (non optical)" bit nor to the prior Ordinary Language argument put in terms of what it *makes sense to say*. Also notice that on the Ordinary Language argument side of the claim that 'it doesn't make sense to talk about a man who was so patriotic he couldn't go to war for his country' or 'a man who was so interested in logic he couldn't study it'. The argument seems to point to a phenomenological content where there is no Linguistic Behavioural claim to justify or a grammatical distinction to attach to the justification. We are naturally led to this by the argument.

The Ordinary Language argument is justifying something, and since there's no distinction at the Linguistic Behavioural level to analyse, this, in fact, is an example of the phenomenological stream running through Ryle underneath some of his Ordinary Language argumentation.

5.3 Exemplar: Ghostographic analysis of Ryle's philosophical practice.

Finally we will do a case study of Ryle's therapy and apply our method of analysis and we shall see how all the parts fall in together. So briefly, if we return to the para-mechanical theory of mind, there are ghostly and non-ghostly thrusts. On the non-ghostly thrust side, Ryle identifies volitions and passions. If we go down the passion side, we can see how Ryle uses Ordinary Language Philosophy as exposited in *The Concept of Mind*.

As we have seen, according to Ryle dispositions that are governed by tendency verbs belong to one or the other of two classes according to whether the tendency can be classified as either *propensities* or *occurrences*. Occurrences contain our sub-structure of the flash-bangs, recall also that the difference between a glow of *warmth* and a glow of *pride* is difficult to demarcate without the phenomenological "feely bit"[357]. This, of course, I will argue points to a theory of consciousness in the next chapter. But here, for the purposes of the exemplar of Ryle's therapy, I note that inclinations, motives and agitations are subsumed under *propensities* proper.

Ryle writes

As the words 'distraction' and 'agitation' themselves indicate, people in these conditions are, to use a hazardous metaphor, subject to opposing forces. The two radical kinds of such conflicts are these, namely when one inclination runs counter to another, and when an inclination is thwarted by the hard facts of the world. A man who wants a country life and wants to hold a position which requires his living in a town is inclined in opposing directions. A man who

---

[357]There is a concern I have been unable to put to rest with regard to David Chlamers. See Chalmers. *The Conscious Mind*. 1996 pg 97, my question is would his phenomenal zombies be able to tell or speak of the difference between a 'glow of pride' and a 'glow of warmth'. Answering the point is made ambiguous because on pg 107 of *The Conscious Mind* there is evidence that Chalmers thinks language can be explained away in a functionalist account and is separable from phenomenal content, but he also takes it that reportability is indicative of conscious experience. The part I struggle with is that while it is easy for us to imagine the zombie with inbuilt temperature sensors, it is less so to imagine him detecting a 'glow of warmth' as warmth has different phenomenal properties to temperature which makes a fundamental difference to the type of thing that warmth is. Even less so for us to imagine him detecting a 'glow of pride' and being able to tell the difference. Even harder to imagine him differentiating between 'flashes of', anger, irritation or insight', and so on.

wants to live and is dying is precluded by the facts from doing what he wants. These instances show an important feature of agitations, namely that they presuppose the existence of inclinations which are not themselves agitations, much as eddies presuppose the existence of currents which are not themselves eddies. An eddy is an interference-condition which requires that there exist, say, two currents, or a current and a rock; an agitation requires that there exist two inclinations or an inclination and a factual impediment. Grief, of one sort, is affection blocked by death; suspense, of one sort, is hope interfered with by fear. To be torn between patriotism and ambition the victim must be both patriotic and ambitious[358]

.

Thus, we can see that when one has a motive, or an inclination they can run ashore on the hard facts of the world and create an agitation. There is something odd about this but we will bring it out later in the paper[359]. For my purposes here note Ryle is saying the following

When a man is described as being both very avaricious and rather fond of gardening, part of what is being said is that the former motive is stronger than the latter, in the sense that much more of his internal and external conduct is directed towards self-enrichment than is directed towards horticulture. Moreover, when situations arise in which a slight financial loss would be accompanied by a major improvement to his garden, he is likely to give up the orchids and to keep the cash[360].

[358]Ryle, *Concept of Mind,* 1983. Pg 90
[359]See 'Chapter Eight : The Occurrences' in this paper.

Thus, we can see, according to Ryle's account, that two contrary inclinations, or two contrary motives, produce agitations. Recall the student who enjoyed symbolic logic, and the question as to how often the desire, need, hankering, or impulse to study logic arose in an hour of deriving propositional proofs. Ryle points out that *it doesn't make sense to talk about* someone who was so agitated by their love or interest in study that they could not study.

Putting all of these things together yields the following argument, or better therapy, applied to Hume's theory of the passions.

> In fact, inclinations and agitations are
> things of different kinds. Agitations can be
> violent or mild, inclinations cannot be
> cither. Inclinations can be relatively strong
> or relatively weak, but this difference is not
> a difference of degree of upsettingness; it is
> a difference of degree of operativeness,
> which is quite a different sort of difference.
> Hume's word 'passion' was being used to
> signify things of *at least two disparate types*[361]
> .

Thus we have the Linguistic Behavioural observation that the sorts of things Hume designates as *passions* with the properties of being mild or violent Ryle claims, pace Hume, actually indicates two things of very different types. The two things of very different types are based on two disparate sets of linguistic behaviour, and that these behaviours are further analysable down into inclinations and motives and when they clash they result in agitations or in the case of commotions, habits. That is the dispositions governed by the behaviour of tendency verbs on the one hand and on the other, the conflict that arises from contrary sets of these tendencies result in clashes that result in agitations.

Agitations are moods. They are of a different order of the propensities from the dispositions governed by occurrences. When dispositions, governed by tendency verbs escalate into moods they

---

[360]Ryle, *Concept of Mind,* 1983. Pg 91
[361]Ryle, *Concept of Mind,* 1983. Pg 91 NB Italics are his.

become episodic and behave in a way governed by episodic verbs, or verbs that behave in ways that are different from dispositional verbs themselves. The day Rod fumed. That time Reg was so agitated he punched the desk. Do you remember when Rod was so agitated he claimed he couldn't see straight?[362]

In the next step we have the Ordinary Language argument to support the linguistic behavioural claims that pick out these distinctions, that is, *it doesn't make sense to talk about* a man who was so interested in symbolic logic he was too agitated to study it. Recall that this is a different sort of argument to that appealed to by asking the question; how many times in an hour of working on logical problems does the motive or inclination arise to study? This latter argument invokes the occult stream of Ryle's argumentation. Alongside this is the phenomenological aspect of the argument. In this strand of the argument we are invited to actually think about a period of study and try to recall any study twinges, or sensations of scholastic feeling, or subset of academic pangs, and so on. Motives and inclinations that incline or drive one to study must be very different things to feelings and occurrences because when we look, or think back, we can't find any feelings or occurrences in our private studies or internal experiences of libraries. But to recognize a 'twinge' or an 'occurrence' requires a phenomenological recognition or sensation. Thus to make the argument about motives and inclinations, and subsequently, the argument on how many times a sensation or academic twinge arises, we must first have the phenomenological insight into what a 'pang of regret', or 'glow of warmth' actually feels like in order to scour our recollections of our

[362] This last claim is one of the ones that may present problems for a Sellarian reading of Ryle. See Sellars Rylean picture in *Empiricism and the Philosophy of Mind*, for Sellars holds that Ryleans cannot have private reports. This is a problematic reading because Ryle has arguments like the Witness/Reader argument and refers to, arguable, non-linguistic or pre-linguistic elements of thought which require or implicate non-linguistic private experience. Sellars' reading of Ryle, I take it, is based on the 'log keeper' cognitive function in Ryle's retrospection, and various species of 'mindologue' that I drew out and listed, and a select number of the fruits of Ryle's linguistic behavioral analysis, most important of which are his 'achievement verbs' which Sellars refers back to. Sellars uses the term 'in foro interno', to refer to the mental narrator in our inner thoughts, which I take it equates with Ryle' log keeper role and species of 'mindologue'. See Sellars, *Empiricism & the Philosophy of Mind*, 1997 pg 91, numbered section 48 for Sellars discussion of a Rylean language. Note, some of the phenomenological strain of arguments in Ryle incorporate elements that would only be accessible from a fully developed Sellarian Report Language. See footnotes 71, 200, 218, 229, 378, 377 and 362 for further clarifications in relation to the Sellars reading of Ryle and Ryle's original arguments in *The Concept of Mind* and the way these relate to the first and third personal positions in the paper.

experiences of the inside of libraries and private studies, to see if in fact there are any there?

The argument asks you to "take a (non-optical) 'look' at what is passing (or passed) in (your) mind" in the same introspective sense Ryle disavows. That is to make the argument you need to grasp a prior act of consciousness by means of a present act of consciousness and see if any of these twinges are or were there. And by a preceding act of introspection and comparison you first have to imagine what these twinges, pangs and states would feel like as they apply to grief, anger, regret or nostalgia so as to compare them to what the argument asks the audience to envisage. So here we find once more an appeal to that rare species of occurrence which I asked you to keep an eye on at the outset of the argument, the flash bang.

Thus we can see that in the process of apprehending Ryle's 'Ordinary Language Therapy' to dissolve Hume's theory of the passions we have Linguistic Behavioural claims about the way certain words behave and the structures of their behaviour. We also see the employment of an Ordinary Language argument making the claim that there is a behavioural distinction on the ordinary language level such that *it doesn't make sense to talk about* a man who was so patriotic he couldn't go to war for his country. However to make such argumentation work we have a Phenomenological argument which requires us to apply an earlier distinction in relation to a sub-set of occurrences, the *flash-bangs*, to differentiate between 'glows of pride' and 'glows of warmth' in order to make the argument that requires one to know what a 'twinge of studiousness' might feel like, in order to see if we have any. That is, to make the argument Ryle requires us to "take a (non-optical) 'look' inside and see if we have any such twinges, thereby invalidating his own strictures. These 'studious twinges' are of course different to the ones he mentions in his sensations and observation distinction since such 'studious twinges' do not rely on a sensory organ like the eye or the skin, but like the "flash bangs" occur on some non-sensory level.

Chapter Six : Midway map of the paper

So far we looked at Ordinary Language arguments and Linguistic Behavioral arguments. I drew my definition of Linguistic Behavioral arguments from the short comings of Chalmers' 'psychological and phenomenal history' and Weitz's 'propositional model' of Logical Behaviorism. The former was inadequate because it ignored Ryle's Ordinary Language justifications for his 'dispositions and associative behaviors' and the latter because its simplified propositional model failed to make distinctions at the level of natural language analysis that Ryle was specifically interested in. The deficiencies of the former allowed us to begin our Ordinary Language classification of Ryle's argumentation. Deficiencies in the latter allowed us to begin our analysis of Ryle's Linguistic Behavourial strand of argumentation.

We also examined determinate/determinable dispositions, along with a species of Rylean emotional flower that I gave the label 'flash-bang' to, and the differing families of dispositions. We looked at occurrences and propensities, Ryle's arguments about hypocrites and charlatans, along with the capacity and tendency verbs which allows Ryle to divide up the dispositions. I demonstrated the difference between ordinary language arguments as a unique criteria for drawing a normative authority from, and compared these with several other types of arguments like Tsai's analysis and critique of Stanley and Williamson's 'true' propositional approach answering questions based on the 'how/that' distinction, or like Ayer's argument that the facts need to be cashable in the native currency of possible experience, which, in the example we used, leads him into problems with statements about the past which went beyond the normative force of ordinary language argumentation since 'it didn't make sense' to talk about past events in terms of future evidence.

The division of verbs into tendency and capacity dispositions along with the limits of the analysis of language are important for deeper philosophical reasons that have to do with Wilfrid Sellars. By moving capacity dispositions out of the psychologistic critique, this will also move Sellars dispositions for fact stating roles in *Empiricism and the Philosophy of Mind,* beyond the critique of this paper. However, there is specific content I want to critique in Robert Brandom. This will become clearer when we reach *Motives, Moods and Reasons* later in this paper. But there are several stages we need to develop before we reach this.

As I just mentioned in a discussion of Dummett's and Ryle's arguments about the meaning inherent in language, I pointed out Dummett's criteria for the Pre-Fregeian view of psychologism which held that language was like a code that thoughts were compounded into. I also revealed that the central tenant of this paper was to argue for a return to a Pre-Fregeian theory of mind on the basis of a push that would come from several sources, one of which, I revealed, could come out of a reading of Ryle which would lead to a 'pre-linguistic how' and a 'non-linguistic how' and which, of the theorists we looked at, perhaps only Tsai and his criticism of Stanley and Williamson's paper, seemed to have picked up on. I also pointed out that what I would be attacking is not some version or inconsistency in this, or those between his polemic on consciousness and cases where he breaks his own strictures, but rather Ryle's direct and strongest claims about the nature of mind on the basis of Frege's maxim it is always better to attack what is philosophically strongest about your opponent's point. Here we removed the capacity dispositions from the capacity-tendency configuration in Ryle's linguistic machinery, because they presented themselves as a strawman.

Most importantly I showed you my methodology for drawing out the phenomenological content in Ordinary Language arguments by looking for a claim that doesn't have a bit of Linguistic Behavioural analysis attached to it. This of course is my mini-methodology of Ghostography, so named, because it is designed to bring out what is haunting *The Concept of Mind*. I further pointed out the rarefied and pure strains of 'naked' phenomenology which exist outside Ordinary Language arguments and couldn't be mapped in any way it '*makes sense to say*'.

We also looked at Brandom and Sartre's position on the apperception of the I and Immanuel Kant. Sartre argued, as a point of historical scholarship, and as his own view, that the 'I' is not present in all of our thoughts. Brandom took the contrary view, that Kant, as a point of scholarship, and as the philosophically right position, held that the I needs to accompany all of our thoughts. We turned to scholarship on this point for consultation and clarification but also noted, that

Brandom had inherited his view with certain analytic insights and ways of interpreting Kant that of course come down from the analytic tradition. In particular, he has inherited insights and views from Wittgenstein and Frege. We then broke off, for a short literature review, owing to a cluster of mistakes that repeatedly turn up in the Anglo-Sartre scholarship, and which I promised would free Sartre up from further discussion of them for the duration of the thesis.

Indeed, when we returned to the argument of the apperception of the I, on philosophical grounds, I argued that I should side with Sartre, on the strength of his phenomenology. This constitutes the first part of a two part general critique of a linguistic treatment of motives and beliefs based in Brandom's philosophy, and, really, the second move in the thesis towards a Pre-Fregeian psychologism. The first, of course, is an attack on the Dummett-Ryle 'implicit position' of meaning.

We now have a taxonomy that can analyse and categorize structural traits of three different types of argument in Ryle. We have the Linguistic Behavioural form of argumentation. These are arguments that consist of grammatical and linguistic types of analyses that offer fruit in the form of specific constructions of verbs, nouns, epithets and other bits of linguistic jargon to make claims about the mind. We saw some of these in the analyses of the short comings of Weitz's critique of Ryle with his classification of Ryle's analyses into specific model sentences and propositional forms. This type of approach lacked the resources for a thorough analyses into the semantic relationships as they actually occur in Ryle's arguments and can not capture the significance of the aspects of mind Ryle is specifically interested in. In Ryle's words Weitz made the mistake of assuming that 'all dispositions have a uniform function' and he is thus unable to differentiate between different sets of dispositions based on the behaviour of specific word groupings. Weitz can't differentiate between capacity and tendency dispositions. He can not differentiate between determinate and determinable forms of these dispositions and he can not differentiate between motives, capacities, tendencies, skills and abilities and what makes the relationships between these interesting. He has failed to capture the exact nature of Ryle's argumentation method. He has failed to capture Linguistic Behavioural analyses.

The second type of argument we find in Ryle is an appeal to the normative value and authority of ordinary language. This is what David Chalmers account of Ryle lacked. Ryle was first and foremost an ordinary language philosopher. He argued and held that all we need to know about the mind is already contained in the language of ordinary men who know how to talk about the mind in the world. The work of philosophers is, as such, to sort through the terminology of these terms and correct their geography. Thus an appeal to ordinary language is an appeal to what it makes sense to say and has an authority based on that linguistic usage. Ryle sees his role as seeing how that usage fits together into a concept of mind. Where the terms *it makes sense to say* or the negation *it does not make sense to say* enters into an argument, line of jargon or analyses of a bit of language, here Gilbert Ryle is making an Ordinary Language claim and drawing on the authority he argues is implicit in ordinary language.

The third strain of argument is that which we have drawn out of Jean Paul Sartre's method of phenomenology to analyse Ryle. Where consciousness appears before itself in a reflective state and we analyse aspects of what Sartre refers to as the 'cortege of the psyche', that is, when Ryle asks us to engage in what Sartre would call a theitic act of consciousness and apprehend the ego complex of the 'I' or the 'me' for the purposes of an argument, here we are engaging the phenomenological strain of argumentation in Ryle. We do this for such arguments as examining our motivations, emotions, the sizes and shapes or lack thereof, of various sensations and what it might take to build instruments to improve our apprehension of them in the way we build telescopes to look at far away objects and microscopes for looking at small ones. The phenomenological strain is particularly important for the developing body of argument based on the emotional content of our thoughts and the 'flash-bang' strain of arguments. These are arguments that we must 'try out' in some way to see the point. We must bring aspects of consciousness to attention in order to see the point of the argument being made. These are arguments like asking us to imagine a light so bright that it hurts our eyes or to try to anticipate our next thought before we have it. It is not until we try such exercises and make consciousness the object of our scrutiny in doing so and we reflect on them that we begin to see the point of them.

Our taxonomy of Ryle's arguments can thus be summarized in this way. Linguistic Behavioural arguments can be typified by examination of verbs, nouns, linguistic phrases and grammatical terminology. Typically they make claims about the mind based upon specific examples of language and the behaviour of different verbs components and configurations. Ordinary Language argumentation engages the normative value of ordinary language usage. It makes claims that involve the term *it makes sense to say*, to forward a claim about the mind or *it does not make sense to say* to negate one. Finally the third strain is made up of arguments that engage phenomenological content, they get us to bring certain aspects of our consciousness before consciousness in order to make a point or get us to engage phenomenological acts in order to understand the argument.

It will be the purpose of this paper to determine which of the three makes the strongest argument and thus offers us the strongest bases for arguing about the nature of mind. The results of making that argument will have long range impacts on 20th Century philosophy of mind in general. For if it turns out that phenomenology is indeed the stronger source and it follows that words are simply codes for our thoughts then psychologism is indeed the correct grounds on which to model a theory of mind. This will have the effect of over turning both forms of anti-psychologism as I have laid out above. Both the implicit anti-psychologism of Ryle and Dummett and the explicit psychologism of Frege, Brandom and Ayer. This indeed is the controversial line I will be arguing through the remainder of the paper. That is, I shall argue that psychologism is the correct approach to a theory of mind. That we must begin with a theory of mind before we can develop a theory of language. Most importantly, that at the heart of an argument for the move towards psychologism is an argument for a theory of consciousness. To this we now turn.

Since Ryle claims both Linguistic Behavioural arguments and phenomenological ones are species of Ordinary Language arguments and that they draw their normative force from Ordinary Language, then they should not be able to contradict each other. That is, if Ordinary Language is a good normative source it should not produce contradictions in what it asserts. I take it that any clash between propositions, premises, predication of existential quantifiers that produces rival propositions that can be reduced to a contradiction is grounds enough for rejecting that normative source.

The goal of this paper is to attack the grounds for accepting Ordinary Language arguments as a unifying source and undermine its normativety by demonstrating one such clash producing a contradiction. The clash between Phenomenology and Linguistic Behaviourism will thus undermine the authority of Ordinary Language as a unifying force, and reveal that the normativety of these arguments does not come from shared knowledge of the linguistic speaker but rather it comes from two underlying rival forms of arguments. That is to say phenomenological arguments and Linguistic Behavioural ones are not happy campers in the Ordinary Language park.

The clash I am preparing the way for and which I wish to draw out with this taxonomy of Ryle's arguments is that between Phenomenology and Linguistic Behaviourism the clash will be drawn out using an obscure Rylean critic, Robert Wolff.

In my view Robert Wolff's paper is important because it pinpoints an insight into problems with Ryle's analyses and use of ordinary language which surface in the Daniel C. Dennett and David J. Chalmers debates over the nature of objectivity and language in relation to the mind. Gilbert Ryle was hugely influential on generations of analytic philosophers, in particular Daniel C. Dennett who is at the forefront of neuroscience and research on the mind. Because of that influence the Wolff article will reveal something insightful to the Chalmers and Dennett debate about the possibility of a First Person Science of the Mind. More important my treatment of Wolff's article will illustrate the problem with using Ordinary Language arguments as an authority in the way that Ryle does. Basically the way I untangle that problem is to show there is a space that arises between two different types of analyses that allows for a contradiction to surface if the two types of analyses are taken as the same type. If the two types of analyses are taken as the same type then they can support rival contradictory claims. This means if we treat them as the same source then we have a normative source that provides us with a contradiction. The contradiction arises via the assumptions inherited from a direct appeal to either first or third personal perspectives and these assumptions are then carried over into an indirect appeal.

The contradiction arises from claims as to what dispositions are and involves the two strains in Ryle I have been at pains to separate. These two strains, of course, are the Linguistic Behavioural arguments and the phenomenological strain as I point out above. This sets the scene for the 'sharp thesis' of this paper which is merely the claim that auto-phenomenology can give us a stronger type of argument than hetero-phenomenological sources like linguistic behaviourism and arguments that use them. Auto-phenomenology is the phenomenological experience from the direct first person. Auto-phenomenology is what we get when we start to drop assumptions inherited from the indirect positions. In *7.3 Dispositions, Phenomenology and Law-Like statements* we will learn that the force in much of Ryle's arguments derives from the indirect assumptions inherited from the mixed domains of first personal position and a third personal position. Auto-phenomenology is the type of first personal phenomenological source we draw upon in Ryle's phenomenological style arguments. Hetero-phenomenology is a combination of different sources accessible from the third personal position. These include Linguistic Behavioural arguments along with physiologistic symptoms and neuro-physiologistic systems. The remainder of the paper will decide the priority and strength of these two positions and build a hierarchy from comparisons of relations between them. I hope that the shape of the thesis should now start to become visible.

If we extract the logical stages from what we have done so far we can see how each stage of the thesis has prepared us with arguments and pieces for the subsequent stages leading to the final insight. The preparation for the division of Ordinary Language arguments on the nature of mind started when I isolated two strains of argument in Ryle. The first strain was made up from the Linguistic Behaviourist strain which I drew against the short falls of Logical Behaviourism in Weitz who failed to examine the behaviour of words and instead tried to abstract to the level of whole propositional models and which gave us the beginning of the Linguistic Behaviourist strain of arguments in Ryle. The second was drawn against the inadequacies of David Chalmers account, which failed to take into account the normative foundation of Ryle's arguments. Ryle was an Ordinary Language Philosopher and his arguments' appeal was to the authority in ordinary language and what it makes sense to say. From Chalmers we started our second taxonomy of Rylean argumentation, the Ordinary Language argument. We then looked at how these related to Dummett's position on language and Ryle's own arguments for an implicit theory of meaning.

I then isolated another strain based on inconsistencies in Ryle's arguments against consciousness and introspection, and cases where Ryle used arguments that either implied the presence of consciousness in carrying them out or that called upon the reader to perform feats of introspective scrutiny. I then clarified these arguments based on Sartre's phenomenology, which I had historical reasons for believing had influenced Ryle and had been active in germinating some of the contradictions and inconsistencies that arise in *The Concept of Mind* between Ryle's Ordinary Language methodology and what he is arguing.

That is Ryle's 'it makes sense to say' arguments that call upon an ordinary language user's intuition contain two types of sub-arguments. Ones that can be pin-pointed by linguistic behaviors and ones that can not. Of the ones that can not, some of them make their appeal at the level of something implicit in ordinary language, but don't bear fruit at the level of linguistic behavior like the flash-bangs and their different types of glow. What I drew out in the previous section, was a careful set of distinctions that give us a fruitful taxonomy of Ryle's arguments. This taxonomy involves (1) Ordinary Language arguments, (2) the Linguistic Behaviorist strain, I identified early on, and (3) the Phenomenological strain. The phenomenological strain has two further strains; the 'occult strain' and the 'pure' phenomenological strain. The occult strain either had ordinary language distinctions 'it made sense to talk about' that were implicit, but could not be identified by linguistic behaviors, or it had linguistic behaviors that had both a linguistic side and a hidden occult side. The pure phenomenological strain were arguments that contained distinctions that could not be pin-pointed by linguistic behaviors like trying to anticipate your next thought, or the reader/witness distinction. These arguments are problematic for a Sellarian reading of *The Concept of Mind*, because they contain phenomenological insight from exercises that are only available at the pre-linguistic level and that can be accessed by a report language. There is a puzzle here and the purposes of this paper is to bring the solution out.

We thus have, so far, two strains in Ryle. A Linguistic Behaviorist strain and a phenomenological strain, the two share Ordinary Language arguments in common, which accounts for the 'occult' strain. The occult strain of phenomenological arguments running through *The Concept of Mind* were arguments like the "Remember When/Remember How" Configuration which can be read as having a Linguistic Behavioural analysis or an introspective phenomenological source of normativity. The pure strain has a number of arguments that are problematic for Sellars' account of a Rylean public language as well as Ryle's own arguments against consciousness and introspection This was because the pure phenomenological arguments make claims based on introspective scrutiny with no Linguistic Behavioural distinctions or appeals to Ordinary Language normativity to justify them. Instead the pure strain contain distinctions that can only be made with insight into consciousness like the Canny Reader in the Reader/Witness argument or exercises like anticipating one's next thought. Later I will introduce a nuance in my reading of Sellars that can account for this by a certain interpretation that favours a close reading of what Sellars actually does with Ryle's achievement verbs.

I will now go onto show how contradictory claims can arise between these two styles of arguments. In such cases, phenomenological arguments that individuate over the authority of another source, I will identify as 'auto-phenomenological claims'. These are claims which take it that first person phenomenology is a stronger type of argument than third person claims like Linguistic Behavorism. But the new rivalry revealed by the break down of Ordinary Language argumentation about the mind also lets us access other third person views like physiologistic and neurological sources. Physiological, Linguistic Behaviourial and Neuro-physiological are all forms of what I call hetero-phenomenology. This is the Sharp Narrow Thesis I pointed out in the introduction. This is supported by the psychologistic thesis that non-linguistic meaning is necessary for talk in some linguistic contexts, specifically reasons, motives and beliefs, as we are attacking the tendency side of the tendency/capacity dispositional configuration I introduced earlier. I shall break each one of Ryle's configurations down systematically. I already started this process with Brandom and Sartre. I will complete it with my critique of Brandom's game of giving and asking for reasons and A. R. White. However prior to this we need to do some ground work with R. S. Peters and Wilfrid Sellars.

This psychologistic thesis has another side. It claims autophenomenology is a better source for arguments that involve elements and individuation on the nature of mind, and specifically for some irreducible aspects of mind. This, of course, is the Broad General Thesis of the paper, which it shares with David Chalmers. Emotions are one type of these irreducible elements of mind. They are the type I am focusing on to break the dominant Cognitive and 'implicit theory of meaning' positions in the philosophy of mind.

This introduces or leads us to the Ordinary Language cum Neuroscientific position, which is a sophisticated version of Ryle's implicit meaning position that allows and takes it for granted that we have some neuroscientific data on the mind and this can be mapped onto language roles and linguistic behaviours. I've drawn this out of my reading of Sellars paper *Philosophy and the Scientific Image of Man*[363]. I argue against this. I argue that autophenomenology to hetrophenomenology is stronger than two heterophenomenological sources. That is to say in cases where neuroscience claims that two mental states are the same it is left up to a third claim, that of introspective scrutiny, for the person to say whether they experience them as the same or different cases. Likewise where neuroscience and Ordinary Language claim that two states are the same, you still have an autophenomenological source that can disagree. That is the same part of the brain might light up on the EEG 'do-hickey' and the person might say they feel a 'flash of anger' and the neuroscientist might tick 'anger' on his chart twice, but it is still open to yet another claim by the person to say the two flashes of anger feel differently even though the 'do-hickey' and the word ticked on the page recorded the same 'anger' in

---

[363]This is a slightly simplified account. In *Empiricism and the Philosophy of Mind* Sellars argues that the scientific conceptual framework eventually comes to replace the one inherited from the manifest image. See Brandom's commentary Brandom, Sellars, *Empiricism & the Philosophy of Mind*, 1997. Pp 177-181

each case. This is the significance of the flash-bang thread I started and told you to keep your eye on. I shall sketch these arguments in more detail when we arrive at them.

Chapter Seven : Motives, Moods and Reasons.

One of Ryle's more interesting followers is the Ordinary Language Psychologist R.S. Peters. In his work *The Concept of Motivation*[364] R. S. Peters aligns himself with Ryle, reaffirming Ryle's position on the properties of dispositions, but limits his own analysis to points at which we encounter their presence in episodes since he thinks that Ryle's account is far too wide. Peters takes a teleological view of dispositions restricting his analysis further to teleological tendency dispositions like motives. A fortori Peters argues, given the status of motives as non-events[365] and affirming their proper role in ordinary language he argues that the non-episodic properties of dispositions are ultimately keyed into sets of reasons which can be revealed in given social circumstances which are episodes.

What is important in Peters' are the social circumstances in which explanations arise. He thinks that efficient causal structures in the forms of sets of reasons are a mistake. Motives as linguistic structures or propositions do not drive actions, rather they are linguistic manifestations of underling purpose and teleological actions which arise in specific social circumstances.

Peters writes

> Motives, in other words, are a particular
> class of reasons, which are distinguished by
> certain logical properties. My thesis is that
> the concept of 'motivation' has developed
> from that of 'motive' by attempting a causal

---

[364]Myers, G. E. "Motives and Wants." *Mind* Vol. 73, no. 290 (1964): Pp. 173-185.

[365]Peters, R. S. "Motives and Causes." *Proceedings of the Aristotelian Society, Supplementary Volumes* 26, (1952): pp. 139-194. Also Peters, R.S. *The Concept of Motivation*. London Lowe & Brydone, 1969. Pg 32

interpretation of the logical force of the term. This is made possible by the failure to distinguish different levels of questions[366].

He then draws out three characteristics of the ways in which motives can be exposed in a layered tier structure. The third characteristic rests upon the first two, which Peters thinks is, essentially, as I stated above, that dispositions are keyed into sets of reasons. The third characteristic is teleological and purposive.

Peters writes

> The third characteristic of motives as a class of reasons for actions (is that) they must be reasons why a person acts. By this is meant that the goal which is quoted to justify a man's action must also be such that reference to it actually explains what a man has done[367].

The class of dispositions behind the teleological structure of motives, the rule following purposive framework, in Peters account, are supervened on by the class of motives that a person gives justifying or explaining their actions in the right social context.

Peters writes.

> In the first place we only ask about a man's motives when we wish, in some way, to hold his conduct up for assessment. The word is used typically in moral or legal discourse where actions have to be justified and not simply explained. We ascribe or impute motives to others and avow or confess them

[366]Peters, *Concept of Motivation*, 1969. Pg 28
[367]Peters, *Concept of Motivation*, 1969. Pg 34

in ourselves. This explains why we often ask for motives when there is a breach of conventional expectations; for it is in just these sorts of contexts that men have to justify their actions[368].

This is interesting because Wilfrid Sellars does something very much like it in *Empiricism and the Philosophy of Mind*. He points out that a report[369] in the presence of a stimulus about that stimulus must have an authority that is recognized by a person. The recognition for the authority of a report ultimately rests upon the social conditions that make such a report possible. The stimuli itself is not enough.

Sellars writes

For we have seen that to be the expression of knowledge, a report must not only have authority, this authority must *in some sense* be recognized by the person whose report it is[370].

Indeed, these social conditions are keyed into a background of standard conditions.

Sellars writes

And this is a steep hurdle indeed. For if the authority of the report "This is green" lies in the fact that the existence of green items appropriately related to the perceiver can be inferred from the occurrence of such reports, it follows that only a person who is able to draw this inference, and therefore who has

[368]Peters, *Concept of Motivation*, 1969. Pg 29
[369]Sellars, *Empiricism & the Philosophy of Mind*, 1997. See pg 72 for the technical specification that Sellars gives for the form for a report.
[370]Sellars, *Empiricism & the Philosophy of Mind*, 1997. Pg 74

not only the concept *green*, but also the concept of uttering "This is green" -- indeed, the concept of certain conditions of perception, those which would correctly be called 'standard conditions' -- could be in a position to token "This is green" in recognition of its authority. [371]

This authority constitutes the second part of Wilfrid Sellars direct refutation of *Konstatierung* statements[372] and the central core of his refutation of Empiricism.

In other words, for a *Konstatierung* "This is green" to "express observational knowledge," not only must it be a *symptom* or *sign* of the presence of a green object in standard conditions, but the perceiver must know that tokens of "This is green" *are* symptoms of the presence of green objects in conditions which are standard for visual perception.[373]

Sellars point is evident. Not only must one have the standard conditions making up the fact stating role, but one must also have the concept of the fact stating role itself. Both rely upon a social background and a set of standard conditions. People are not geared up like Price's thermometer to simply state 'this is green' in conditions in which green

---

[371]Sellars, *Empiricism & the Philosophy of Mind*, 1997. Pg 74
[372] These of course are Moritz Schlick's, even though Sellars doesn't reference them. See Willem A. DeVires, Timm Triplett. *Knowledge, Mind and the Given*. Indianapolis: Hackett Publishing Company Inc, 2000. Pp 72- 77.
[373]Sellars, *Empiricism & the Philosophy of Mind*, 1997. Pg 76 Again I take it that Sellars is referring to the 'role' of an achievement verb out of Ryle. This is important to the reading. See Triplett and Devires discussion of William Aston on this point. Triplett, DeVires, *Knowledge, Mind and the Given*. 2000. Pg 84-86. I think once we take Sellars position from the perspective of Ryle's achievement verbs it clears up the alleged ambiguity in Sellars claim. Wilfrid Sellars specific type of claim is linguistic about the use of achievement verbs sanctioned by a community and recognized by the user rather than an epistemic claim about the conditions of knowledge.

obtains as the proximal stimulant on the visual organs of the human sensory apparatus in the way mercury reacts to heat bulb of a thermometer. Rather, the act of a report must come from the specific practice in a linguistic community[374]. And I think this is a good argument by Sellars, since it could be argued, prima facie, that one could conceive of communities that don't possess the social history which leads them to invent simple fact stating roles of the behaviourist empirical schools of philosophy. They might, for instance, possess some sort of exotic social history that leads them to talk about everything in terms of spirits, or to reduce fact stating descriptions to explanatory statements in terms of numerical ratios[375].

Sellars writes

> As we have already noticed, the correctness of a report does not have to be construed as the rightness of an *action*. A report can be correct as being an instance of a general mode of behavior which, in a given linguistic community, it is reasonable to sanction and support[376].

In order to keep the point I'm about to make clear, let us relate this all back to Ryle's original framework[377]. Peters is making an argument

---

[374] See O'Shea, *Wilfrid Sellars*, 2007. Pg 75-81 for a discussion on this point.

[375] See Wilfrid Sellars' "Truth and 'Correspondence'." In *Science, Perception and Reality*, Pp 197 - 224. California: Ridgeview, 1991. Pg 203 for Sellars position on fact stating roles in a language and their relationship with forms of 'correspondence'. Sellars treats a report qua 'fact stating role' merely as one 'use' or 'role', role being his preferred terminology, in a language. However he thinks that a role and role aspects change in different languages, and they don't all 'correspond'. One example of this might be the classic Whorf hypothesis that reports made of geological features by Native American speakers use a complex vocabulary of geometrical figures, compared with simile and metaphor which plays a larger role in English descriptions of Geological features. Another could be the deep structures that arise in English use of metaphor about containers, impersonalization, personification and deeper principles of coherence in structuring metaphors; see G. Lakoff, M. Johnson. *Metaphors We Live By*. Chicago: University of Chicago Press, 1980. Pp 29 -30, 41-46, 49 – 51, 126 – 128. See also O'Shea, *Wilfrid Sellars*, 2007 and the discussion of Linguistic Communities pp 75-81, 102-105.

[376] Sellars, *Empiricism & the Philosophy of Mind*, 1997. Pg 76

[377] I am aware that this point is contentious since Ryle's original framework was based on the linguistic behavior of 'how' and 'that' clauses. It is debatable whether Ryle on purely linguistic grounds would classify Sellars' dispositions in fact stating roles as capacities or tendencies. I think

about dispositions, in particular tendency dispositions, which reveal themselves in episodic actions. These have attached social conditions that allow us to pry into these motives for a justification. The motives of the disposition behind the episodes, when revealed by social practice, are given as sets of reasons. In contrast Sellars is forwarding a position on linguistic reports given in the presence of certain stimuli. In this case Sellars is attaching social conditions onto what is essentially a fact stating role. One must first have the concept of the fact stating role and this concept must come out of social practice or a social linguistic context that allows one to acquire the concept of a fact stating role. This is an epistemic claim and unlike Peters position would in the original Rylean framework be moved over to the capacities[378], abilities and skills. One must know how to make a fact statement in the presence of the stimuli. That is to say Peters is on the side of the tendencies and Sellars is on the side of the capacities. This is important because as I pointed out in my earlier section on Tsai, Stanley and Williamson, dealing with capacities as part of a psycholohistic critique of anti-psychologism is a strawman, thus *Empiricism and the*

---

that he would once he understood Sellars' sophistication even if a Sellarian disposition as a fact stating role behaves like a tendency dispositional verb or a belief. I think it might trouble Ryle but I think he'd endorse Sellars' point of view, firstly because a Sellarian disposition behaves like a sophisticated achievement verb and secondly one must acquire the concept of a fact stating role, which means that one must know *how* to make it, even though they use a 'that' linguistic structure, they invoke a 'how' form of knowledge.

[378]Following on from the above footnote, the development of a language of the emotions, in my reading of Sellars, is actually a development of achievement verbs the success of which come from learning Jones' behavioral vocabulary in linguistic practice as part of a linguistic community. For instance, one day Rod hits Reg, and Jones develops the term 'red-hit' to describe Rod's face when he hits Reg. Red-hit is the ancestor to 'anger'. Further down the track, Bob, when asked by Jones what happened, describes hitting Rob, and how he felt as 'Red-hit'. This is the shift from an observation language to a report language. The ability to apply and describe his feelings as part of a report language is thus a further development on a capacity disposition, and itself, a new capacity. This accounts, as Sellars puts it, for the 'intersubjective achievement' of language, See Sellars, *Empiricism & the Philosophy of Mind,* 1997. Pg 107. My reading of Sellars is actually a cogent one. I'm going to argue at the end of this paper that 'red-hit' or 'anger' is not in the word, the word is like an internal disposition in the sense of an achievement-observation acquisition. This, I argue, is a codification of non-linguistic information accessible to the person. They don't think 'anger' as a word spoken in the mind or displayed as verbal imagery to an inner eye, rather, it is how they describe what is happening at a non-linguistic level. My reason for this reading of Sellars is simply that there must be something that is happening in Rod's thinking on that first day when he hits Reg, even if he doesn't have a name or a word for describing it yet. The codification or self description of dispositionally acquired emotional vocabulary follows the achievement verb structure like the scientist who learns to see hooked vapor trails in the Wilson Cloud Chamber. This puts it on the side of the capacities, even when it becomes a report language.

Philosophy of Mind is outside of this paper's critique of anti-psychologism. I pointed this out in *The Midway Map of the Paper*.

However, Brandom picks up on Sellars point about the social context. Brandom calls this the 'space of reasons' and endorses the position that the concept of the authority of a report is tied into discourse within the space of reasons. Brandom however moves over from the capacities and into the tendencies with Peters and the game of giving and asking for reasons.

The difference between Brandom and Peters is that Brandom maintains the presence of the I in the propositional structure of judgements as the minimum unit of thought. Peters thinks that language is something that happens to underlying purposive and teleological actions in given circumstances. Peters doesn't think motives in the form of utterable speech acts are what drive behaviour, rather motives as speech acts are what people give in the right social circumstance when a person's reasons for doing something are brought into question. This subtle difference will become important later in the paper when we look at internal and external reasons. For the moment let us return to Sellars to draw out a distinction that will be fruitful in our analysis of Brandom and will lay the scene for Hetero-Phenomenological sources.

Initially, when Wilfred Sellars scrutinizes[379] perception in terms of dispositionally based responses and abilities in *Empiricism and the Philosophy of Mind*, it is to the end goal of three moves based on the propositional and descriptive content that Logical Positivists took to be sense data, that appears in structures based around 'looks' talk. Initially he divides the treatment of looks talk between a sense datum and an appearing in a syllogistic disjunction[380]. The appearing he argues is ultimate and irreducible and he dismisses it for reasons he doesn't completely give over in the paper, but, which, are ultimately tied into what's bothering him in *Philosophy and the Scientific Image of Man*[381] and the Thomistic structure of the possible intellect in *Being*

---

[379] I'm trying to avoid using the word 'analyses' in my exposition of Wilfrid Sellars because it acquires a highly technical use with specific applications in given contexts in relation to the body of his philosophy. See Sellars, *Empiricism & the Philosophy of Mind*, 1997. Pg 33, 86, pp 53-53
[380] Sellars, *Empiricism & the Philosophy of Mind*, 1997. Pg 35
[381] Sellars, *Philosophy and the Scientific Image of Man*, 1991. NB The division of the Scientific Image, from the Manifest and the Empirical.

and Being Known[382] where he constructs a vocabulary of the senses[383]. However in *Empiricism and the Philosophy of Mind*, the sense data, he argues, can be broken down between an explanation and an analysis[384].

What is interesting is that Sellars introduces his final position early on[385]. In section Nine after considering the problems raised by treating sense data theories as an enriched code flagged[386] from ordinary language expressions, in his critique of Ayer[387] he raises the possibility that sense data are actually theoretical entities, but seemingly dismisses it in a slight of hand, on account of 'no one thinking of them this way'. Sellars then argues that sense data theories are a mismatching of two ideas. Firstly, what Brandom will come to refer to as 'sentience' that is, there are certain inner episodes without which it

---

[382] The senses informed by the character of the possible intellect. See the Tractarian criticism in *Being and Being Known*. Sellars, Wilfrid "Being and Being Known." In *Science, Perception and Reality* Pp 41 - 59. California: Ridgeview, 1991. Pg 49

[383] Sellars, *Being and Being Known*, 1991. Essentially the difference between 'picking' and 'signifying'; the former pertaining to the real order and the later to the logical order in his critique of Wittgenstein's Tractatus. The core of his concern involves the move from 'first act' in *Being and Being Known* which is dispositional, Pg 43, in which someone gains the concept and the second act. The second act being the ability to think of something as an instance of that concept. The move involves being informed by the 'thing's' nature. This in turn involves an isomorphism between the knower and the known, involving the intellect and the senses. Pg 41-48. For Sellars picking involves more than just the logical order of significance it requires an isomorphism of the second act.

[384] Sellars, *Empiricism & the Philosophy of Mind*, 1997. Pg 34

[385] Sellars, *Empiricism & the Philosophy of Mind*, 1997. Pg 31

[386] I am using flagged here in the highly technical sense Sellars introduces on pg 27, Sellars, *Empiricism & the Philosophy of Mind*, 1997

[387]. Triplett, DeVires, *Knowledge, Mind and the Given*. 2000. Pg 104-106 Comparatively, see also Triplett and DeVires summary. However I think it is problematic that they introduce their vocabularly. Many of the terms are irrelevant to explaining Sellars argument, like locutions, which is Austin's vocabulary, not Sellars, nor is it Ryle's which Sellars is using. See Lecture IX Austin, J. L. *How to Do Things with Words*. Massachusetts: Harvard University, 1975. Pp 109-120 for Austin's locutionary, illocutionary and perlocutionary acts. Triplett and DeVires also do things like changing Sellars other highly technical terminology which puts the explanation of *Empiricism and the Philosophy of Mind* they develop out of synch with Sellars other papers. They replace Sellars technical term 'report' which is a specific type of 'language role' with 'beliefs', and although it is perhaps influenced by a Brandom reading it is nonetheless problematic. 'Report' has specific applications in *Empiricism and the Philosophy of Mind*, as well as his other papers. Moreover it is problematic because it is not obvious that 'language roles' and 'reports' are synonymous given Sellars developments in Sellars other papers like "Naming and Saying." In *Science, Perception and Reality*, Pp 225 - 246. California: Ridgeview, 1991, and Sellars, *Truth and Correspondence*, 1991. Sellars, *Language of Theories*, 1991. Considerations about maintaining and preserving the consistency of Sellars' highly developed technical vocabulary aside there are other problems. On a naïve view, the general philosophical problem with treating 'reports' as synonymous with 'beliefs' as Triplett and DeVires do is that there is an interpretive 'gap' between reports and beliefs. One can take a beliefs as de dicto of the report or de re of the object the report is about. For a more sophisticated position on de dicto and de re see Devitt's work in *Thoughts and Their Ascription*, 1984.

would be impossible to hear musical notes or see a three sided patch of colour[388]. That is to say, that such inner episodes are necessary, for what you might call 'higher functions' like recognizing and knowing that a certain musical tone is the note C#. These at the basic level are shared by man and beast alike. The second idea is that there are certain inferential 'knowings that'. These are what Sellars reveals later as the fact stating roles of the subjunctive conditional structure revealed in looks talk between the withholding of a statement of fact and the residue of the descriptive content[389].

My reading of Sellars presents a closer analysis of the 'residue' and fact stating role, in order to avoid Jerry Fodor's criticism of the 'New-look psychology' and 'the assumption all perception is boundlessly theory laden'. Fodor's point is that if this were the case then we would learn to see the Muller Lyre arrowed lines at an equal length[390].

Fodor writes

> The very same subject who can tell you that the Muller-Lyre arrows are identical in length, who indeed has seen them measured, still finds one looking longer than the other. In such cases it is hard to

---

[388]Sellars, *Empiricism & the Philosophy of Mind*, 1997. Pg 33

[389]What I am defending Sellars against in this reading is the charge that he advocates a 'boundlessly theory laden view of perception' and a criticism that his view is subject to critique by 'bullheadedness'. See Fodor, *The Modularity of Mind*, 1983. Pg 70. NB This 'bull-headedness' is a particular property of encapsulation that makes faculties, a fortiori, vertical. Vertical domains are modular by hypothesis and domain specific by definition, pg 101. Horizontal faculties are widely distinguishable across multiple content domains, pg 13. Fodor gives examples of these, and they are generally things like attention span, memory, perception, the multiplicity of imagination, pg 10-11. On this view a thoroughly horizontal faculty, functionally individuated, is one that may access mental content in other domains at one time or another. This horizontal and vertical architecture allows Fodor to distinguish input systems, which are encapsulated, modular and vertical, from central systems which are non-modular, un-encapsulated and involved in long term processes of review and chained reasoning like belief fixation. In-put systems, on this view, would deal with the residual descriptive content of looks-claims, while central systems would deal with factual claims. The incompatibility isn't in the architecture of Fodor's system, it's in the nature of specific elements of his claims against those of Sellars like categorization that relies on the most abstract members of implication that subtend objects of similar appearance. See pg 95 of Fodor's *Modularity of Mind*. Sellars just wouldn't think these are phenomenologically given. His residium allows for descriptive content but not categorization. To insist on categorization over descriptive content, would be to go too far, and to commit the myth of the given.

[390]Fodor, *The Modularity of Mind*, 1983. Pg 66

see an alternative to the view that at least
some of the background information at the
subject's disposal is inaccessible to at least
some of his perceptual mechanisms[391].

This, I think is a very important point that has been overlooked in
much of the literature. However, I argue that there is a distinction in
Sellars that takes account of this, and that distinction is strong enough
to bear up to Fodor's criticism as it parallels Fodor's own distinction
between input and central systems and can give an account of
information encapsulation at the perceptual level[392]. The distinction in
Sellars arises from the subjunctive conditional structure in Sellars
exploration of 'looks' language which can be revealed by a close and
careful reading of *Empiricism and the Philosophy of Mind*. The
subjunctive conditional structure is revealed by carefully scrutinizing
the difference between the existential distinction and the qualitative
distinction. The existential and qualitative distinction is revealed by
distinctions in three separate cases. The three cases are

     a.   Seeing that x, over there, is red
     b.   Its looking to one that x, over there, is red
     c.   Its looking to one as though there were a red object
over there[393]

---

[391]Fodor, *The Modularity of Mind*, 1983. Pg 66 NB Fodor's criticism is aimed in general at what he
calls the 'new-look psychology' to which Sellars only features as a small part.Other examples are
Bruner. "On Perceptual Readiness." *Psychological Review* 64, (1957): 123-152., (Fodor's in-text cites
a 1973 article, but his works cited list only has a 1957 article by the author of the same name.)
Goodman, Nelson. *Ways of World Making*. Indianapolis: Hackett Publishing, 1954., Schank, Abelson.
"Scripts, Plans and Knowledge." In *Proceedings of the 4th international joint conference on Artificial
intelligence* 1 Pages 151-157 San Francisco: Morgan Kaufmann Publishers Inc.1975.
[392] Where the key clash occurs is between Fodor's 'vertical' and 'horizontal' architecture and Sellars
'residuum' and 'precept' and bottoms out in Fodor's basic categorizations, in particular (g) the
'giveness' of the categorization which Fodor draws from the evidence of high frequency counts and
associations in natural descriptions on pg 95 Fodor, *The Modularity of Mind*, 1983. However Sellars
would argue that these are acquired as part of a linguistic community. The residuum doesn't change
but our fact stating roles and linguistic practices do. Categorization in the sense Fodor means it in
terms of information encapsulation as the essence of modularity involves fact stating roles and
recognition, at least on the level of 'perceptual encoding'. To insist that these are given basic
categorizations, as such that they cannot be changed as part of linguistic community's set of roles or
linguistic practices would be to commit the 'myth of the given'. See Fodor, *The Modularity of Mind*,
1983, pg 66, 71, 94 – 97.

As Sellars points out

> (a) is so formulated as to involve an
> endorsement of the idea that x, over there,
> is red, whereas in (b) this idea is only
> partially endorsed, and in (c) not at all. Let
> us refer to the idea *that x, over there, is red*
> as the *common propositional content* of
> these three situations[394].

That is, (a) agrees with this propositional content (b) partially agrees
with it, and (c) is the existential case that disagrees with what is
common between (a) and (b) namely the existence of x. The difference
between (a) and (b) is qualitative, which he can explain away with a
story about standard conditions. (b), the existential case is more
troublesome. The propositional content for (b) involves a claim about an
object that isn't really there. This is what leads him to his final position.
But the propositional content is only the first part of the story. If the
propositional content was the only part of the story then Sellars would
be subject to Fodor's criticism of the New-Look psychology school.
However on Sellars account all three share a residual descriptive
content as well as the propositional content

Sellars writes

> The propositional content of these three
> experiences is, of course, a part of that to
> which we are logically committed by
> characterizing them as situations of these
> three kinds. Of the remainder, as we have
> seen, part is a matter of the extent to which

[393]Sellars, *Empiricism & the Philosophy of Mind*, 1997. Pg 50.
[394]Sellars, *Empiricism & the Philosophy of Mind*, 1997 Pg 50 – 51. His italics.

this propositional content is endorsed. It is the residue with which we are now concerned. Let us call this residue the *descriptive content*[395].

By separating the propositional content from the descriptive content of the 'residue' he can thus implement a distinction between the two within the subjunctive conditional structure that a denial in a 'looks' claim takes[396].This is like the ordinary language subjunctive conditional that Lewis analyzed except its linguistic behavior has a mechanism that works on two levels. Instead of false antecedent, the protasis contains a denial, or a withholding of propositional assent to the descriptive residue, which is attached to social and standardized conditions. It uses the subjunctive conditional, but it is not a proper counterfactual. See the discussion in the appendix for Lewis' distinction between an English subjunctive conditional and a formal logical counter-factual one.

Sellars writes

Now, and this is the decisive point, in characterizing these three experiences as, respectively, a *seeing that x, over there, is red, its looking to one as though x, over there, were red*, and *its looking to one as*

---

[395]Sellars, *Empiricism & the Philosophy of Mind,* 1997 Pg51, his italics.
[396] My treatment of Sellars here closely parallels Triplett and DeVires, t. *Knowledge, Mind and the Given,* 2000.Pp 24 – 28. My reading runs along similar lines to DeVires and Triplett's distinction between ascribing and endorsing a claim, except I follow what I consider to be a closer reading of the text and proceeded under the assumption that Sellars himself isn't using 'locutions', which is Austin's vocabulary, but 'achievement' verbs which is Ryle's, like Sellars says in Section 16. See Sellars, *Empiricism & the Philosophy of Mind,* 1997 Pg 40,Triplett, DeVires, *Knowledge, Mind and the Given.* 2000. Pg 223 of the DeVires and Triplett reproduction of *Empiricism & the Philosophy of Mind* for the same section. Both state "I pointed out above that when we use the word "see" as in "S sees that the tree is green" we are not only ascribing a claim to the experience, but endorsing it. It is this endorsement which Ryle has in mind when he refers to *seeing that something is thus and so* as an *achievement,* and to "sees" as an *achievement word.* "I prefer to call it a 'so it is' or 'just so' word, for the root idea is that of *truth.*"

though there were a red object over there, we do not specify this common *descriptive* content save *indirectly,* by implying that *if the common propositional content were true[397]* , then all these three situations would be cases of *seeing* that x, over there, is red. Both existential and qualitative lookings are experiences that would be *seeings* if their propositional contents were true[398].

This subjunctive conditional structure, the one Sellars is identifying with the indirect mechanism of assent in 'looks talk', is different, of course, from that which I outline in my discussion of David Lewis' Counterfactuals and the ordinary language usage. Rather, it is concerned with the set of propositional statements that one would attach to descriptive contents given in a stipulation in which the social linguistic and standardized conditions of the linguistic community are obtained. The Muller-Lyre lines, on the reading I am purposing, would appear in the residue of the descriptive content. In this way, I argue, that the 'residue' or 'residuum' of the descriptive content, in Sellars argument, is flexible enough it can account for modulation in the form of information encapsulation of the sort that Fodor's criticism of the New-look school of psychology makes. Further I argue it does so without making Sellars himself subject to the 'giveness' of the sense data schools he criticizes. This is because, in Sellars account, as I pointed out, the confusion in sense data theories originally begins with the mix matching of a brute component with the 'non-inferential knowings that'.

Sellars writes

---

[397]Note that his 1963 amendments actually support my reading. *"and if the subject knew that the circumstances were normal."* Added to his 1963 version, NB see
http://www.ditext.com/sellars/epm.html in Sellars, Wilfrid. *Empiricism and the Philosophy of Mind.* . 1963 ed. Electronic Text. 1963 Amendments, edited by Andrew Chrucky.
http://www.ditext.com/sellars/epm.html 1995.
[398]Sellars, *Empiricism & the Philosophy of Mind,* 1997. Pg 51 The italics are his however I should like to emphasize them as well, as they reveal the subjunctive conditional structure of the residuum and the fact stating role.

> (1) The idea that there are certain "inner episodes," e.g. the sensation of a red triangle or of a C# sound, which occur to human beings and brutes without any prior process of learning or concept formation, and without which it would -- in *some* sense -- be impossible to see, for example, that the facing surface of a physical object is red and triangular, or *hear* that a certain physical sound is C#. [399]

And also

> (2) The idea that there are certain "inner episodes" which are the non-inferential knowings that, for example, a certain item is red and triangular, or, in the case of sounds, C#, which inner episodes are the necessary conditions of empirical knowledge as providing the evidence for all other empirical propositions. [400]

Now the split between sensory qualities and the non-inferential *knowings that*, are ultimately based on a 'faculty-dualism'[401] he's left with at the end of *Philosophy and the Scientific Image of Man* between firstly thinking processes which he thinks can be dealt with as analogous to sentences and secondly sensations which he is unsure of what to do with at the end of that paper.

The root of this faculty dualism begins with Descartes distinction between different levels of thinking. There are the low-grade levels which comprise of sensations, images and feelings and the high grade conceptual elements of thinking which lays at the foundation of

---

[399]Sellars, *Empiricism & the Philosophy of Mind,* 1997. Pg 21

[400]Sellars, *Empiricism & the Philosophy of Mind,* 1997. Pg 22

[401] My term used in a sense derivative of Fodor's use of 'faculty' in *The Modularity of Mind.* By it I mean a dualism like that in Sellars, between thoughts qua thinking and sensations qua sensory faculties. This dualism, for Sellars, originates in Descartes distinction between 'conceptual thinking' and sensory correspondence of perception, See Sellars, *Philosophy and the Scientific Image of Man,* 1991. Pg 29 – 31 following the discussion of Eddington's two tables.

the Cartesian attempt to incorporate the two in an account. Sellars sees the Cartesian project as representative of an early attempt at integrating what Sellars refers to as the manifest and scientific images.

Sellars writes

> Let us consider in more detail the Cartesian attempt to integrate the manifest and the scientific images. Here the interesting thing to note is that Descartes took for granted (in a promissory-note-ish kind of way) that the scientific image would include items which would be the counterparts of the sensations, images, and feelings of the manifest framework. These counterparts would be complex states of the brain which, obeying purely physical laws, would resemble and differ from one another in a way which corresponded to the resemblances and differences between the conscious states with which they were correlated. Yet, as is well-known, he denied that there were brain states which were, in the same sense, the cerebral counterparts of conceptual thinking[402]
>
> .

Basically Sellars has worked out what to do with higher cognitive states, that is, he treats them as analogous to speech which is how they are found in the manifest image and allows one to ignore their introspective qualities[403]

> Thus our concept of 'what thoughts are' might, like our concept of what a castling is in chess, be abstract in the sense that it does not concern itself with the *intrinsic*

---

[402]Sellars, *Philosophy and the Scientific Image of Man*, 1991. Pg 29 - 30
[403] The problem of Introspection, Sellars, *Philosophy and the Scientific Image of Man*, 1991. Pg 31

character of thoughts, *save as items which can occur in patterns of relationships which are analogous to the way in which sentences are related to one another and to the contexts in which they are used*[404].

This allows him to map conceptual thinking as analogous to speech processes, picking out the role they play and identifying them with neurophysiolocial processes[405]. This, in turn would allow him to merge the scientific image with the manifest without a clash.

Now if thoughts are items which are conceived in terms of the roles they play, then there is no barrier *in principle* to the identification of conceptual thinking with neurophysiological process. There would be no 'qualitative' remainder to be accounted for. The identification curiously enough, would be even more straightforward than the identification of the physical things in the manifest image with complex systems of physical particles. And in this key, if not decisive, respect, the respect in which both images are concerned with conceptual thinking (which is the distinctive trait of man), *the manifest and scientific images*

[404]Sellars, *Philosophy and the Scientific Image of Man*, 1991. Pg 34

[405] This is an over-simplification of the paper and his general position is a lot more complex than I have room to exposit here. In Willfrid Sellars see for example "Truth and 'Correspondence'." In *Science, Perception and Reality*, Pp 197 - 224. California: Ridgeview, 1991. Pg 197 – 207 where he distinguishes between thought as acts of thinking, and thoughts as that which is thought. Linguistic utterances can express an act of thinking if it is the culmination of a process in which the initial stage is the act of thinking. On the other side of this distinction he points out in Sellars, *Language of Theories*, 1991 not all linguistic roles are conceptual. See his footnote on page 115. The position offered by this paper supports the earlier close reading, where an emotional vocabulary is inter-subjective because it is learnt as an achievement verb which becomes part of a report language after it shifts from an observation language where it starts as a capacity disposition.

could merge without clash in the synoptic view[406].

However, at the end of the paper integrating the sensory qualities of the manifest image still eludes him. We will return to his treatment of conceptual thinking as analogous to language later in the paper. For the moment let's take a closer look at how he resolves the problem of sensory qualities.

In *The Language of Theories* Sellars argues that the division between the theoretical elements like postulates, theorems and various calculi and the non-theoretical elements, conveys what seems to be an ontological dualism between observables and non-observables which is bothering him in *Philosophy and the Scientific Image of Man*. The difference between observables and non-observables is what he characterizes as the main difference between the Manifest Image and the Scientific Image[407]. Here, however he deals with it by collapsing first the theoretical framework into the correspondence rules, and then the correspondence rules with the observational framework[408] giving us the first sight of his solution in *Empiricism and the Philosophy of Mind* of unifying ordinary language discourse with science, and the argument that theoretical and observational frameworks can shift over time. The way they shift is what Brandom will come to call 'the space of reasons' and the subjunctive conditional structure of withholding assent is important to this solution. Sellars solution of course, as Brandom points out[409], will be to argue the difference between 'theoreticals' and 'observables' is methodological and not ontological.

The relationship that develops between' analyses' and 'explanation' is interesting in this regard. Both embody elements of the faculty dualism that I mentioned has roots in issues arising from *Philosophy and the Scientific Image of Man*. In *Empiricism and the Philosophy of Mind* Sellars holds that if the 'looks' part of the sense datum is analysed then we find it hard to disbelieve the analyses, in

---

[406]Sellars, *Philosophy and the Scientific Image of Man*, 1991. Pg 34
[407]Sellars, *Philosophy and the Scientific Image of Man*, 1991, Pg 18-19
[408]Sellars, *Language of Theories*, 1991. Pg 106 - 109
[409]Brandom, Sellars, *Empiricism & the Philosophy of Mind*, 1997. See Brandom's commentary on Section 44, Pg163

this case, treating a car or B. C. Broad's penny, say, as merely an oblong or elliptical bit of colour. However, if the sense datum in the way something *looks* is explained away we can believe otherwise because as Sellars points out one can accept a fact without accepting the explanation.

Eventually Sellars does away with 'sense data analyses'in *Empiricism and the Philosophy of Mind*. He argues that it is merely a spatio-logical bit of sophistication that is related to our framework, but does not belong to it[410]. Sellars argument here is that it is a spatial-conceptual sophistication to treat an object like a car as an oblong bit of colour. Here, his argument seems to be drawing on Ryle's success verbs[411] except that Sellars is taking Ryle both in the strictly normative ordinary language sense and a naive phenomenal sense. But what Sellars is, of course, doing is developing specific grounds for the qualitative and existential distinction behind statements about the way something 'looks'.

This then leaves him with what he holds as the only two remaining possibilities that the 'sense datum' in looks talk can be explained by, which, of course, is his third and final move[412]. Either we treat impressions or immediate experience as theoretical entities, or we make the discovery that the sense datum contains impressions or immediate experience as components. The later he's already ruled out with the existential case since there is no object[413]. For the former he needs to make a case for treating impressions and immediate experience as theoretical entities. He then, of course, goes on to provide a story about a Rylean community to do this. The community begins

---

[410]Sellars, *Empiricism & the Philosophy of Mind*, 1997. Pg 52 -53

[411]Sellars, *Empiricism & the Philosophy of Mind*, 1997. Pg 40 Sellars tells us that these correspond with Ryle's success verbs in particular Ryle's class of 'see' achievement verbs, which also have applications in a linguistic behavioural analysis that Ryle applies to the way sensations and observations linguistically behave. Ryle, of course, lacks the resources to make a number of the distinctions that he makes in order to uphold this distinction since he denies the existence of consciousness and introspection. Sellars' position is pragmatic. He doesn't deny the existence of consciousness, rather he thinks it is problematic for the Scientific Image of Man. For Ryle's Observations and Sensations distinction see Ryle, *Concept of Mind,* 1983. Pg 190 – 200. In particular, see pg 196 for his argument on telescopes and confusion between sensations and observations. Observations have objects. I assume this is why Sellars' fully developed Rylean language contains both public properties and public objects. Sellars, *Empiricism & the Philosophy of Mind*, 1997 Pg 91, 87

[412]Sellars, *Empiricism & the Philosophy of Mind*, 1997. Pg 86

[413] (a) and (b) vs (c)

with a Rylean language which lacks the ability to talk about thoughts and develops that ability first through descriptions of behaviour and secondly the ability to come to see theoretical terms in observational descriptions. The point of the story is to make credible treating thoughts as theoretical entities which in turn gives an account for what seemed so puzzling about the existential case.

Now the mechanics inside of the sense data of a bit of 'looks' talk is operated by a set of reasons that change the structure of a causally keyed-in disposition. This is Brandom's space of reasons. He pulls this off by dividing the propositional content of the disposition in a claim, from that of the descriptive content[414]. The descriptive content is the residue at the back of the propositional content which posses a subjunctive conditional structure. We briefly looked at this a moment ago with Fodor's criticism. This is the structural element that were the propositional part true, then the descriptive part would be an accurate description. This is an important piece of Linguistic Behavioural analysis. For instance, 'x seems to be the case' states, counterfactually that 'x isn't the case' but that it 'looks' to be. The 'looks' of course refers to the descriptive content. Similarly, 'x looks y to z' states that x is not y, but that it carries the descriptive content of y, the residuum, for z. This is why he explicitly rejects treating 'looks' statements as sets of relations[415].

To explain this division between a descriptive and propositional content with a counterfactual structure, Sellars introduces a short piece of fiction about a tie shop[416]. I only mention it briefly as it illuminates the division of the descriptive content from the fact stating role in the

---

[414]Sellars, *Empiricism & the Philosophy of Mind*, 1997. Pg 50
[415]Sellars, *Empiricism & the Philosophy of Mind*, 1997. Pg 36, Section 13.
[416] For those unfamiliar with it: during the course of his career as a tie salesman, lighting is introduced to the neighbourhood that John, the tie salesmen, works in. At first this puts John in to a bit of a confusion as a tie which looks green suddenly appears blue. 'I know it is blue, even though it looks green.' John might say. The 'looks' green is the part where John withholds his acquiesce from the propositional content and simply states the descriptive residuum. However, were one to convince John that electric lighting is a better medium to judge the visible band of electromagnetic radiation by than the old kerosene lighting, then John would come to see the residuum, the descriptive part that looks green, as if it *were* green. The standard conditions by which John takes to be true the propositional content will have shifted. In this way, Sellars is able to solve the earlier problem of the stereoscopic image between the Scientific Image and it's methodological feeding upon the Manifest Image, in *Philosophy and the Scientific Image of Man*, and demonstrate how theoretical entities become visible.

subjunctive of the conditional and the way reasons can be keyed into dispositional statements. This is important because Brandom, who I'm going to critique, picks up on this.

Brandom in his commentary calls this move 'entering the space of reasons'[417]. Brandom takes Sellars to be arguing that (a) 'looks' talk is parasitic on 'is' talk, that is, the nature of the descriptive residue associated with a dispositionally keyed bit of observation is parasitic upon standard conditions, and (b) that reasons, such as an argument that the reliability of electric lighting is more loyal to the discernment of the visible band of electro-magnetic radiation, may cause the person to shift from withholding assent to a propositional content by saying "x 'looks' y", with the subjunctive conditional structure, to a propositional affirmation of the descriptive residue associated with the statement, to, of course 'x *is* y'.

Recall now the move between the descriptive and the fact stating roll is whether someone uses 'looks' or 'is' in their claim. x 'looks' or 'is' in their claim. "x looks to be the case" is descriptive while "x is the case" is fact stating.

Interestingly enough, to reinforce Sellars argument Brandom in his commentary on *Empiricism and the Philosophy of Mind* introduces an Ordinary Language argument with a Linguistic Behavioural analyses, that doesn't appear to be in the original text. Brandom points out that 'looks talk' doesn't iterate, that is, we can not withhold our assent more than once, because the withholding has already been done in the original structure of the subjunctive conditional. That is, *it doesn't make sense to say* 'it looks as though x looks y' or *it does not make sense to say* 'it seems like x seems to look like y' and so on, through various iterations[418].

There is a background argument to explain where I think Brandom may have lifted this argument from or at least which can align Brandom's introduction with a central thesis of *Empiricism and the Philosophy of Mind*. As I pointed out above, Sellars thinks that a report language itself is quite a sophisticated piece of linguistic behaviour that rests upon a foundation which brings with it a lot of

---

[417]Brandom, Sellars, *Empiricism & the Philosophy of Mind*, 1997. See Brandom's commentary pg 123
[418]Brandom, Sellars, *Empiricism & the Philosophy of Mind*, 1997. Pg 142, Brandom's commentary.

social 'baggage'. Not only must a perceiver know the standard conditions, he must also know the socio-linguistic conditions of appropriateness[419]. Before someone can make observations they need the ability to make an endorsement based in those conditions. This ability rests on linguistic competence and linguistic competence itself rests upon having the concepts[420].

Now, as I pointed out earlier, in section forty of *Empiricism and the Philosophy of Mind*[421] Sellars argues, pace Ryle, that science follows the pre-scientific stage of language, and that failure to accept this will result in failure to understand ordinary language[422]. Thus, the affirmation that 'looks' talk can be based in the Linguistic Behaviourist analyses of iteration and applied to a normative Ordinary Language argument can be made on Brandom's behalf. That is the propositional shift of the qualitative distinction of the residue in the subjunctive conditional structure of the 'looks' statement revealed by a Linguistic Behavioural style argument, which of course is the inability to iterate, as a bit of linguistic behaviour is supported by an Ordinary Language argument. But that Ordinary Language argument itself ultimately rests on the continuity between ordinary language and science and the position that Sellars takes in *Empiricism and the Philosophy of Mind*.

The moral of the Brandom-Sellars position I've been at pains to draw out carefully is that Sellars originally introduced the distinction between the propositional and descriptive content in the subjunctive conditional structure of *looks* talk, as a means of explaining the shift in propositional statements, causally keyed into dispositions on recognition of social grounds, and those social grounds were made in relation to standard conditions. The difference between the dispositions of Price's thermostat to display symptoms of a temperature or a bit of iron and its disposition to co-vary with its environment displaying temperature in the presence of heat, or rust in the presence of wet, is that people have propositional structures that are keyed into reasons which change according to what they accept as the standard conditions upon which such dispositions rest. It is the major part of a significant move Sellars makes from *Philosophy and the Scientific Image of Man*,

---

[419]Sellars, *Empiricism & the Philosophy of Mind*, 1997. Pg 73
[420]Sellars, *Empiricism & the Philosophy of Mind*, 1997. Pg 75
[421]Sellars, *Empiricism & the Philosophy of Mind*, 1997. Pg 80
[422]Sellars, *Empiricism & the Philosophy of Mind*, 1997. Pg 80

where theoretical entities, 'imperceivables' separate the 'Scientific Image' from the 'Manifest Image' and present themselves as a genuine rival to the manifest qualities of the 'everyday world' to the central thesis of *Empiricism and the Philosophy of Man*, where theoretical entities become visible, via a shift in the fact stating component of the subjunctive conditional that 'pivots' from 'looks' to fact stating affirmation, or 'alignment' with the descriptive residuum; on the basis of a set of standard conditions that the person accepts. That's as far as Sellars goes in *Empiricism and the Philosophy of Mind*.

Brandom, however, picks up on this shift and describes it as a move into the space of reasons. That is, the standard conditions for the descriptive content withheld in a 'looks' statement, shifts, if adequate reasons are given. The part Brandom thinks is important is the domain of reasons, which he bases his sapience and sentience distinction on in the commentary for *Empiricism and the Philosophy of Mind* along with other pieces of writing. In *A Social Route from Reasoning to Representation*[423] for example, Brandom expands on this dispositional structure, and attributes reasons to behaviour with linguistic conditions of utterance and propositional statements attached. This is like the R.S. Peters position in relation to tendency dispositions and motives, but on analytic steroids and grounded in the propositional structure of judgements and the defacto presence of the I and he minimum unit of thought.

Here Brandom moves from the capacity dispositions in the Rylean framework I introduced, which is where Sellars position on fact stating rolls is situated and over into the tendency dispositions with Peters.

Now note, before we move into Brandom's criteria, that indeed he says '*sometimes*'. The argument I'm going to be forwarding is a pessimistic one and it is part of the overall push towards a reconsideration of psychologism that I'm making in this paper. To put it blatantly, I think that Brandom is being optimistic on two accounts; firstly, the affirmation that reasons poses a propositional structure, and two, that giving linguistic utterances *as reasons* just aren't all that important to people out in the world. That's the pessimistic part. The

---

[423] In Brandom, *Articulating Reasons*, 2001.

move in my argument towards a pre-Fregeian psychologism is that, in fact, some behaviours actively resist being made intelligible in any linguistic propositional structure. We just can't imagine what the reasons for their behaviour would be like as a bit of linguistic utterance. It is this property that makes them philosophically interesting. This allows me to suggest an internal/external framework to get around the difficulty of some behaviours. A reason can only serve as a motive if it is capable of being an internal reason and not an external one. External reasons are justifications after the fact or explanations given from a third person point of view. The former typically arises in moral discourse, the latter typically in theoretical explanations like Darwinism, Scientology or Freudian psychotherapy. Internal reasons, however, are reasons that the person themselves might give, which figure in their behaviour. This will allow me to argue the position in which I claim the driving 'motives' cannot be justified or explained in the game of giving and asking for reasons without invoking non-linguistic knowledge. I define this non-linguistic knowledge as knowledge one must arrive at the game of giving and asking for reasons with in order to play, it is not knowledge that can be set up within the propositional structure of the judgements and beliefs within the linguistic frame work of the game itself.

7.1 The Game of Giving and Asking for Reasons

Earlier we began a critique of the position that beliefs must be linguistically formulated. We noted that Brandom held that the minimum unit of thought was the judgement and that the "I" must accompany our thoughts as a de facto condition. Further we saw that Sartre objected to this and I argued that Sartre's view would be that Brandom would be making the same mistake, (both philosophically, and in terms of Kant scholarship as Brochard). More significantly we saw on Sartre's view that the 'I' need only accompany thought as a de jure possibility for reflection on a prior act of consciousness brought before the present one and not as a de facto reality.

We saw that Sartre's de facto conditions for consciousness occupies the second level of Hoffe's analysis of the stages between Kant's Original Synthesis and Apperception of the I, while Brandom occupies the third level. The third level of course is the level at which the I is present and indicative of judgements. My position is that Sartre's de facto position is the correct one, however it is up to the introspective scrutiny of the reader to decide whether they agree with me or not. This may take a moment.

Now, the critique of this section has two parts, and the first part has two stages. The two parts correspond to the two sides in reason-linguistic utterance frameworks for intentional explanation. On one side, and the first part, I am going to argue against conditions that obtain to a rough approximation that one can only entertain 'beliefs' if they are or can be formulated as linguistic utterances by the belief holder. The other side will concern the 'wants', 'needs', 'desires', 'wishes' and so forth, and an insight I am willing to share.

For the first stage of the first part I am going to argue that beliefs are not de facto linguistic representations. That is, we do not carry around with us, de facto; lists, scrolls, memorandum, mental logs, or an itinerary of propositions formulated as linguistically expressible utterances or written statements containing our beliefs. That is, de facto, we do not carry belief statements in our thoughts. That's the first stage of the first part of the critique. The second stage of the first part is that I am going to argue that beliefs do not have to be de jure expressible in a language to be held.

Now this is different from Sartre's de jure possibility for reflection. I want to avoid the possibility of a confusion arising in what I am getting at. Sartre's conditions for the 'I' to appear in an act of consciousness as the third level of Hoffe's schematic of Kant's stages, is different in the way that it might 'appear' to the way that Brandom thinks the 'I' inhabits the third level. Brandom's is a linguistic formulation, as we will see, that utilizes conditions of propositional state-ability to correspond to belief conditions based on the 'taking to be true' which Brandom argues characterize Sapients.

The difference between how we should see Brandom's occupation of Hoffe's third level, to Sartre's, is like the difference between the phenomenological and Linguistic-Behaviourist side of Ryle's 'Remember-when/remember-how' configuration. Brandom's occupation of Hoffe's third level is like the Linguistic-Behaviourist side, except he attaches propositional statements and structures based on believability and whether the statement can serve as a premise in an argument or not, while Sartre's position is like the act of recalling when uncle Reg tripped over in the dining room last Christmas except that in Sartre's position the I appears as the presence of the prior act of consciousness that arises in the reflected sense of horror and shock of watching Reg trip over and upsetting the Christmas pudding, recalled in the present act of consciousness in the act of reflecting. To put it sharply Brandom's occupation on Hoffe's third level is a linguistic one and Sartre's is a phenomenological one. To be clear, what I am arguing against are the de facto conditions of belief ascription that Brandom holds. That is, the first stage, of the first part of my critique is an argument that people don't go around formulating and holding sets of linguistically structured and expressible belief ascriptions attached to an I, as a de facto condition of consciousness. This view of Brandom's is like a formalized version of Ryle's log keeper who records everything in affirmations given in a propositional structure with an I attached to them since this is the minimum unit of thought.

The second stage of the first part will be to argue against a position that beliefs, de jure, need to be expressible linguistically as a condition of holding them, and tied into that is an assumption I want to clear out and attack which has it that truth and falsity can only enter the scene once we have a language. The idea behind this, I take it, is you need a form of representation before you can have false representation. The position, I argue, is implicitly implied by Brandom's position, and explicit in Ayer's. I disagree with both of them that language is a pre-condition for false thought and being mistaken. I think you only need consciousness. I think that a conscious being without a language can be mistaken.

This is why Brandom and Peters are different. Both represent a dispositional account of motives within the Tendency Configuration of Ryle's original framework. However Brandom adopts a Pro-Fregeian position on thought, while Peters is closer to what we might characterize as a borderline implicit position.

To put this in perspective, Brandom and Ayer are on an anti-psychologistic position on the next level up from Ryle and Dummett. They are Pro-Fregeian explicit-cum-anti-implicit about a theory of meaning. Brandom has inherited his assumptions directly from Frege, which he freely admits, while Ayer is openly explicit in his position on meaning. Brandom and Ayer, however, differ in their view on language in that Ayer is Pre-Sellarian while Brandom has inherited Sellars framework.

The above is important because my position is that introspection can argue away the de facto conditions of propositional apperception, but my argument against the de jure conditions for entertaining a belief which Brandom maintains are that it need be expressible, are a little harder. My argument rests upon an analogical structure not unlike that which I have been critiquing in Ryle. This is a weak argument. I am not unaware of this. But my point in arguing it is to show that it rests, merely, on a technicality. Moreover, since the being in question posses no language, there is no possibility for a Linguistic Behavourial vs Phenomenological rivalry that can allow a contradiction to arise from an analogical structure in an indirect domain. Thus there is no rivalry, and thus no second source that can produce a contradictory claim. All we have as a form of analysis is a weak third person argument. However, I am willing to give this position up if need be and it can be shown that some sort of problematic thesis follows from it. These two stages, thus, constitute the first part of my critique on the game of giving and asking for reasons.

The second part is to attack what is implicit in reason-linguistic utterance frameworks about motives. I am going to attack the 'rationality' of intentional behaviour. I think 'rationality' is an illusion of these types of models, that comes from implicit acceptance of the 'wants', 'desires', 'urges', and 'needs' that feature in these specific types of reason-linguistic utterance frameworks. I think the commonality leads to a complicity about the types of wants and needs which acts as a camouflage to something deeper that goes beyond the game of giving and asking for reasons, and that once we can see our way out of this illusion, the ordinary language position breaks down, and we are left with some sort of appeal to psychologism.

However, this won't be a psychologism with appeals to things like Freudian or Evolutionary paradigms rather this will be a psychologism that precludes such lecture room explanations in the very nature of its inception. This will be a psychologism based on an appeal regulated by an internal/external distinction in the types of things that count as a feature or as part of our explanations and towards meaning in language. I think this internal/external distinction is important, and can be seen as evident in Ryle's original insight and his 'ex hypothesi' argument for an implicit ordinary language position against the Freudian and psychological schools[424]. But I think that in his attempt to formulate these in a way accessible from the perspective of ordinary language, and his two radical insights[425]: it lead him to category mistakes that resulted in him modelling the mind from the outside, and taking the behaviour and the way we talk about mental phenomena for the mental phenomena itself. I will return to this point following my discussion of A. R. White. With this in mind, let us, now, turn to Brandom and the game of giving and asking for reasons.

Brandom states

> Sapients act as though reasons matter to them. They are rational agents in that their behaviour can be made intelligible, at least, sometimes by attributing to them the capability to make practical inferences concerning how to get what they want, and theoretical inferences concerning what follows from what[426].

And he develops this into a philosophically interesting idea of 'score keepers' who can keep track of inferences and beliefswith sets of instructions and a criteria. Behaviour can be explained through intention, and intention can be explained by belief and desire. What serves as a belief must be able to be taken as true, since taking to be true is what Sapients do and thus belief must be propositional. For Brandom, this implies that belief must be capable of serving as a premise or a conclusion in an argument. Furthermore it must be capable of being given in the form of a spoken utterance. Like R. S. Peters, the motive behind the behaviour must be able to be given in the

---

[424] This is the second problem with Chalmers reading of Ryle.
[425] See '3.2 Linguistic Behaviorism: Episodes and dispositions. What exemplifies the linguistic behaviorist's claims?' in 'Chapter Three : Towards a Methodology' in this paper.
[426] Brandom, *Articulating Reasons*, 2001. Pg 158

form of a reason. Unlike Peters Brandom maintains the structure of the propositional form as the minimum unit of thought with the de facto presence of the I.

Brandom argues

> Making behaviour intelligible for this model is taking the individual to act for reasons. This is what lies behind Dennett's slogan "Rationality is the mother of intention". The role of belief in imputed pieces of practical reasoning, leading from beliefs and desires to the formation of practical intentions is essential to intentional explanation – and so is reasoning in which both premise and conclusion have the form of believables[427].

This is simply to say that the person, when practically reasoning, must be able to believe the premises or conclusions of their own reasoning as a precondition to the explanation of intention.

Ryle would have a problem with this sort of approach. What Brandom calls taking to be 'true' Ryle calls propounding[428] and Ryle thinks this is an unnatural form of thinking. Ryle thinks people don't normally think or speak in this way, nor do they seem to solve problems in this way, and especially not as a necessary pre-condition to getting what they want. The peculiar properties of 'propounding' as a species of speech or thought, that is, the propositional linguistic form of such thought, Ryle argues, only arises in artificial conditions as when, for instance someone is forced to give reports, or to explain a situation, or to offer the grounds for the solution to a problem in a formal context[429]. Or, in particular, Ryle thinks logical structures as a linguistic trait are germane to the method of transmitting theoretical knowledge, and this, Ryle argues, is why it has acquired such a 'sophisticated air'. But Ryle

---

[427]Brandom, *Articulating Reasons,* 2001. Pg 101.

[428]Ryle, *Concept of Mind,* 1983. Pg 174

[429]Ryle, *Concept of Mind,* 1983. For 'propounding' see pg 174, for reasoning pg 46 and the inheritance conditions of reasoning related to propounding see pg 32.

thinks this isn't the natural de facto way one thinks and reasons while one is engaged in problem solving. Ryle does not think people form propositional structures which they affirm or deny in their day to day business in the world. Ryle's internal log keeper is different to Brandom's propositional judgement keeper. Indeed Ryle is emphatic on this point. Ryle thinks logically structured reasoning only surfaces in formal contexts as part of the process of presentation and justification. Thus practical bits of reasoning, on Ryle's view, do not, nor need not behave like premises or conclusions in an argument to be believable. Ryle thinks people don't actually do their thinking in it. At best, Ryle should argue, it would be a de jure condition of possibility for representation, not a de facto representation of natural thinking or reasoning. It is important to separate these two strands and I don't think Brandom entirely does that since one of his conditions, as I pointed out, is that the 'I think' must accompany thoughts, de facto, as it is the minimum unit required for judgement.

Ryle writes

> The detective was, perhaps, given certain clues on Tuesday and at some moment on Wednesday he says to himself for the first time 'it could not have been the poacher, so it was the gamekeeper who killed the squire'. But when reporting his results to his superiors he need not say in the past tense 'On Wednesday afternoon I argued that the gamekeeper killed the squire'; he may say 'From these clues I conclude that the gamekeeper killed the squire', or 'From these clues it follows that he was the murderer', or 'The poacher did not, so the gamekeeper did kill the squire'. He may say this several times to his slow-witted superior, and later say it again several times in Court. Each time he is using his argument, drawing his conclusion, or making his inference[430].

Further, the premise, conclusion and argument structure that Brandom argues defines what can serve as a reason, Ryle argues could *not* come naturally to the way we conduct our natural reasoning[431]because it is not a natural property of people's thinking. Rather, as a property of 'inner speech' it is unusual and we acquire it only by it being introduced to our thought patterns through training and education[432]. This, rather interesting view on the conditions of formal logic and classical reasoning, as being an unnatural form of reasoning, with origins as a linguistic trait associated with the transmission of knowledge, and presentation of facts, invites the question where did the academics get it from? While this is an interesting topic, it, of course, goes beyond the scope of this paper.

Ryle's polemic against this kind of approach is perhaps most explicit in his short article *Thinking and Language*[433] where he argues that while thinking resembles language, it is of a fundamentally different sort from the kind of propositional statement Brandom should describe thoughts as. For while Ryle builds into the majority of *The Concept of Mind* and *On Thinking* the thesis that natural thought follows language, he holds that natural language is of a fundamentally different sort to the type of thinking that something is the case. For Ryle thinking, rather than being an articulated structure as Brandom argues, rather for Ryle it is a meandering and unplanned process. For Ryle thinking is ra 'polymorphous family resemblance activity'[434] which Ryle maintains occurs as inner chatter or as part of an internal log keeper process linked to his concept of retrospection and the ability to give a status report[435].

What Ryle is objecting to, we might characterize as one half of Brandom's criteria, which is that of the de jure necessity and de facto

---

[430]Ryle, *Concept of Mind*, 1983. Pg 81 -81
[431] Ryle, *Concept of Mind*, 1983. Pg 267
[432] Ryle, *Concept of Mind*, 1983. See pp 291 – 295, section titled Saying and Teaching.
[433]Ryle, Gilbert. "Thinking and Language." In *Collected Essays 1929-1968*, edited by Julia Tannery, II, 269-283. Oxon: Routledge, 2009.
[434] Ryle, *Thinking and Language*, 2009. Pg 269
[435] Except in cases of untangling a skein of wool or doing a jigsaw puzzle as we discussed earlier. We disagreed with Ryle about chess.

reality of the propositional structures of statements that can be taken by Sapients as true. Indeed, Ayer raises a similar position to what troubles me about Brandom, in Ayer's work *The Problem of Knowledge*. Ayer thinks that truth and error become possible once language is on the scene since propositions, beliefs and opinions need a language to be expressible and thus true or false. We might even characterize the part in Brandom that Ryle is concerned with, the part where Brandom thinks that reasons must be able to serve as premises or conclusions in an argument with Ayer's position on language as a de jure precondition for the de facto expression of truth or falsity. That is, for Brandom since Sapients are takers to be true, what serves as a reason must have a propositional content to which the I, which is the minimum unit of thought, as a de facto reality of thinking and expression can be seen to share factors with Ayer's position on truth values, since for something to be believable it must be capable of being true.

Now I don't think that Brandom comes right out and says that language is necessary for truth or falsity, but I think that this is implied in his account. According to Brandom reasons must be utterable and contain judgements and beliefs. Since judgements must be propositionally accompanied by an 'I think' as a de facto reality and beliefs must be represented as either premises or conclusions, which, if they are believable, must be capable of being taken as true or false since this is what Sapients do. The judgment is the minimum unit of thought possible and it must be in some way represented. Truth and falsity are the conditions for believability, in Brandom's account, and since beliefs must be presented, such that they might serve as conditions or premises, and thoughts must be accompanied de facto by the I, since judgments are the minimum unit of thought, then it seems one cannot posses beliefs without representing them as propositions, de facto, and the possibility for linguistic expressions, de jure. This would for Brandom, I take it, make the possibility of expression in a language a de jure necessity of entertaining a belief, which can only be a belief if it is capable of being true or false, since it must be believable, and as such, the conditions holding a true or false belief is the de jure possibility of expression in a language, and the de facto reality that the I must accompany the thought. On Brandom's account, this, I take it, would be regardless of whether someone entertained their beliefs as linguistic propositions or not. I take it they would, on this account, since the 'I think' must accompany the thought in the propositional context of a judgment since this is the minimum unit it is possible to think.

However, even if this is not the case, I would like to block off Ayer's position that language is a pre-condition for entertaining true or false beliefs. The reason for doing so is part of my general strategy for undermining the assumption that beliefs need to be expressible or capable of being formulated in a language as a de jure condition for formulating them, based on the claim that truth or falsity become possible only once you have a language.

Let's relate this back to an approximation of Ryle's original framework. Ayer's position is like Williamson and Stanley's syntactically embedded answers to the how-question but it relates to the 'that' side of the 'how-that' configuration and concerns the propositional structures expressible in a language on the that-side. For Ryle motives and beliefs, of course, are what characteristically exhibit the linguistic behavioural properties of 'that' clauses. In a moment I am going to argue the counter to this view. I am going to argue that a conscious being does not need a language to have true or false beliefs. I think that this position is implicit in Brandom and explicit in Ayer, and is indicative of a much larger mistake and a costly one that impedes our progress towards uncovering the nature of mind[436].

Ayer writes

> (I)t is only with the use of language that truth and error, certainty and uncertainty come fully upon the scene. It is only such things as statements or propositions, or

---

[436]Stitch, S. *From Folk Psychology to Cognitive Science*. . Cambridge: The M.I.T. Press, 1983. Pg 231. I am aware of the parallels between mine and Steven Stitch's position, only mine is slightly more formalized and aimed specifically at attacking the implicit ordinary language position in the philosophy of mind rather than folk psychology in general. Stitch holds that there are two systems, which he describes as 'sets of books', underlying vague belief states. One for verbal reporting and one for non-verbal behavior, and what conceals this is that often they agree. Where they disagree the person does one thing and says another. My argument is that the verbal utterance is a different sort of thing than what the person is thinking. It's a codification, that depends on assumptions that carries meaning from pre-linguistic knowledge using analogical structures and third person descriptions. Once the knowledge becomes linguistic it is no longer at the pre-personal level of the first person position, but rather, now involves analogical structures that make explicit connections to descriptions from the third personal point of view. See below, my footnote on Dennett and Chalmers.

beliefs or opinions, which are expressible in language, that are capable of being true or false, certain or doubtful. Our experiences themselves are neither certain, or uncertain; they simply occur[437].

And it rather seems that way on first appearance.

However, anyone who has ever fed the seagulls at the seaside from a bag of fresh hot chips might have played the game of pretending to throw the chip, and watching the bird react. Indeed, the very same experiment might be made with a dog, who like the bird; it seems posses no special powers of language or any propositional statement affirming ability.

Indeed I, as a boy, discovered many hours of amusement in playing this exact same game of feeding the dog by tossing a small morsel of food a short distance, several times, then waiting for the dog to return, then tossing another small morsel of food, and waiting for the dog to return. Then, of course, pretending to toss a small morsel of food, and then finding a satisfying sense of amusement as the dog searches around for the morsel of food in a puzzled way.

Indeed, the dog, to me, seems entirely certain that the food has been tossed. He's going over there to find it and nibble it up. Having tossed no food, I, on the contrary, know he is entirely mistaken in that certainty. Yet, indeed, not a single word has passed between us.

This insight, of course, is what underlies Norman Malcolm's attack on the Cartesian position that animals can't feel or think in his paper *Thoughtless Brutes*[438]. He makes that argument in the comparison of his friend patting his pockets for the set of keys which Norman knows are hidden in the glove box and Norman compares this with the behaviour of the dog who thinks the cat is hidden up the tree. The dog, like his friend, acts 'as if'.

---

[437]Ayer, *The Problem of Knowledge,* 1956. Pg52
[438]Malcolm, Norman. "Thoughtless Brutes." In *The Nature of Mind*, edited by Rosenthal, Pp 454-461. Oxford: Oxford, 1991.

The core of the propositional kernel, of language statements, as affirmations of mental states, Malcolm contends, after a short, shrewd and powerful bit of scholarship, lays in the assumption made on the behalf of Descartes. Norman maintains that Descartes argues that animals can't think on the basis that they can't use language, and thus, thought cannot be proved.

Norman Malcolm writes

> (Descartes) said it could not be proved either that animals do or that they do not have thoughts "hidden in their bodies" thus conceding it to be possible that they do have thoughts after all. But the idea that we cannot determine whether dogs have thoughts in them is a dreadful confusion. Suppose we did know everything that is hidden in their bodies: How could we tell which of these things were thoughts?[439]

Now I think this presents a worthwhile and very interesting starting point of a criticism to Brandom's model with its propositional constraints, however, my argument, a fortori, is a return to psychologism on the basis that thoughtful behaviour is irreducible to language. This attack on beliefs as expressible in language is not the strongest thesis of this paper. It's actually an argument based on a third person assumption and on par with physiologoistic and Linguistic Behavioural claims[440]. There is something interesting here and I'll bring it up in my conclusion and afterword. However the role of the argument is to soften up the linguistic behavioirual side of beliefs. The central attack will be on motives themselves that are found in the 'wants' and 'needs' of the explanation. The argument I'm going to be forwarding,

---

[439]Malcolm, *Thoughtless Brutes,* 1991. Pg459
[440]Chalmers. *The Conscious Mind.* 1996. Pg 225. Initially Chalmers defines awareness as the state where some information is available for consciousness, either for action or report, here he backtracks by suggesting a modification such that availability for report is not indicative of awareness or required for conscious experience since he concedes that it is possible mammals are conscious though unable to report.

here, is the position that a linguistic utterance model expressing reasons just can not account for human action without falling back into a form of psychologism. Ultimately I'm pessimistic not just about the form such an analysis must take, but about how much behaviour can be analyzed on the basis of a concern for reasons and made subject to the conditions of a linguistic utterance system of rationale or explanation. I am even more pessimistic as to how much concern people have for making their behaviours accountable to a set of reasons presentable to a linguistic utterance system of rationale or explanation, prior to the behaviour. I think such a criterion is unreasonable, except in cases where a person, perhaps, is doing something they shouldn't, and may have gone to the trouble of preparing a set of reasons beforehand, in which case, commendable as such a story might in fact be, these aren't the actual reasons for why the person was doing what they've done[441].

I think some behaviours, in fact, I think rather a lot of them are intentional, but they're not 'transparent' to linguistic analysis, which is to say they don't convert into the currency of reasons given as linguistic utterance. I base that on the fact that we just can't imagine what the reasons might actually be like as linguistic utterances for any number of behaviours we encounter in the world. Even if we make allowances for these behaviours we still need to assume knowledge prior to and fundamental for the game of giving and asking for reasons. The central argument I am going to make is that there is knowledge that precedes the game of giving and asking for reasons, knowledge which can't be established or justified by a set of reasons in the game on linguistic grounds, but it is necessary for someone to have this knowledge to play the game. It's a psychologistic move because the 'knowledge' I'm proposing one needs is non-linguistic, but it is also prior and necessary to a theory of meaning in the game of giving and asking for reasons.

---

[441]There are strong logical grounds for this type of argument. See Milligan, David. *Reasoning and the Explanation of Human Action*. New Jersey: Humanities Press, 1980. Pp 124-128, for Miligan's discussion of Abraham Irving Melden's argument that motives must either be causes, or effects of actions. If they are effects then they can't be causes of our actions on pain of turning a cause into an effect. However, we might argue that if a motive figures as part of an explanation, then we might distinguish between a justification which contains a formulation of the person's reasons and a motive that may be re-presented in an explanation. To do the latter will avoid Melden's contradiction, since I'm going to argue that it bottoms out in non-linguistic knowledge, once we understand what drives the game of giving and asking for reasons. The linguistic motive is not the motive that drives the person, rather it is something that happens at a pre-linguistic level which then becomes a 'motive' in the game of giving and asking for reasons.

Motives as reasons or linguistic utterances are meaningful only on the assumption of this non-linguistic knowledge.

Now, rather than get out the 'psychological empirical studies and surveys' and argue numbers, figures and statistical significance, basically the argument I'm making is a pessimistic one[442]. Where Brandom thinks that the intentional criteria he describes is the rule, I, on the contrary think that such concern for representation in the form of linguistic utterances that can account for behaviour by 'Sapients', or rather ordinary everyday people is the exception. Indeed, it may only be limited to philosophers and crooks, the former because we wouldn't be having this discussion otherwise, and the latter I gave in an example, just now. Indeed, I just don't think people care all that much about reasons, beyond the domain of crooks and philosophers, and if they do then perhaps less as a precondition for rational action, and that can be offered as a justification after the fact when hindsight distorts all too typically.

Now to uphold this, as promised, I'm going to argue that there are just some behaviours that defy rationalization in terms of a linguistic utterance or propositional structure and the evidence for this pessimistic view is that we encounter behaviour out in the world which actively resists the sort of criteria for intentional action that Brandom argues it should have. Indeed, this is what makes up the first part of the central push towards a Pre-Fregian psychologism in this section.

Brandom argues that the premises or conclusions themselves have the afore mentioned constraint of being give-able in the form of speech acts in utterance. This is like Ayer's argument about truth, certainty, beliefs and opinions, and his insistence they enter the 'scene' only once you have a language up and running. Brandom, ultimately, if you recall the discussion at the start of this paper and with Sartre and the apperception of the I. Kant, thinks that judgements incorporating I are the minimum unit of thought. This is the propositional form of the 'taking to be true' which Ryle would object to, and together with 'desire' and stipulated conditions for utterance, makes up Brandom's criteria for intentionality.

---

[442]See for instance Stitch, Steven. *The Fragmentation of Reason: Preface to a Pragmatic Theory of Cognitive Evaluation* Massachusetts: The M.I.T. Press, 1993.

Brandom states

> On the side of propositionally contentful
> speech acts, paradigmatically assertion, the
> essential inferential articulation of the
> propositional is manifested by the fact that
> the core of specifically linguistic practice is
> the game of giving and asking for reasons.
> Claiming or asserting is what one must do
> in order to give a reason, and it is as a
> speech act that a reason can be demanded
> for.[443]

As I said, this idea of score keepers is philosophically interesting and, perhaps, reminiscent of both Leibniz's attempts to develop a calculus for arguments, and as one might imagine, the early development of marriage counselling. However, after much reflection, and as I also stated above, I'm inclined not to be quite as optimistic as Brandom.

There are any number of cases or incidents which seem to me just don't fit Brandom's optimistic view of human behaviour and linguistic utterances. I have selected five philosophically interesting and pessimistic counter-examples from my own experience.

Case 1

Now imagine you are driving along a country stretch, and a car starts to trail you.

In your re-view mirror you catch sight or it, then it disappears.

You look again, and just catch a glance of it, sitting, less than a meter behind you, in your blind spot. You travel along for another few minutes and rather annoyingly the car won't move.

So you speed up a little.

It speeds up.

You go up to ten kilometres over the speed limit to try and put some distance between you and it and rather annoyingly it does the same.

---

[443]Brandom, *Articulating Reasons,* 2001. Pg161, 162

You then slow down, thinking, after all, it might just want to pass. It slows down.

You slow down further and the car in your blind spot you keep glimpsing, slows down further.

You continue to slow down and it continues to slow down, until it gets to the point of being ridiculous, where you're now travelling at between ten and twenty kilometres an hour.

Finally, travelling at under twenty kilometres an hour, you arrive at the first set of lights on the city limits and they're red. Your curiosity suddenly gets the better of you, since you've experienced this strange and annoying phenomenon of blind spot drivers a number of times before and you want to know why people do it. So you get out of the car, walk over, and tap on the little old lady's window. She winds it down and you ask her what she was doing.

Suppose this isn't the first time you've done this, and like the other two times, the lady can't explain what she was doing. At first she doesn't know what to say, then, she denies it. You point out, that both cars were going 5km an hour, you know they both were, because you slowed down, on purpose to let her pass, and were watching the speed dial the whole time. Again she denies it, and claims you don't know what you are talking about. The light abruptly changes and she speeds off.

You stand there, puzzled.

Case 2

Consider the case of a woman who sleeps with other women's husbands.

A string of them.

Every time she does it seems she goes out of her way to make sure the wife finds out at some stage, by leaving finger nail marks on the man's back, calling his house, sending him messages, or leaving her perfume scent or lipstick stain on his collar. Suppose you've witnessed a handful of occasions where someone's wife has confronted this woman, at a football game, or a parent-teacher meeting, usually with the evidence, and once with a very sorry looking husband in tow. Yet the woman, in the face of the wife's accusations, and silent husband, either has nothing to say or goes on to openly deny it. Yet, it seems she wanted the wife to know what she was doing because she plants the tell tale evidence that the wife shows up with each time.

Case 3

Consider the case of an art teacher who victimizes an otherwise talented art student, to the point where the entire class see what he is doing to her. Then one day, one of the class members asks him, outright, in front of the class, why he is doing it, and the teacher loses his temper and throws a tantrum in the hallway, smashing student canvases and clay models, rather than answering the allegation.

Case 4

Consider the girl who, by sly and quick moves at the bar of a night club, leads two men to fight over her, but takes off before either, as the victor, can claim her. She comes back and does it the next night. Then the following weekend. She's done this three times, and here she is, a fourth night, doing it again.

Case 5

Consider the father who, most should expect to express some sentiment of pride, but instead throws his son's acceptance letter to a university in the bin. Then when confronted by the son over why he did it, flies into a rage, then kicks the boy out.

These examples have a few things in common.

Firstly the person seemingly can't or won't explain the behaviour, and when pressed, will usually deny it, fly into a rage, or do something strange. Secondly, these sorts of cases are familiar enough when pointed out and we recognize them on some sort of intuitive level, but we don't tend to notice them until they are pointed out. They seem to hover at the point where we are aware of them, but only notice them after they've been pointed out. Eric Bernie, for instance, in *The Games People Play*[444] points out two similar cases; the friend who fidgets with something each time he comes over until he breaks something and the mechanic who always makes his customers wait for their car but seems

---

[444]Steiner, Claude M. *Scripts People Live: Transactional Analysis of Life Scripts*. New York: Bantam Books, 1982. And Berne, Eric. *Games People Play*. New York: Random House, 2004. Also see Rorty, Amelie. *Explaining Emotions*. Los Angeles: University of California Press, 1980. For interesting strategies and Rorty, Amelie. "Enough Already with Theories of Emotion." In *Thinking About Feelings: Philosophers on Emotion*, edited by Robert C. Solomon, 21-33. New York: Oxford University Press.

to get annoyed with them in his work shop. There are any number of similar cases one could think of but I will restrain discussion to the five laid out. Thirdly, and this is the interesting part, it is hard for us to imagine what a reason-style explanation for any one of these behaviours would be like, or to try and eke anything more than a mere description, as something substantial that goes on below the level of an observation language. We just don't know how to explain the sorts of things these people do. We can't imagine what an explanation in the form of a set of linguistic utterances or reasons might actually be like.

What could the little old lady who travels around in people's blind spots actually say? What might the woman who sleeps with other women's husbands, plants the evidence, then shows up to be caught out, what might she say? Presumably she is entirely aware of what she is doing, since it takes considerable planning, manipulation and skill, and likewise with the woman who started fights at the bar the past two weekends, but what might a justification or an utterance; what might be a reason they might give to explain the motives behind such behaviours?

It seems there's something in the behaviour that actively resists linguistic description or explanation on the level that Brandom is looking to determine.

My own further pensive analysis suggests it's not because these might be taboo behaviours in the community, like theft, incest or murder, which we can think of and apply linguistic terms to, but rather, the behaviours themselves seem to resist classification within a linguistic utterance structure. There is something odd about them. In a moment I'm going to suggest it is not because the actions are 'irrational' or 'unreasoned' in any sense. Far from it. They, variously, contain reasoning, systematic planning, deliberate steps and the employment of careful manipulation. Rather it is what drives these behaviours that makes them so interesting.

Case two isn't merely the result of coveting another woman's husband because she doesn't restrict her coveting to the one man. We might imagine in that situation something like 'I really do love him' or 'he's truly the one for me' for a propositionally contentful reason or a linguistic utterance. But her behaviour is a bit stranger than that, because there are a string of other women's husbands. Moreover, it seems she nearly always wants the wife to find out because of the evidence that gets spat out by the wife during the denial and confrontation. Whatever the woman in case two is after it is a bit stranger than the typical love story.

I chose these examples for this reason: the 'typical story' has a way of blinding us to the nature of the 'wants' themselves. The 'typical everyday things people want' like 'love', 'trust', 'the warmth of a good friendship' we relate to these and as such we don't tend to question them. Their 'commonality' and 'relatablness' is part of the illusion that camouflages them to analysis. It is only when we encounter cases like the woman in case two, which we find a little harder to relate to. I suggest we take this insight, and when we think back to the things we want and normal cases that we can begin to think our way out of this illusion and see our way around the illusive nature of Brandom's reason-explanation model. From there we can take systematic steps out of the ordinary language model and towards an insight that is on offer but first we need an internal/external reasons structure to clarify what is at the heart of the insight that is on offer and to deal with the parts of Brandom's and Peters' model, which I have listed above, that are most problematic.

Now admittedly cases two and four may resist our attempts at formulating a linguistic explanation or justification for the behaviour. It is too hard to justify what the person is doing within the external framework we have available to us from the outside. We might put the distinction like this. External reasons in ordinary language are often difficult, we don't quite have the motive words nor do we find it easy to imagine what justification these behaviours might have. Someone might see this gap and argue that a Freudian theory, or argue that some sort of evolutionary paradigm might work in these cases but these would be external reasons.

By external I mean something like Bob Hearn's concern with Freudian therapy in Norman Mailer's anti-war novel.

> "If you saw my analyst. . ."
> "The hell with that. If I'm afraid of having
> my (. . .) cut off or something like that I
> don't care to know it. That's not a cure, it's a
> humiliation, it's a deus ex machina. I find
> out what's wrong and bango I'm happy and

go back to Chicago and spawn children and terrorize ten thousand people in whatever factory my father decides to give me[445]."

Bob's point is that he is unaware of whatever motives the psychologist might impute him with and moreover he doesn't want to know them. Thus they can't serve as reasons in his motives. They can not be given as reasons and utterances in a Brandom sense, nor can they be given as parts of the agent's "being alive to what he is doing" in Ryle's. These motives seem to fall short of Ryle's special status reports. It is hard to imagine what the old lady is thinking as she is driving around in people's blind spots. We might imagine ourselves in her shoes, behind her windscreen, watching ourselves or another driver up-front through her eyes, and feeling some sense of mirth or enjoyment at the irritation we are causing. Similarly we might attempt to imagine ourselves somehow oblivious to the situation as we trundle along driving in people's blind-spots without realizing it and can be chalked up to a bad habit. But when we try to put these into a framework like Brandom suggests we start to run into problems with artificiality. Part of this 'artificiality' comes from Brandom's conditions for the 'I' and the premises-propositional structure that the reason must take, like I brought out with Ryle above, but part of it, I argue, also comes from trying to justify her actions from our own standpoint. Thus, external reasons, we might suggest to get around this, are those that are not available to the person in terms of reasons that can be given for these motives from an external view that is not part of the person's own cogitations.

Thus if we take case four, because on first appearance it is the simplest of five, we can build a framework for what an internal explanation might be like. We might suggest, for instance, that a reason is an internal reason if it is accessible to the person and can figure in their explanation. We might say that the girl in case three is starting fights at the bar because she wants or needs reassurance about her desirability.

Someone may respond that this isn't her actual reason, that her actual reason is she is doing it out of boredom, a joy of mischief, entertainment, or for as elusive a reason as simply for something to do

---

[445]Mailer, Norman. *The Naked and the Dead*. London: Harperperennial, 2006. Pg 353

on a Saturday night. They might even argue she is doing it out of something altogether too difficult to put into words or linguistic terms.

We should then respond to them by saying that an internal reason need not be the reason they are doing it, but it can function as one, suggesting that desirability is of a class of similar such reasons. Here, of course, we are not quite breaking Peters condition that a reason must be the reason why the person who did it, did it, but simply allowing for a class of reasons to stand in place of *the* reason[446].Further, while we may not agree with her motives, that is, her need for reassurance or her sense of mischief and boredom, we may, nonetheless respond on Brandom's behalf by pointing out that in obtaining what she wants, whether it be reassurance or relief from boredom, she does reason in such a way as to get what she wants. We might even be able to argue that her reassurance could be represented, with some ingenuity, in a reason-explanation utterance framework with practical steps and inferences like Brandom suggests.

So wouldn't Brandom be right? That is, one might argue, even though we don't agree with what she wants or her wants may appear irrational or be difficult to describe, like in the case of the little old lady who drives around hovering in people's blind spots, or case two, the serial adulteress, they are still following Brandom's framework which involves making deliberate actions with premeditated steps and utilizing complex forms of manipulation.

However, in response to this, let's examine the 'want'. The 'want' itself doesn't have reasons to justify it. If we search for them we are driven to the point where the want is on the edge of the expressible. The want can't be justified by a reasons-explanation framework by the person. The 'want' of desirability by the woman, that in turn, leads her to dress up, go to the bar, pick out one of the males, flirt, then turn around, pick another male, flirt and so on; the 'want' that drives these steps, is itself beyond the game of giving and asking for reasons.

Now someone might respond with a story about her father, or evolutionary paradigms and say she evolved in such a way as to require reassurance about her desirability, but this would be an external reason. This was Bob Hearn's concern about going to see the

---

[446]See the part dealing with Peters earlier in this section. '7.1 The Game of Giving and Asking for Reasons.'

psychologist. The reason must be one that she gives, and it must be capable of serving as her reason. It must be internal.

We are at the point where we need non-linguistic knowledge, or at least the girl does, in order to know what the meaning of her wants and desires are, in order that she might make practical inferences and represent them as premises in the game of giving and asking for reasons. That is, she needs knowledge that precedes the limits of the game of giving and asking for reasons because the wants and needs from which a person makes practical inferences about getting what they want are themselves not justifiable in the game. Which is simply to say that to play the game of offering reason-explanation or reason based utterances one must arrive with a special knowledge, the knowledge of what one wants which must in itself precede the limits of the game because it can't be justified by the game.

There are two cases relevant to the problem of utterance explanation left along with a case reported by the psychologist Janet which Sartre analyses and will allow us to bring all of these threads together.

The first is the art teacher. Seemingly his job and passion is to encourage talent in his students, to push them to new aptitudes and foster their creativity, and yet, he runs down this one student, and when confronted about it, flies off the handle.

The second and last remaining case is the father of Case 5. The father's reaction to the son's acceptance letter, likewise, seems plainly irrational, and more so because he flies into a rage when confronted, and kicks the boy out of home. Instead of offering a reason, such as, perhaps, 'boy, I wish you really would follow in my footsteps' or 'son, listen, I really think you'd be better off with a trade than some fancy good-for-all education' he flies into a rage.

These situations are philosophically interesting, because it seems as though in the case of the adulteress, the woman at the bar who has started fights the past two weekends, and the little old lady who drives around in people's blind spots, linguistic utterances as reasons are *conceivably difficult*, and what seems most conceivably difficult are the wants themselves. Once we strip these wants of the conditions that obscure them in the game and include a distinction between those which might serve as internal reasons and those which may be external, it can be seen that these 'wants' reveal to us something that exists beyond the game of giving and asking for reasons. Furthermore, the 'wants' reveal something that cannot be established once the game is set up. Rather one must arrive at the game of giving and asking for reasons with some sort of pre-linguistic knowledge. That is, if one wants to play the game of giving and asking for reasons one must already possess knowledge of what one wants, and this knowledge is of a type that can't be justified and is not reducible at the level of reasons and justifications as linguistic utterances. These last two cases in particular are reminiscent of something that Sartre picks up on in the case work of the French Psychologist Janet. It is to this we now turn.

7.2 Psychasthenia, Dispositions and the Meta-Concern.

An interesting case that Sartre introduces in *A Sketch for a theory of the Emotions* involves a woman who cries each time she goes to see her psychologist, and this crying, in turn, prevents her from revealing or confessing any information about herself. Sartre wonders if this is an emotional reaction that has a foundation in part of the primitive emotional response circuit that Janet speculates on. The theory of Janet's circuit is beyond the scope of the paper, however the way it functions is analogous to what Ryle would call a disposition, but this is closer to an emotional disposition, that is, a disposition keyed into an emotional response and this is how we will treat it. If we refer back to Ryle's Occurrence and Propensity distinction that what I am investigating is a dispositional occurrence which in Ryle is impossible since occurrence are episodic, however if we insisted on a Rylean classification, then we would treat we could treat it as a dispositional mood, since moods are propensities, here, similar to an agitation. Since what triggers the mood is a 'flash bang' there is something important in what happens next. The question then is whether she has a disposition to be overwhelmed by her emotions in certain contexts, or if, in fact this is something else, something a little bit more interesting?

> The Patient who comes to see Janet entrusts
> him with the secret of her troubles and a
> minute description of her obsessions. But
> she cannot; this is social behaviour that is

too difficult for her. Then she bursts into
tears. But is she weeping because she can
say nothing? Is her sobbing in a vain effort
to do so, a diffuse upheaval that represents
the decomposition of the behaviour she has
found too difficult? Or rather is she crying
precisely in order not to say anything?[447]

Clearly what intrigues Sartre about this case is that all important
distinction. Is the woman crying because she is overwhelmed by the
task of confessing to her psychologist, or is she crying in order *not to* say
anything to her psychologist? This last case is precisely reminiscent of
the father in case five and the art teacher we looked at in case two.

Like the father and the art teacher, Janet's patient is either
crying because (a) she can't give a reason, which branches off into (a: i)
because her dispositions incline her towards crying and not giving
reasons, or (a:ii) because there is something inherently un-utterable
about her case. That is to say her case is not the sort of thing you can
give reasons or utterances about. Or is it (b) she is crying in order not to
give reasons.

(a:i) might be taken in either a linguistic predicative sense with
a certain logical force (c:i) or a mechanistic and physiological sense (c:ii).
The first I strip of much of its power in the next section, where I side
with Robert Wolff and argue that dispositions in ordinary language are
a third person descriptive phenomena. The first person strain, the Ryle
strain, survives, but only as a form of auto-phenomenology and
concludes with the account I give of 'flash-bangs' at the end of the
paper. I argue against (c:ii), the physiological strain of symptoms on the
grounding that individuation on the basis of phenomenological insight
gives us a stronger type of argument. But something of this account
survives in the problem of "the 'Roid Rager in the void" I introduce at
the end of this paper. The difference is in that version the problem of
the "Roid Rager" takes into consideration an account of consciousness[448]

---

[447]Sartre, Jean-Paul. *Sketch for a Theory of the Emotions.* Translated by Philip Mairet. New York:
Routledge Classics, 2002. Pg 21

[448] The benefit of Ryle's implicit ordinary language position on the mind is it can avoid these sorts of
causal problems by focusing on the way people use language rather than on physiologistic and
neurophysiologistic symptoms and causal issues. If we dissolve the implicit ordinary language
position on the mind or we weaken it, as we shall see, one of the effects is that we reintroduce the

. The significance of the (a:i) premise in either its (c:i) linguistic sense or (c:ii) mechanistic sense is that both try to give an account of the mind that leaves out consciousness.

On the other hand, (b), it will be reflected, implies a theory of consciousness. This is because the central difference between the two, I argue, is that (a:i) claims it can give an account of the mind without a theory of consciousness using either some linguistic behavioural analysis of dispositions that can tell us everything we want to know about whether she is going to cry or not, or a mechanistic sense in which physiological symptoms can offer us that same account instead of ordinary language analysis. Both the linguistic behavioural and the physiological accounts leave out the first person position on consciousness. The reason why (b) implies a theory of consciousness is that it implies that she must be conscious of the fact she does not want to give a detailed confession since she cries in order not to give a detailed account. She is making a move in Brandom's game or rather she is making a move that avoids giving a reason in Brandom's game. If we are to have an account of why she is crying it must include the fact she is conscious that she is making this move. On this view crying is a move one might make in order to avoid giving reasons, and like the woman's need for reassurance in case four, while we might not agree with the reason why she 'wants' to make this move, or the 'want' of the reasons she is concealing by making it, she is still using reason to determine her moves. If it can be argued that this view is right, in at least some cases, then it puts her, the art teacher and the father into the same category as the other three cases in the prior chapter and offers us grounds for moving forward our psychologistic thesis for non-linguistic knowledge.

Prima facie (b) does not rule out (a:ii), but I will argue it does preclude (a:i). If (b) is true and 'she is crying in order not to give reasons' then (a:i) must be false, since (b) implies she has a choice in the matter and that implies an element that can not be grasped from a third person perspective. Let us pause for a moment to consider what this might mean since it relates directly to my fall back position and illuminates the nature of the disjunction I am building.

Let us take the case of a liar. Let us take it that there is a fool proof method of constructing a lie-detector test. This method can tell us

___

problem of causation.

absolutely when the person is lying and when they are not lying purely from a third person perspective based entirely on neuro-physiological information. Is this all that there is to know about lying? I argue that it is not. I argue that to give an account of the 'lie' one must also give an account of why the person is lying. Whether this can be done from the third person perspective or not is debatable. It leaves us with the problem of internal and external reasons. If a neuro-physiological account can give an entirely satisfactory account of lying it must be able to, from a neurophysiological position, also give us an account of the person's choice to lie without introducing an analogical structure in either the form of an auto-phenomenological argument or a concealed phenomenological assumption like a sympathetic or empathetic argument[449] in some disguised form like 'if I was Jones and I murdered my wife, I would lie about it too'.

If the neurological account falls back on an analogical structure hidden as an assumption with insights or information from an autophenomenological source, or phenomenological argument then the foundation is psychologistic and implies a theory of consciousness. For instance, if our concept of what anger is, is ultimately based on our own experiences of anger then the root of our understanding of anger is auto-penomenological. If we apply our own understanding of anger which is based on auto-phenomenological roots of our own experiences of anger to another person then our understanding and application of anger to that person has an analogical structure hidden as an assumption. This includes when we use the term in a research sense. This is important to hetro-phenomenological sources. Does our use of the heterophenomenological source, whether it is descriptive language, an EEG or a Polygraph, contain any hidden assumptions.

---

[449] This is an analogical structure between a first and third person position, which ultimately, derives its appeal from a first person introspective act of scrutiny, or phenomenological reflection. If we think Johnny did it out of jealousy, for instance, because his girlfriend was having an affair and we base this on our own experiences and what this might have felt like in the past, then this is an internal reason and applied on an analogical basis and the argument is ultimately theitic and analogical. If we've never been jealous, or find it hard to imagine so, but we know that this is what causes some people to murder others, then it appears as either one of Hearn's 'deus ex machina' and the sort of thing that sounds good in a lecture hall but doesn't actually get applied in people's reasoning as part of their behaviour; or it is descriptive in the Wolffian sense from the third personal point of view, but has no actual causal basis in regulating the person's behaviour. Indeed we find this is quite common, men often don't understand what women are talking about or viceversa since in general men have no experience of being women and in general women do not have any experience of being men.

The polygraph must tell us *why* someone lies through some configuration of lights and squiggles not just *when*. Further it must give the decision making process itself and in so doing give us the reason why it went that way as an internal reason. Lying has special conditions that preclude it from being dispositional in a purely reactionary sense and can reveal something interesting about hetero-pheneomenolgical sources. Since lying is a conscious decision to deceive the person must make the decision to lie. It follows that a lie detector that can tell you when they lie is not mimicking the behaviour of the liar. The person lying is making a conscious choice to deceive, the machine is the one being dispositional, not the liar. Thus a hetero-phenomenoloigcal account of lying must give an account of the reason why the person is lying in order not simply to be a demonstration of how the machine works. Moreover it must do so in such a way as not to fall back on analogical structures like in the case of anger.

To relate it back to Janet's case (a:i) as either (c:i) or (c:ii) either case must, like the liar, tell us why she cries without a blatant or occult argument that makes a phenomenological move.

I argue that a move that bottoms out in the person offering an internal reason is sooner or later through a process of inquiry going to lead back to non-linguistic knowledge, or knowledge which it is necessary to posses for the game of giving and asking for reasons. Similarly, as we shall see, my argument on the emotions claims that the emotion and its linguistic expression are not the same thing. There is something more to the emotion than a linguistic behavioural account or a physiologistic story. The basic underlying message of this paper is that physiology and linguistic behavioural accounts cannot give us the full story on the nature of mind. We need an account of consciousness.

There is another class (a.iii) which are cases where the person does not give a reason, because they are too tired, they haven't considered the case and similar reasons, or they were distracted and made a snap decision, but we might dismiss these with the argument that if they were pressed (a.iii), the undetermined case, would collapse into (a.i) or (b). But there is a side related case to (a.iii) which are people who deliberately avoid choosing through some subterfuge. We might represent this with a "refraining/ failing to" account. In a sense, the former, the person who chooses to withhold assent in some manner that involves bad faith or subterfuge, like for example, going to work and working really hard, and avoiding thinking about the situation. In the latter case, the 'failing to consider' they simply react. We might represent this distinction as analogous to Aristotle's teleological and efficient causes. In the latter case, the person reacts without thinking, because they might be tired, or simply haven't considered the situation. This latter reaction is similar to Aristotle's efficient cause. Thus, in this sense it is purely a reaction, x happens, the person reacts with y. But surely this would fall into (a.i), and conform with a dispositional account, if not on the account of (c.i) and (c.ii) then in some vague sense still applicable to (a.i), possibly defined as (c.iii) or (c.iv) or the like. It depends on whether such an account can be explained without an appeal to consciousness. If so, then it belongs in the case of (a.i). If not, then it either fits the final account I give for flash-bangs or as some form of epiphenomenalism[450] or non-reductive functionalism[451]. The earlier case might be likened to Aristotle's teleological causes. On this view the person is employing a subterfuge in order not to give an answer and on this view going to work, crying, getting angry are all done in order *not to* give a reason. This disjunction is precisely what happened when the lights changed and the little old lady in case one drove off. If she drove off not to give a reason it fits clearly with case (b). If she drove off because the lights changed and one might argue that the lights changing are like (a.i), we can give an account of why she drove off like Price's thermostat. It can then be seen that we are able to argue for a reduction in the above cases to either (a.i) or case (b). This

---

[450] See Carter, Rita. *Mapping the Mind*. Los Angeles: University of California Press, 2010. Pg 190 There is a curious form of this in the Benjamin Libet experiments that came out of the University of California in the mid to late 1980s. Subjects were fitted with EEG sensors and told to move their hand, at any moment, as a self willed action. While doing this they were told to watch the clock and note the precise moment when they chose to move their hand. The curious result was that activity in the brain was found to occur *before* the person had reported the conscious experience of deciding to move their hand. Now this is an interesting claim, but there is an interesting counter-position which leads me to shy away from epiphenomenalism and it is this. Assume *this* is consciousness and your decision to keep reading is an act of consciousness. What you are experiencing as my little voice inside your head as you read is a consequence of your decision to keep reading. Now, using *this* awareness, pick a number between one and ten. Now, count backwards from that number, and raise your right hand. Good. Consciousness and conscious decision making has just preceded the act of carrying it out.
[451] See Chalmers. *The Conscious Mind*. 1996. Pg 229 for his account of functional organization as an abstract pattern of causal interactions. Pg 241, and the discussion of Nagel's bat and 'phenomenological insight'. See also footnotes in the succeeding sections on his notion of a bridging principle. I shall bring out some of the implications for a Neural Correlate of Consciousness and a method for collecting formalisms at the end of this paper in the conclusion.

reductive move affects the legitimacy of whether we can give an account of these within an internal first person position and the efficacy of the options that are open to us from a third person perspective.

Further analysis of (a:ii) will show that as well as being compatible with (b) it is also compatible with (a:i) since it is imaginable that Janet's patient may have a disposition to cry instead of giving reasons, and that the reasons she might have gave, or rather attempted to give, would have something un-utterable about them. In addition (a:ii) is supported, as its own class, at least prima facie, by the case of the old lady who drives in people's blind spots, the girl who starts fights at the bar and the repeat adulterer.

There is something difficult to articulate about these cases. The person clearly knows what they are up to. They are using reasoned moves and deliberate actions, but it is hard to make their 'wants' and 'desires' transparent to articulation. Moreover, as we have seen, when we do, like the girl who starts fights between men at the bar, and which, we might characterize as the need for reassurance of her desirability, we find that what the person wants can not be justified or explained in the game of giving and asking for reasons. It depends on non-linguistic knowledge they arrive at the game with. This is the hard problem of psychologism. How to access, talk about and gain systematic insight into these non-linguistic forms of knowledge, either without breaking strictures against, or reverting to external reasons. This I argue is the genuinely difficult problem that we face in uncovering the nature of mind. We arrive at this position in the philosophy of mind if we take our point of departure from mainstream analytic philosophy starting with an implicit ordinary language position on meaning and the nature of the mind, like I laid out at the start of this paper with Dummett and Ryle, and we then come to reject that position on the basis of the arguments I have advanced.

Thus it can be argued that the final disjunction is between a conjunction of (a:i) and (a:ii), and (b) and the conjunction with (a:ii)[452]. Now note the shape of the argument. It looks like a syllogistic disjunction with a shared middle premise, minus the deciding premise, which, of course, would be the negation of one side or the other[453]. That

---

[452] (a:i) and (a:ii) or (b) and (a:ii)

is, to complete the syllogistic disjunction, since (a:ii) is compatible with either side, the conjunction of a true and a false, of course being a false, negating (a:i) or (b) which would then negate that side of the argument. That is to say, we need a not (b) 'she is crying in order not to give reasons', or a not (a:i) 'her dispositions incline her towards crying rather than giving reasons' in the structure of our argument.

I have already argued that my auto-phenomenological thesis will deal directly with refuting the (c:ii) case of (a:i) on the basis of the strength of phenomenological arguments as a normative source to individuate elements of the mind. This will open the way to my final account of 'flash-bangs'. My refutation of (c.i) will come from a cogent reading of Wolff's linguistic behaviourism which will strip much of Ryle's ability to offer an account of the mind in terms of linguistic dispositions once it is shown that the power of Ryle's dispositions rests upon a confused sense of first and third person positions and it can be seen that the correct role of dispositions is descriptive and holistic, not predicative and 'forceful' or 'compelling'. The element that is compelling will become the grounds for the forcefulness of phenomenological arguments in the negation of (c:ii).

The negation of (b) I take it would be a vindication of a dispositional account either as (c:i) a mechanistic account or (c:ii) an affirmation of dispositions as law-like statements that can be derived entirely from a third person's perspective without an appeal on phenomenological grounds. The latter is the Ryle, or Ryle-Peters position, the former the mechanistic position that Ryle is opposed to[454] but is the way Ryle has been read in a lot of cases, and can be characterized by a special reading of Sartre's reading of William James and a type of argument that Sartre uses for negating it[455] where Sartre

---

1) [(A . C) v (B . C)]   (a:I) = A
2) - A
          (B . C)   (b) = B

3) ⌐ - (B . C)   Ass   (a:ii) = C
4) | (A . C)   SD 3, 1
5) | A   AD 4
6) | (A . - A)   CD 2, 4

          (B . C) RAA Ass

453
[454]Ryle, *Concept of Mind,* 1983. Pg 74-80 NB See 'the bogey of mechanism'.
[455]See '8.1 Occurrences: Moods, States, Acts and Feelings' in 'Chapter Eight : The Occurences' for

argues that anger is not a form of ultra-joy even though the symptoms of anger are magnified or a more intense manifestation of the observable physiologic symptoms of joy. The structure of this argument is important. The argument appeals to phenomenological insight into anger and joy to argue one is not the other no matter how similar the two phenomena are in terms of physiological symptoms. This, as I indicated earlier in the paper, I call an auto-phenomenological argument, in contrast to Dennett's 'hetero-phenomenology' because it individuates on the authority of an appeal to phenomenological insight and introspective scrutiny. I argue this is a stronger source for individuating mental elements and presenting arguments about the mind than either (c:i) the linguistic or ordinary language account or (c:ii) the mechanistic account.

If we lay out a sort of structure behind some positions it might help to make my argument more clear. Ryle, once freed of the phenomenological strain, is arguing that the nature of mind can be discovered from an analysis of ordinary language. That is, implicit use of language contains all the information we need to discover the nature of mind and this can be done using a linguistic behavioural approach. Sellars, if we simplify his position in certain regards for the purposes of clarity[456] is arguing that ordinary language can be aligned with scientific information of a certain sort to give us a view about the mind. That is, I read Sellars position as advocating that by treating sentences as analogous to thoughts, with special considerations, we can map 'linguistic roles' in ordinary language as 'conceptual thinking' onto neurophysiologistical information as it becomes available and thus merge the manifest image, which contains the ordinary language position, with the scientific image which contains the developing strains of neurophysiologic information. In a clash between these two positions I should argue that Sellars basic position would be to favour the scientific image over the ordinary language position, since science is continuous with processes in ordinary language, and given the shape of the argument in *Empiricism and the Philosophy of Mind*, science will eventually replace our manifest framework of the mind. I take it he would argue scientists will come to see new information as part of the structure inside the observation framework once rationale has been

Sartre's 'anger is not ultrajoy' argument.

[456]Sellars, *Philosophy and the Scientific Image of Man*, 1991

given signalling a move into the space of reasons, and from this learn to identify mental phenomena from a report language.

There are two other key positions. One is that partially of Dennett and specifically the Churchlands that is hetero-phenomenological scientific sources are the correct position regardless of phenomenological or ordinary language sources[457]. I disagree with these arguments as I think that they usually contain occult appeals to ordinary language sources or phenomenological assumptions and appeals. While here, in the body of the text, there isn't room to go into specific types of claims[458] it is worth noting, in relation to our inquiry,

---

[457]David J. Chalmers "How Can We Construct a Science of Consciousness." In *The Character of Consciousness* 2004. Pg 55. Chalmers is right to distinguish critical phenomenology from hetero-phenomenology, and I also argue Chalmers is clearly right in arguing that Dennett often confuses the two. Critical phenomenology depends upon comparing a phenomenological account with a third person position. Hetero-phenomenology takes it that the first person position is either reducible to or equivalent to the third person position. I hold a position similar to this latter view, that is the reductive or equivalence and prevarication, can be argued for language but not for consciousness itself. A good example of this are Simons experiments. See, for instance, Dawkins, discussion in *The Greatest Show on Earth*. Random House: London, 2009. Pp 14-16.Daniel J. Simons experiments focus around 'inattentional' blindness, and compare their first person account as a form of critical phenomenology to a third person position. In these experiments he had people concentrate on tasks like how many times a ball was tossed, and had a man dressed in a monkey suit walk past. When they compared the accounts the people concentrating on the tasks didn't see the man in the monkey suit. Chalmers in *How Can We Construct a Science of Consciousness*, 2004. Pg 42-43 discusses another form of critical phenomenology where researches were able to contrast 'explicit memory' with 'implicit memory', by examining cases where people learned something and were or were not conscious of what they learnt when they went on to perform that action again. Conscious recollection and application was labelled 'explicit', and 'implicit' being, by contrast, the case where they were unaware they were using the new skill. I can see both Dennett's position, coming from Ryle, and Chalmers position coming from his research and *The Conscious Mind*. The solution, I propose, is in the puzzle of *the Concept of Mind*. See the footnote below for further clarification.

[458]See Stitch, *From Folk Psychology to Cognitive Science,* 1983. Pg 210 -214.Dennett, Consciousness Explained, 1993. As well as Chalmers, David J. "How Can We Construct a Science of Consciousness." In *The Character of Consciousness* 2004. In particular Chalmers, on pg 40, distinguishes between a claim that there is no first person data, like Ryle's implicit ordinary language position in the philosophy of mind, and a less extreme version of that where first person data is treated as equivalent to third person data because it is given as verbal reports. This is like Dennett's position on hetero-phenomenology. What I am arguing is that Dennett is right in the sense that verbal reports are equivalent to third person reports if taken from a third personal perspective. This is because the first personal perspective is a pre-linguistic codification of non-linguistic sources. If Psychologism is the right position to adopt, which I think it is, then reports are something that happen to our thoughts. The meaning of those reports depends upon assumptions, analogical structures and third person descriptions known to the ordinary language speaker making those reports. The meaning isn't generic or implicit to the medium of the language, but rather, is constructed from what happens prior to the linguistic codification, while the codification depends on non-linguistic knowledge present in assumptions. These assumptions involve analogical structures which involve the third person position once the thought becomes linguistic. Hence why linguistic thought implicates the third personal view. In regards to the second type, the pure hetero-

that they are to the radical right wing of Sellars and Ryle's position. And then my position is that phenomenology and phenomenological insight is a stronger source of appeal for discovering the nature of mind and individuating elements than either ordinary language or ordinary language and science. Now notice that the form of the argument I'm putting forward isn't making a metaphysical claim about the mind, but rather it is about the strength of the type of argument. That is I am claiming that phenomenological arguments or arguments that offer some sort of phenomenological insight are stronger sorts of arguments than those that individuate the mind and its nature on the basis of ordinary language, science or both. The reason why this might be so is my claim that non-linguistic knowledge is necessary for establishing linguistic meaning. Thus if ordinary language requires pre-linguistic knowledge for meaning then a theory of the mind can not stop at

phenomenological sources, like those Paul and Partricia Churchland argue for, my general argument against these follows similar lines to my argument against individuating on physiological data alone. For instance, if scientists discover that a certain part of the brain is active in, and only in certain states of depression, and they come to classify activity in that part of the brain as depressive state x, and thus, can from that point claim somebody suffers from depressive state x, if and only if they have activity in that part of the brain, without any insight or appeal into the patient's state of mind, then the original classification of brain state x contains within it a concealed linguistic or phenomenal argument, namely the original classification itself. For this reason I don't think there can be a pure form of neuroscience based on physiology and brain states. I think arguments that try to do this, in general have some sort of hidden basis or assumption for the nature of the individuation. I think once this point is grasped, the question then becomes what is the strongest source for individuating the mind. Ordinary language arguments about the mind are different. What makes them interesting is that unlike purely physiological arguments, prima facie, they *appear* to offer the possibility of giving us a theory of mind without appeal to another source purely on the authority of an intrinsic understanding of language. But, as this paper shows, this is only an appearance. Once we analyse the types of claims that ordinary language arguments posses, as this paper has attempted to show, most of them will simply break down into phenomenological style arguments and linguistic behavioural arguments. See also Chalmers, David J. "First-Person Methods in the Science of Consciousness." *Arizona Consciousness Bulletin*, (1999). For Chalmers original version of "How Can We Construct a Science of Consciousness." Professor Velmans unpublished diologue with Daniel Dennett which was both inspirational and instrumental in forming the above view, Velmans, Max. "Heterophenomenology Versus Critical Phenomenology: A Dialogue with Dan Dennett." 'Unpublished', deposited at " http://cogprints.org/1795/, Febuary 2006, (2001). David Chalmers and Dennett's debate "The Fantasy of First-Person Science." In *Daniel C. Dennett, David J. Chalmers.* Transcript at http://ase.tufts.edu/cogstud/papers/chalmersdeb3dft.htm, 2001. Video at http://catcomcon.blogspot.com.au/2012/09/dennett-d-unpublished-fantasy-of-first.html. Ratcliffe, Mathew. "Phenomenology, Neuroscience and Intersubjectivity " In *A Companion to Phenomenology and Existentialism*, edited by Hubert L. Dreyfus and Mark A Wrathall,. http://www.blackwellreference.com/public/tocnode?id=g9781405110778_chunk_g978140511077826, 2006. It is my view that European phenomenological practice in the main supports Chalmers position on developing a first person science, critical phenomenology, and his claim in the earlier paper that first and third person positions are separate positions. Seen from this perspective, my concern merely relates to some ambiguities in the details of his treatment of language.

ordinary language. We must go beyond ordinary language and into a psychologistic domain to understand the mind.

My argument thus is that phenomenological arguments are stronger arguments because they can appeal to insight that draws upon non-linguistic knowledge like in the case where you have to recollect being the witness of a race or a football game from a perspective up in the stadium and compare that with the freedom of the reader of the race to imagine his perspectives in order to see the distinction. You have to have some sort of pre-linguistic knowledge about what pride feels like and what warmth feels like in order to distinguish between the two types of glow in linguistic contexts. You have to imagine thunder so loud it hurts your ears, or lightening so bright it hurts your eyes to see the difference between 'see' and see. These arguments that offer phenomenological insights are stronger types of claims about the nature of mind than ones that trade on the currency of mere grammatical differences and linguistic behaviours. That's the first part of the auto-phenomenological strain. The second part of my auto-phenomenological argument arises from Sartre's argument about ultra-joy.

I would, however, be prepared to give this up, if it can be shown that (a:i) is in some sense the correct position on the basis of either (c:i) which I take it, given the considerations following the discussion of Wolff seems weak and unlikely, or (c:ii) which, given Sartre's argument that anger is not ultra joy, from which I take it, it seems it is almost as unlikely. Either or both would do, though it seems intuitively likely that one is incompatible with the other without some sort of bridging principle that takes into account arguments raised by Ryle[459]. That is, (c:ii) implies not (c:i), and (c:i) implies not (c:ii). As I stated, of the two (c:i) I should argue is the weakest source and least likely to give us an account of the mind. Once it has been shown that Ryle's dispositions are a confusion of first and third person positions, then third person linguistic behavioural arguments lose much of their power. They are external in the sense I gave in the prior chapter and in my final account I shall argue that they are a linguistic code for talking about something that happens at a non-linguistic level. The meaningfulness of this code depends upon non-linguistic knowledge.

---

[459]Ryle, *Concept of Mind*, 1983. See Ryle's 'governed but not ordained argument'. Pp 74-80

My fall-back position, thus, would be an account that favours the authority of arguments that call upon (c:ii) physiologistic symptoms on the basis of a link with phenomenological data. To make this argument one would have to show why physiological symptoms offer a stronger foundation to individuate the nature of the mind than phenomenological sources and thus are better than and stronger than auto-phenomenological arguments. That is, one would have to show firstly why anger is a form of ultra-joy and secondly how a physiologistic account can give an account of the mind that can tell us why the person sitting the polygraph is lying on purely physiological grounds. It's not enough to tell us that the person is lying it must also give an account of the person's choice to lie[460]. This would, in effect, block my re-routing these cases, like Janet's patient and the two cases in the prior section from Brandom's game of giving and asking for reasons. This is a critical move in the argument because it allows me to raise the argument of the prior section. The argument of the prior section, was, of course, that someone must arrive at the game of giving and asking for reasons with non-linguistic internal knowledge, that is, knowledge that can not be set up or justified in the game but is necessary for playing it. Re-routing these cases would be a step, but it would only be a first step. The argument must do so in such a way as to convince me it applies to all cases. That is, it can give an account of decisions, either as physiological or linguistic dispositions, without an account of consciousness. Consciousness would then be an after effect. In this sense physiologistic and neurophysiologistic claims would have authority over phenomenology, and what we would be left with is a form of scientific epiphenomenalism rather than auto-phenomenology. Essentially, one that advocates a correspondence between neuroscientific information and the introspective qualities of thoughts, and that favours the authority of neuroscientific information over phenomenological introspection.

This fall-back position is the position that Sellars eschews in *Philosophy and the Scientific Image of Man* on account of the problem of introspection. The problem of introspection is what leads him to

---

[460]Carter, *Mapping the Mind*, 2010. Pg 24-25 This is different to the recent discovery that activity in the orbito-frontal cortex is linked to a person's ability to choose, self awareness and consciousness based activities, or that damage to this area results in the person being unable to make decisions like in the oft cited case of Phineas Gage.

bridging an ordinary language perspective, that is, mapping the roles in ordinary language on to neuroscieintific data. I think if Sellars had been more familiar with some of the details in the phenomenological style arguments that Jean-Paul Sartre was developing, that he may have started to go the other way. In a very real sense this paper is the culmination of this conflict and the product of a debate that I should argue, should have occurred in 1956, between Wilfred Sellars, Gilbert Ryle and Jean-Paul Sartre on the nature of mind. It has been my intention to bring this debate out in full. I hope that now all the pieces have been set that this can be seen in the way the paper has unfolded[461]

.

We will now turn to the breakup of ordinary language arguments into the phenomenological and linguistic behavioural strains. This will set the scene for the remainder of the paper in the manner I have indicated above.

.

7.3 Dispositions, Phenomenology and Law-Like Statements.

Robert Wolff, in a very short paper[462] attacks Ryle's dispositional account as a piece of speculative metaphysics. Robert Wolff accuses Ryle of metaphysical postulation and hypostatization in much the same spirit as Empedocles indulges with love and strife and the positing of grand metaphysical substances. One might almost say, that Wolff's criticism, rather amusingly makes Ryle, himself, the subject of Carnap's satire of Ryle's own polemical position[463] as that of indulging in a bit of Fidoism[464].Indeed, one could almost posit the debate in the form of a satisfyingly philosophical Ouroboros which is of course our discipline's native animal, except that in Ryle's case, Wolff argues, these

---

[461] Explaining this three way conflict before this point would have been hopelessly complex and technical. Each stage must be built up and understood before the nature of the conflict becomes evident as a pivotal Twentieth Century debate over the nature of mind between three foundational thinkers on the nature of mind.

[462] Wolff, Robert. "Professor Ryle's Discussion of Agitations." *Mind* Vol. 63, no. No. 250 , (1954): Pp. 239-241.

[463]Carnap, Rudolf. *Meaning and Necessity: A Study in Semantics and Modal Logic*. London: Phoenix Books; The University of Chicago Press, 1958. See pg 216 in'Supplement A. Empiricism, Semantics and Ontology' Pp 205-248

[464]Ryle, Gilbert. "Systematically Misleading Expressions." In *Collected Essays 1929-1968*, edited by Julia Tannery, II. New York: Routledge, 2009.

metaphysical substances underlie the structure of the dispositions and not the dispositions themselves.

As such, I suppose, a Caduceus will have to do.

Wolff writes

> The attempt to explain law like statements about the physical world has often led to the postulation of some sort of "substance" or "stuff" which endured through the many alterations of the world and hence accounted for the continuity and order of those alterations. In the same way, a dispositional account of mental concepts runs the risk of hypostatizing the patterns of behaviour either as "Faculties" and "Ideas" or as Dispositions.

Here, of course, Wolff is accusing Ryle of creating ontological distinctions on the basis of ordinary language foundations. As Wolff points out even though Ryle "commits this error, he would undoubtedly repudiate it if confronted with it explicitly[465]" And indeed, Ryle, would passionately object to such an attribution. In fact, it is this very same type of argument which Ryle uses in an early paper where he develops an interesting Ordinary Language critique of Platonism. In *Systematically Misleading Expressions*[466] Ryle points out that sentences which have as their subjects non-existent entities, or as their predicates the claim that the subject does not exist, present a paradoxical problem for the philosophy of language. If the subject is a non-existent, or the predicate denies the existence of the subject, then it raises the two part question; what is the predicate referring to and what is the sentence actually about?[467]

Ryle takes this concern in a different direction.

---

[465] Wolff, *Professor Ryle's Discussion of Agitations*, 1954.

[466]See Ryle, *Systematically Misleading Expressions*, 2009.

[467]See Ryle, *Systematically Misleading Expressions*, 2009.

His concern is that 'terms couched in grammatical or syntactical terms' which are perfectly useable and understood in every day ordinary language use by the natural language speaker, become 'monsters' when philosophers begin to take them too seriously. Their 'syntactic elements' make them problematic when philosophers apply truth conditions, search for ontological hints, or begin to analyse them in ways in which the terms were never meant to be used[468].

Formally, Ryle thinks that all quasi-ontological statements are systematically misleading and end in a sort of layman's Platonism. For instance, taking a bit of natural language like ' 'honesty compels me' to mean that there is a Platonic force called honesty that literally compels someone is to fall into the illusion created from the expression and be misled by it. In a moment we will see that Ryle commits a similar error with his dispositions although it arises, not from a Platonic force, but rather a confusion between first and third person perspectives and the two types of argument he makes.

Nonetheless Ryle would most likely respond that Wolff has committed the error of reading *The Concept of Mind* too literally, and missed the point of Ryle's original project. Indeed, I would argue Ryle's case against Wolff , in the sense he has indeed met such objections, if not directly in *The Concept of Mind*, then, most directly, elsewhere.

Nevertheless there is something at the base of Wolff's criticism that I want to dig out.

Wolff writes

> One of the most interesting examples of this hypostatisation is the discussion of agitations in the chapter entitled the Emotions. An analysis of the argument will illustrate the way in which the error is committed and the care which must be exercised to avoid objectifying dispositions, tendencies and other pseudo-substantives. . . .Motives are simply the dispositions and inclinations which he has previously analysed; pride, vanity, avarice, patriotism, laziness and so forth. "Feelings are the sorts

---

[468]Ryle, *Systematically Misleading Expressions*, 2009. See Pg 44

of things people often describe as thrills, twinges, pangs". . . . Quite different to these are agitations or commotions[469].

Again

> (Ryle maintains) agitations are conflicts or interferences between two motives, or between a motive and the world (factual impediment) they are frustrations, shocks, anxieties, and distractions. . . For example if a man is patriotic and cowardly he will be torn between a desire to serve his country and a fear of being wounded or killed[470].

Wolff makes the following argument

> As soon as we speak of two motives, or inclinations as opposing and interfering with one another, we get into trouble. For "patriotic" and "cowardly" are descriptions of the man's behaviour and therefore *the description of what he would do when confronted by conflicting interests must necessarily be a part of that self-same pattern*[471].

Wolff contends thus

> Part of saying that this particular man is patriotic is saying that when offered a chance to serve his country, he does so unless there is danger involved. Likewise, to describe him as cowardly is to say that he shies away from danger, *although on*

---

[469] Wolff, *Professor Ryle's Discussion of Agitations*, 1954. Pg 240

[470] Wolff, *Professor Ryle's Discussion of Agitations*, 1954. Pg 240 brackets are mine

[471] Wolff, *Professor Ryle's Discussion of Agitations*, 1954. Pg 240 Italics are his

occasion he will risk danger for the sake of
his country.[472]

Wolff's criticism of Ryle's dispositions, in its essence, is simply that they are far too narrowly formulated, and that the way dispositions behave when they go together in ordinary language usage by the natural language speaker, Wolff thinks, is to form a general holistic description of a person's characteristic nature.

And indeed, that sounds about right.

But, given that, surely Ryle is onto something.

Surely there are cases like the following which we can relate to.

Plato, as Socrates, writes

> Well, I said, there is a story which I remember to have heard, and in which I put faith. The story is, that Leontius, the son of Aglaion, coming up one day from the Piraeus, under the north wall on the outside, observed some dead bodies lying on the ground at the place of execution. He felt a desire to see them, and also a dread and abhorrence of them; for a time he struggled and covered his eyes, but at length the desire got the better of him; and forcing them open, he ran up to the dead bodies, saying, Look, ye wretches, take your fill of the fair sight[473].

I think everyone can relate to this type of turmoil and self conflict.

Perhaps, if not in that specific context, then the more general agitation one feels when one wants to do one thing, and feels an inhibition not to or a compulsion to do something else. Notice the space we have at last moved into. In the course of formatting an Ordinary Language argument in terms of a Linguistic Behavioural analysis we have arrived in the space between a holistic descriptive analysis of dispositions as Wolff argues for in natural language usage, and a

---

[472] Wolff, *Professor Ryle's Discussion of Agitations,* 1954. Ibid
[473] Plato. *The Republic.* Translated by Desmond Lee. Victoria: Penguin, 2003. Pp 147-148

narrowed interpretation of dispositions as Ryle argues for and to which we can find our own sympathises. That is we've moved into a direct conflict between introspective scrutiny and ordinary language usage in a piece of linguistic behavioural natural language analysis. On the one side we have the Plato-Leontis-Ryle position, that is, a direct appeal made to the first personal perspective about what it is like to have competing impulses. This direct appeal to the first personal perspective carries over into an indirect appeal to the third personal perspective via an assumption that the Ryle-Plato argument makes on behalf of ordinary language when we think someone else is agitated. These assumptions make up the third person indirect. Similarly we have a direct third person appeal in the Wolffian form of a linguistic behaviourist analysis about the holistic behaviour of dispositions expressed as descriptions as exemplified by the cowardly but patriotic man who "shies away from danger, although on occasion he will risk danger for the sake of his country". This incorporates an implicit indirect appeal to the first person domain of language enabling Wolff's negation of Ryle's own position for that is in fact how such descriptions work.

The two indirect appeals, the indirect first personal appeal, and the indirect third personal appeal taken together with the direct appeal might constitute an inchoate normative source for the claimed authority of the linguistic understanding of the natural language speaker. From these one might ground an Ordinary Language claim like that which I gave in the earlier example of Ryle's critical censure of the 'volitions'. Ryle advances the claim that nobody actually uses 'volitions' in natural language descriptions and so concludes against them. He effectively uses this domain as a normative body to advance his own arguments and negate others. In the present case, however, Wolff out-Ryles Ryle. Wolff derails Ryle's attempt at developing an 'agitational calculus' based on an occult appeal to a first person perspective. Wolff does this, firstly, by pointing out that dispositions, inclinations and motives in the third person work holistically as descriptions in Ordinary Language use. Secondly, Wolff has revealed that Ryle's 'agitational calculus' is actually based on an occult appeal to a first personal perspective. The allegedly occult first personal perspective is what I have called the direct first personal perspective understood phenomenologically. The argument works by sympathy. We see someone behave in a certain way, and we introspect and apply our own recollections or memories of a prior consciousness when we found ourselves in a similar agitational state. We might see the same domain involved in the form of an appeal for the difference between, for example, a glow of warmth or pride where we cannot locate a Linguistic Behavioural distinction for such an analysis. This, of course, is the sphere from which I've distinguished theitic arguments as an occult subset of Ryle's Ordinary Language arguments.

The problem concerns how we understand dispositions. We have introspective motives on the one side and descriptions attributing them on the other. The introspective source, the theitic act, is entirely opposed to the bit of natural language analysis as Wolff's holistic linguistic behavioural argument about ordinary language motive talk shows. The two direct sources say different things, and the indirect sources that they ground, consequently, say correspondingly different things. The linguistic behavioural analysis encourages us to go one way and maps the ordinary language claim at the source in that direction. The direct third personal perspective in turn encourages an indirect first personal perspective view of dispositions as holistic entities. The introspective scrutiny that arises from a theitic act, however, which has our sympathies in the form of a direct first personal perspective, such as the case I quoted from Plato's *Republic*, encourages us to go in another direction and this in turn maps the indirect third person assumption that dispositions can conflict with each other, i.e. a man can not be both patriotic and cowardly or he'll suffer from an agitation. We sympathize with this agitation from the first personal direct point of view which gives a force to the indirect third person positional perspective and ends with ascribing conflicting motives to another.

It seems that there is a direct contradiction in what a disposition is as understood in Ordinary Language: from the directly first personal point of view motives can conflict; from the directly third personal point of view they can not.

Having now arrived at this point, I want to take it back in a certain sense. That is, though so called "Ordinary Language arguments" concerning motives as dispositions lead us to contradictory positions about them, I wish to diagnose the contradiction. The cost, however, will be to undermine the authority of Ordinary Language arguments, by revealing two underlying sources of analyses which can rival one another and in the case in point, produce contradictory claims. The contradiction between these claims seems inevitable until we realize what is going on in the phenomenology of the theitic act of reflection that we make implicitly and in the Linguistic Behavioural analysis that Wolff offers about the holistic way that dispositional descriptions fit together. If the data of the direct first personal perspective is irreducible to the data of the direct third personal perspective then that suggests we should expect the possibility for inconsistency between some claims which purportedly make their claim to authority by appeal to a shared common source and that shared common source is the normative force of ordinary language usage. Ryle seems to get away with it because of an inconsistency connected to his claim that consciousness does not exist. Ryle uses unacknowledged phenomenological arguments that rely on introspective scrutiny, but he pretends that he doesn't, hence the clash between the normative authority in Linguistic Behavioural arguments and phenomenology is not obvious. Indeed this is the fault line running through *The Concept of Mind*, that I pointed out at the start of this paper in the introduction.

What we have identified is a contradiction between a source disclosing the normative force of a phenomenological claim based on the introspective consciousness of a theitic act, and another based on analysis of a bit of natural language. The source of the tension is readily identified. Each notion of dispositional motives maps a rival source of normativity. This complicates a straight forward division of perspective into the first and third person. The contradiction is made serious by the assumption they both are aspects of the domain of knowledge possessed by the ordinary language user and this is where both claims are drawing their normative force from. The Linguistic Behaviourist analysis lodged by Wolff makes an appeal to the behaviour of language based on the direct knowledge of the third personal use of language. Likewise, the appeal to our own introspective scrutiny, in the form of a theitic act of sympathy, of recalling a moment of self conflict is also a direct appeal. The problem of the contradiction persists for as long as we think of 'ordinary language' as one unified source. This problem is solved even if the contradiction is not dissolved once we recognise that indirect knowledge of the third person is reducible to direct knowledge of the first person. Indirect experience of the first person is reducible to direct experience of the third person. Neither direct knowledge of the first person, nor direct knowledge of the third person, is reducible to the other. If we insist that they are then we do so on pain of admitting a contradiction. That is to say if we insist things are reducible by the way Ryle treats them, then Ordinary Language arguments are bad arguments because they produce contradictions.

The domain they would have shared is indirect. That is, the indirect domain posses any number of assumptions carried over from bits of analysis, as the domain we used to map onto the area of knowledge possessed by the natural language user, as a normative source. This domain, under scrutiny, has now disappeared into the first and third person direct views. These are simply rival perspectives from which to approach the nature of mind.

The impact of the argument should now become apparent. Ordinary Language usage can't be a normative source for mapping arguments on pain of admitting a contradiction from distinct sources, such as a Linguistic Behavioural analysis, or from introspective scrutiny in the form of phenomenological arguments. In short order, the domain of knowledge marked out by the 'ordinary language' user is no good as a source for justifying arguments concerning the nature of mind.

However though this argument tells us the problem with using Ordinary Language Claims as a normative source and how these rival perspectives conflict, they don't tell us whether motive dispositions actually exist. To be more precise, they factorize the rival claims made by Wolff and Plato-Ryle-Leontius into direct claims in the first and second person. It is precisely for this reason, that I differentiated two strands early on, namely the meta-philosophical concern and an argument for a theory of consciousness. The latter argument is obsfucated until we achieve greater clarity about the former.

It is to this second thread that we now turn.

7.4 Motives that are neither dispositions nor propensities.

A. R. White in *The Language of Motives*[474] makes an interesting argument against Ryle, headed by the claim that motives are a category mistake. For while White agrees with Ryle that motives are neither states, acts, or feelings, or any of the other occurrences, he thinks Ryle is clearly mistaken about them being propensities.

A. R. White argues

> In Chapter IV of the Concept of Mind Professor Ryle has argued that motives are 'propensities, not acts or states', that they are traits of character, or trends, not happenings or occurrences and that motive words are used 'to signify tendencies or propensities[475].

And that

> I shall try to show that while he is right to deny that motives are acts or states or feelings he is wrong to say they are propensities.

And the reason is

> unless I am making an enormous howler, (Ryle) is guilty of a 'category' mistake'[476].

The basis of this category mistake has to do with the sort of things that White argues that motives are.

> Motives. . . are not kinds of things which feature in our explanations of human conduct, they are kinds of explanation. To ask for the motive from which some action

---

[474]White, A. R. "The Language of Motives." *Mind* LXVII, no. 266 (1958): Pp 258 - 263.
[475]White, *The Language of Motives*, 1958. Pg 258
[476]White, *The Language of Motives*, 1958. Pg 258

was done is to ask for the reason why it was
done[477].

On first appearance it looks as though he is adopting the Peters-Ryle
position, which assumes there are dispositions which make their
presence known in episodes and which we can reveal in certain
contexts, or the Brandom-Sellars position, which assumes there is a
'space of reasons' which we enter into, and which in turn can explain,
and modify our dispositions. However, A. R. White is not arguing this at
all, rather he is arguing that motives are *not* dispositions. They are
something very different. They are a class to which reasons can apply.
On White's reading dispositions can serve as motives but they are not
motives.

> Motives are a class of reasons they are not a
> class of dispositions. Being of a certain
> disposition or having a certain propensity
> may be the reason why a man does what he
> does, may explain his conduct, may furnish
> the motives of his action, but the motive is
> not itself a disposition or a propensity.

The confusion White thinks Ryle is guilty of is between motives and
the sorts of things that can be motives, and it is similar to the basic
type of category mistake that Ryle points out at the start of *The
Concept of Mind* as trying to place both the Cricket team and the team
spirit into the same grammatical categories[478] except here in White's
description of a Category Mistake the appropriation of the properties of
the team membership, in the form of the properties we can ascribe to
the team as individuals gets mistakenly appropriated by the sorts of
properties we can predicate team members qua team members with
and is what lay at the cause of the mistake. For instance Joe, our
batsman may be good at bating, but Joe qua Joe is also very selfish.
His selfishness may clash with his role as batsman and his capacity as
a team member.

---

[477]White, *The Language of Motives*, 1958. Pg 258
[478]White, *The Language of Motives*, 1958. Ibid.

To clarify, White gives the analogy

> Members of Parliament are not as such to
> be characterized as barristers or clergymen,
> university lecturers or civil servants, over
> twenty one or minors, although the people
> who are eligible for membership of
> Parliament can have some of these
> characteristics and not others. Members of
> Parliament as such are backbenchers or
> Ministers, in the Cabinet or not, members of
> the Government or of the Opposition,
> representatives of this or that constituency.
> To examine the characteristics of members
> of Parliament, although since members of
> Parliament are people they have also the
> characteristics of people.

So while a university professor can serve as a Member of Parliament, a
university professor of anarchism qua professor of anarchism, perhaps
at the Université de Vincennes[479]for instance ,in his role as a professor
of anarchism, may find instances where his role clashes with his duties
as a Member of Parliament qua Member of Parliament. White's
argument is that motives qua motives are not the sorts of things that
can be assigned the properties belonging to things that can serve as
motives. Indeed, if White is correct the logical aspects or the properties
of things that serve as motives might clash with motives qua motives.
This would be like trying to classify soft drink cans as red and white
merely on the basis that those are the colours used by a brand of cola.

On White's view once we recognize this category mistake fear
then, for instance, is something that can be a motive, but the properties
of fear are not those of motives, since motives are a specific subclass of
reasons. Fear can explain an action or a bit of behaviour, as a motive,
but the motive itself, qua motive, is not the fear. Now, notice how this is
different to Brandom's position. The fear would serve as an internal
reason but if treated as 'fear' it would need to be from the direct first
person. Fear would only serve as an internal reason if it was articulated

---

[479] I can't imagine where else.

310

in Brandom's framework. But the articulation itself is not the fear. It is a proposition about the fear to be sure, but the 'something else' must serve in the capacity of the motive qua motive than the mere articulation. Since the fear is not a linguistic utterance, from the first person it accords more closely with Robert Wolff's treatment of dispositions as holistic descriptions, but unlike Wolff's position this isn't a descriptive dispositional account of 'fear' and 'fearful' from a third person direct view. Also note that unlike Peters' position the motive qua motive is not a disposition that makes its presence known in episodes as a linguistic manifestation of an underlying teleological drive or purposive action. This is important.

Recall now, that White stated at the start of his paper that he agreed with Ryle that denying that motives are acts, states or feelings, generally one of the occurrences and specifically one of the 'flash-bangs' was the right thing to do. Recall also that White argued that Ryle was wrong in claiming they were a propensity.

> I shall try to show that while (Ryle) is right
> to deny that motives are acts or states or
> feelings he is wrong to say they are
> propensities[480].

One could read a contradiction into this. By making motives a class that can contain acts, states or feelings as a form of explanation, he's opened himself up to a 'shallow reading'. The contradiction arises since the occurrences, say, feelings, can be motives. If feelings can be motives then he can't agree with Ryle, when Ryle denies that feelings can be motives without a contradiction. The contradiction is as follows: Ryle says feelings can not be motives, White then agrees with Ryle that feelings can not be motives. It follows that White's position is that feelings can not be motives, White then argues that feelings can serve as motives.

White could argue for the category mistake again, and reinforce that fears, states, feelings, and propensities are categorically different to motives qua motives in the way that soft drink cans qua soft drink cans are not red and white, but this particular soft drink can happens to be red and white. That is, motives, once understood as motives qua

---

[480]White, *The Language of Motives*, 1958. Ibid.

motives, are the sorts of things that can be explained by feelings, acts and propensities, the same way they explain inclinations, but that motives are not these things and I think this is the correct reading. That is, I take it that his argument, once we remove the possibility of a contradiction from a 'shallow reading' is the sophisticated position that motives are not 'flash bangs' and they are not reasons, but reasons and 'flash-bangs' might serve as motives.

This would explain his position that reasons qua reasons are not motives

> (m)any things can be reasons, perhaps only
> a few can be motives, but reasons have not
> the logical characteristics of the things
> which can be reasons nor motives of the
> things that can be motives[481]

This is to say that reasons might serve as motives but they are not motives qua motives. There is room for a clash between the logical characteristics of motives and the logical characteristics of reasons. Hence some reasons would not be motives. External reasons, for an example, may not serve as motives since it could be argued that the person is not aware or unaffected by them. External reasons might be reasons for the psychologist but they are not reasons for Bob Hearn[482].

However he leaves himself open, since, now, this raises the question, what individuates the motives themselves? Unless he can individuate the types of things that can be motives, from motives, he faces the problem that his category mistake can't uphold its own distinction, and the class of motives will just collapse into the objects that can serve as motives. This is the point of the specific type of claim he makes against Ryle's category mistake. That is, if his claim is that motives qua motives are not reasons qua reasons because motives do not have the logical characteristics of reasons then he needs a theory about what those logical characteristics are.

There is, of course, evidence laying around in White's paper that he is not unaware of this and has tried to manoeuvre his way out of the

---

[481]White, *The Language of Motives,* 1958. Pg 259
[482]The distinction taken from Norman Mailer's Anti-war novel, *The Naked and the Dead,* 2006 earlier in this paper.

former horn of the dilemma by rather, unconvincingly, trying to claim that motives have as their properties 'life traits'. However, the more interesting prospect is to reaffirm the claim he makes for a category mistake as a type of distinction about the sorts of things that can be motives, and I think this is not only the correct reading, but it is the correct course to take. But it opens up questions as to what sorts of things can serve as motives, what sort of things are motives and how do we individuate the sorts of things that can fit into explanations of motives? What is it about the chill of fear that makes it part of the picture that allows us to explain motives? To this, and thus over from propensities and onto the occurrences, the other side of Ryle's distinction, which I begun this chapter with, I will now turn.

Chapter Eight: The Occurrences

8.1 Occurrences; states, acts and feelings.

If you recall, back to a previous chapter on methodology[483] I isolated one of Ryle's bits of terminology, which he labelled 'feelings proper' and I applied the handle of 'flash-bangs' so as not to lose them in the various discussions and debates. These of course, in the section immediately prior, are the sorts of things, it turns out, which can be explained as motives. One of the immediate consequences of accepting this point was that Ryle's terminology began to break down. What A. R. White effectively did, if we follow the horn of the dilemma that doesn't end in a contradiction, was to create a new class called motives qua motives, to which fears, inclinations, feelings proper or flash-bangs, states and acts can belong to. But we also saw that White did not have a theory that could give us the logical characteristics for motives, only that some things cannot serve as them.

It follows from this reading that White thinks that dispositions can serve as motives but they aren't motives. Moreover to come up with a theory of motives to give motives logical characteristics we can't use 'dispositions', because 'dispositions' as Ryle used the term was broken down in the chapter prior to that with Wolff's Linguistic Behaviourism. Dispositions in Ryle's original sense arose from two types of claim: the direct first personal perspective which used theitic acts, which contributed phenomenological information in the form of assumptions to the third person indirect perspective with hidden and occult arguments like an appeal to sympathy; and the direct third person perspective which used linguistic behaviourist descriptions which when properly analysed are holistic descriptions of what a person is like. Thus Ryle's dispositions were actually a confusion of two different types

---

[483]See '3.5 Linguistic Behaviorist Claims, the Occult Stream of Consciousness, Propensities, Occurences and the Agitational Calculus.' In ' Chapter Three : Towards a Methodology'. There is a footnote in this chapter which outlays a map of the arguments for 'flash-bangs' and how that strain weaves through the various stages of the argument in this paper.

of argument taken from two different points of view but presented as an authority in one unified source.

Prior to that we saw that Brandom's game of giving and asking for reasons broke down into 'wants' and 'needs' that could not be justified in the game of giving and asking for reasons and referred us, once we had made the internal and external distinctions, to non-linguistic sources of knowledge. One must arrive with non-linguistic knowledge at the game in order to play it. That is, while someone might argue that the actions someone undertakes, might, with some difficulty, be represented in the game once we have distinguished the game of asking and giving reasons from the context of thinking, and certain other artificial constrains then the 'motives' as 'wants' and 'desires' could not be set up or justified in the game itself.

As such an appeal to internal reasons might give White, or a position like his, a theory about the logical characteristics of motives but this would be an incomplete theory. The reason why it would be incomplete is that internal reasons, as we saw in the section on Brandom, point us towards non-linguistic forms of knowledge. For while the woman in Case 4 who instigates fights at the bar between two males and her 'need' for reassurance from which she makes reasoned moves and might even be articulated and represented in the game in the way Brandom suggests, the need itself is not justifiable within the game, subject to the conditions for internal reasons. Her 'need' points to something non-linguistic. Motives qua motives are internal reasons, motives qua things that can serve as motives are something else. What White has picked up on here are some of the logical structural aspects of consciousness. But he is missing an important piece of the puzzle.

Now that we have seen how the propensities systematically break down under analysis, let us move to the other side of Ryle's original frame work and take a look at the occurrences. If you recall [484] the occurrences contained 'feelings proper' or 'flash-bangs which are still 'at large'. These are things like fear, anger, joy, sadness, hankerings, pride, that come in flashes, glows, thrills and pangs but we are short of an explanation as to what these things are. We don't have a criterion that can individuate them.

What exactly is the thrill of anticipation?

---

[484] Earlier in the 'game of giving and asking for reasons' and the problem of language utterances.

What are those chills of fear?
What exactly is that shock of horror?
What are those bursts of anger?
What is that glow of warmth?

And in fact, it is this last sort of flash-bang, or rather, pseudo-flash bang that caused the original problem, because Ryle claimed there was a linguistic difference between the glow of warmth, and a glow of pride. That chill of fear and that chill of cold. The shock of horror and the shock of amazement. What exactly is the 'shock' part and how does it differ in each case? Why is a shock of horror different to a shock of amazement? If they are not different, save for the horror and the amazement part, then, of course, what does the 'shock' refer to?

The answer I'm going to be selling in this part of the paper is that 'flash-bangs' like Brandom's reasons, refer us to non-linguistic knowledge. This is knowledge that we need in order to differentiate between a glow of pride and a glow of warmth. We thus need a theory of consciousness and the reason why is that many emotions have an object as their content, that is they are 'about something'. We can't differentiate between a 'glow of pride' and a 'glow of warmth' on a specifically linguistic basis because a 'glow of pride' involves apprehension of an object, the object of pride. The glow is a glow about something the person is proud of. This is what White was picking up on. Linguistic explanations contain gaps. The linguistic behaviours do not quite match what we know about the things they represent. This is because the linguistic behaviours belong to the ways we represent and communicate. Their meaning depends upon a pre-linguistic domain of knowledge they take for granted. We need a different type of argument for looking at and comparing non-linguistic knowledge.

That is to say we need an auto-phenomenological argument. By auto-phenomenological I mean specifically, an argument that trades on the authority of phenomenological style insight over another source to individuate some element that can explain something about the nature of mind. In this sense I'm going to move beyond the two strains in Ryle, and argue against both the implicitly of a Dummett-Ryle position on language and advance an auto-phenomenological strain of psychologism against physiological and physiological style arguments as well. That is, I'm going to be arguing that individuation on the basis of an appeal to 'psychic phenomena' or insight into emotional 'qualia' is a stronger argument than an appeal to physiology or linguistic behaviourism.

If we take an overt physiological position we can see very quickly the problems that develop from a sort of naive distinction between the 'psychic' or 'qualia' and the purely physical symptoms of an emotion. This, in fact, is what Sartre does as one of his opening arguments in *The Sketch for a Theory of Emotions*.

Sartre writes

> The physiological modifications which correspond to anger differ only by their intensity from those that accompany joy (such as) somewhat quicker respiratory rhythm, slight augmentation of muscular tone, increase of biochemical exchanges. . . For all that, anger is not a greater intensity of joy, it is something else, at least as it presents itself to consciousness. . . The idiot who has become angry, is not ultra joyful[485].

Sartre's argument is that anger is irreducible to joy on the basis of the physiological symptoms they have in common because anger and joy present themselves as totally different sorts of things in consciousness even though they may have similar or even, one might suppose, identical physiological manifestations[486].

Prima facie, this should seem the right position at least on an intuitive level. We often find it very hard to tell the way someone feels about something from their visual physiological symptoms. Sometimes it is easy, like when the person is yelling, or crying, but sometimes it can be very hard. People are often described as smiling on the outside and crying on the inside.

Let us look at a similar claim that trades on the authority of an implicit ordinary language linguistic behaviourist position. In *The Concept of Mind* Ryle claims that the difference between frowning

---

[485]Sartre, *Sketch for a Theory of the Emotions*, 2002. Pp 15 -56

[486] See Stanley Schacter and Jerome Singer experiments of the 1960s for a good example of this. Subjects were injected with identical levels of epinephrine, subjects interpreted the effects either as anger or euphoria depending on the types of situations they were in and the way they apprehended them. There's a good discussion of this early on in Sousa, *Emotion*, 2013.

intentionally and frowning *unintentionally* is in the 'inheritance conditions' of an act. The difference occurs at the level of the adverb 'intentionally' and 'unintentionally' although the two frowns may be photographically and physiologically the same.

Ryle writes

> In particular, it is not to bring about a frown on one's forehead by first bringing about a frown-causing exertion of some occult non-muscle. 'He frowned intentionally' does not report the occurrence of two episodes. It reports the occurrence of one episode, but one of a very different character from that reported by 'he frowned involuntarily', though the frowns might be photographically as similar as you please[487].

Both arguments, Ryle's and Sartre's present two cases which are identical, and both differentiate between the two cases they present on the basis of a third aspect. In Ryle's argument it is the 'adverbials' or 'inheritance condition' that differentiate between two types of smile. In Sartre's argument it is an appeal to introspective scrutiny on some aspect of emotional qualia that can be discerned by consciousness. Ryle's normative source is the behaviour of linguistic expression, found in the difference between the adverbial use of 'intentional' and 'unintentional'. In Sartre's argument it is an appeal to reflective consciousness to differentiate between anger and joy.

Now there is a question here how Ryle might maintain such a distinction given that 'the frowns might be photographically as similar as you please'. Ryle's position I argue simply can not bear up under scrutiny. It is either an argument from the first personal perspective or the third personal perspective. The term 'intentional' is either given as a third person description or it is a first person linguistic codification for something going on at a non-linguistic level.

If 'intentional' is taken from the third personal view, following on from the arguments we looked at in the section on Wolff, then the adverb is descriptive. If the two smiles are identical then we can not

---

[487]Ryle, *Concept of Mind,* 1983. Pg 72

differentiate between them on the basis of physiological symptoms. Thus a third personal perspective cannot differentiate between the smiles from a third person perspective. The only way Ryle can make that differentiation is from a first person perspective. This is a naive position that becomes more complex when we introduce neurophysiological information. Ryle's position doesn't have the complexity to deal with this development, but we will look at Sellars, who has come up with a way of dealing with this complexity, in the next section.

If it is an argument from the first personal perspective then I argue it is a theitic argument, that is, the person knows it is an 'intentional' smile and that describing it as 'intentional' is a linguistic codification of something that happens on a 'non-linguistic' level that is only open to them from a first person point of view. That is, the person's linguistic knowledge that the smile is 'intentional' depends on non-linguistic elements.

So to be clear, in Sartre's case, what Sartre is offering here is an argument that has two parts. One involves a justification or an appeal much like our 'thetic' act of 'introspective scrutiny'. This introspective scrutiny is what one might call the 'method' of the argument, that is what it appeals to for justification as a normative source and the second is the content which is what the argument seeks to establish. In this case the argument's content is that anger is not joy. Its method is an appeal to introspection about what anger and joy feel like.

Now note that this is a different sort of argument to whether the presences of physiological symptoms caused the emotional state, or whether the emotional state caused the physiological symptoms. The latter is an argument about the causal order and takes it for granted that we need a theory of consciousness in our account. I will address the problem of the causal relation between physiology and consciousness in the problem of the "Roid Rager In The Void'. The former argument, the one above by Sartre, is an argument about the identity of the emotional states in consciousness. The causal argument and the identity argument are two different arguments. That is to say, they argue for two different things. One argues on the causal order of emotions and physiological symptoms, the other argues about the identity of emotions to consciousness on the basis of some appeal. The latter, as I have pointed out, can be made on the basis of a normative source, citing, firstly Sartre's appeal to consciousness in the form of introspective scrutiny that anger is not 'ultra-joy'. Secondly, we have Ryle's 'linguistic behaviourist argument' which is either a third person direct appeal on the behaviour of a bit of natural language which Ryle thinks rests upon the intuitional semantics in the difference between the adverbial description 'intentional' and its negation, 'unintentional' or as I've argued above can be broken down into a first personal perspective in which case it is a thetic argument. If it is from a third personal perspective then physiology is not enough to establish whether it is intentional or unintentional. This in turn shows why autophenomenological arguments are stronger than linguistic behavioural arguments, since linguistic behavioural arguments either break down into third person descriptions or first person linguistic codification. Next I am going to show why the theitic argument has more normative force than the linguistic behaviourist argument applied to neurophysiologistic grounds.

8.2 The Normative Appeal of an Argument.

Recent work done in neuro-physiology reveals interesting insights into the behaviour of fear. Ledoux argues from the body of his neurological research that there is a direct route and indirect route of neural behaviour in a fear response [488]. From what we know, the first, the direct response, occurs because the amygdala has laid down a fear triggered response associated with an event, situation or pattern of recognition, that was initiated during an episode of past traumatic stress and will be triggered when the thalamus receives a similar stimulus. The amygadala sits in the temporal lobe where it sifts through information passed through and encoded by the thalamus[489]. The pattern of this direct route has come to be associated with Post Traumatic Stress Syndrome[490].

In contrast to the direct route, the indirect route, being the route that a pattern of neural behaviour might take as a reaction to a fear response, passes back through the somatosensory cortex and into the

[488]Carter, *Mapping the Mind*, 2010. Pg 96
[489]Carter, *Mapping the Mind*, 2010. Pg 90
[490]Carter, *Mapping the Mind*, 2010. Pg 82

frontal cortex where it is experienced as an emotion, after being hormonally encoded by the hippocampus[491] and sent into the body.

Thus, according to a consensus recently achieved within the field, the difference between the indirect and direct route that a fear response might take rests upon whether the stimulus, after being received by the thalamus, is recognized by the pattern laid down in association with a past event by the amygdale. If this is the case then it 'short circuits' the hormonal encoding by the hippocampus and skips past sending a response through the body, and back up via the somatosensory cortex into the frontal cortex. In short order the difference between the two fear responses is that while the indirect route passes through the somatosensory cortex on its way to the frontal cortex, the direct route doesn't.

But how might we differentiate between the two forms of fear? What sort of an appeal might we use to claim that the two are in fact different types of fear and not the same? One might, on the one hand, make a claim on the basis of some distinction in language, that is, it makes sense to talk about fear in this way which it also makes sense to talk about the symptoms of the indirect route, while talking about fear in this other way, corresponds more closely with the set of linguistic behaviours in use when we talk about terror. 'Terror' might be the type of fear we associate with Post Traumatic Stress syndrome. One might likewise make an argument that the two types of fear, individuated by neuroscience, don't have anything to distinguish them, in the domain of a direct third person appeal at the level of a linguistic behaviourist argument. There is nothing in the way people speak about or use terms in relation to fear that could uphold a distinction between the direct and indirect route that a fear response might take. That is, we dismiss the appeal of introspective scrutiny, in order to lodge an appeal on the basis of linguistic data.

This is more or less what Sellars does[492]. Except that Sellars would favour the neurophysiological information over the linguistic roles in the ordinary language position if the two should disagree.

In *Philosophy and the Scientific Image of Man*, Wilfrid Sellars thinks he can solve the problem of higher brain states and the problem of introspection by ignoring the intrinsic properties of thoughts, and

---

[491]Carter, *Mapping the Mind*, 2010. Pg 90
[492]Sellars, *Philosophy and the Scientific Image of Man*, 1991

just focusing on neurological data and language by 'mapping concepts on to roles'. The problem of introspection originally arises from Sellars concern over a given reading of Descartes[493] where the person is aware of themselves, as being aware of a cognitive role, like believing, choosing and wondering. In the same way, that the perceptible qualities of the manifest image are challenged by radical advances in science and the development of a Scientific Image, so too, the intrinsic problem of what thoughts are like, come to be challenged by neurological data.

Sellars writes

> But to this proposal the obvious objection occurs, that just as the claim that 'physical objects are complexes of imperceptible particles' left us with the problem of accounting for the status of the perceptible qualities of manifest objects, so the claim that 'thoughts, etc., are complex neurophysiological processes' leaves us with the problems of accounting for the status of the *introspectable qualities* of thoughts. And it would seem obvious that there is a vicious regress in the claim that these qualities exist in introspective awareness of the thoughts which seem to have them, but not in the thoughts themselves. For, the argument would run, surely introspection is itself a form of thinking.[494]

There is a viscous regress here and Sellars is entirely aware of the problem. That regress of introspective qualities, that is, the introspective qualities that introspective thoughts have, is problematic, because it can change the nature of what we are thinking about. In a sense, this will be my final position in the paper. I will argue that

---

[493]Sellars, *Philosophy and the Scientific Image of Man*, 1991. Pg 30-34
[494]Sellars, *Philosophy and the Scientific Image of Man*, 1991. Pg 31

something happens to some forms of thought when we introspect and talk about them. The problem that precedes this is the same that bothered Dummett. How do we talk about and represent what thoughts are *like?* This type of problem about what is communicable is similar to Ryle's point about psychologists and what lead him to an ordinary language position on the mind[495]. This problem becomes more complex and acute when we introduce neurophysiological information. Higher cognitive brain states, like, for instance, the body of consensus we've developed out of research and work done with fMIR since it became available in the early nineties[496] are really just blotches of colour on a screen that display blood flow which neurologists take to be indicative of brain activity and a number of other similar techniques[497].

In Sellars day, he was probably thinking in terms of pencil graph printings on log paper, and hieroglyphic style print outs. Indeed, our own position isn't that radically different, we can see, for instance, the different bits of the brain light up on a computer monitor from readings given by the machine, but we can't know what it is like for the person to experience those bits lighting up unless we apply analogical structures. Similarly, spikes, graphs, pie graphs or cartoon sketches won't tell us this either.

For his purposes in *Philosophy and the Scientific Image of Man* treating thoughts as analogous to verbal utterances allows Sellars to

---

[495]Ryle, *Concept of Mind,* 1983. Pg 52

[496]Carter, *Mapping the Mind*, 2010. Pg 26

[497]See David Chalmers, Chalmers. *The Conscious Mind.* 1996 pp 284 – 287, Chalmers, *How Can We Construct a Science of Consciousness*, 2004. Andy Clark, David J. Chalmers. "The Extended Mind." In *The Extended Mind*, edited by Richard Menary, 27 - 41. Massachusetts: The M.I.T. Press, 2010. NB. There is a deep problem arising from the Extended Mind Hypothesis and Chalmers project of building or establishing a bridging principle related to the above. Since cognition in terms of the Extended Mind hypothesis is established on the parity principle as part of a functionalist criteria then there are questions about what happens to the phenomenal or conscious aspects of mind when we extend the bounds of cognition explicitly beyond an implicit assumption, that is, where do we draw the line in terms of cognition and consciousness? If every time I tap a hammer on a pipe in the next room you experience a headache then does this make your mind and the pipe part of the cognitive process on some elemental basis of the extended mind hypothesis? What about if the pipe is in the next street, or New York? There is a genuine problem here with Chalmers' program for (non)reductive functionalism and the extended mind hypothesis that revolves around the bounds of cognition, a bridging principle between consciousness and physical systems and the parity principle, but I'm not entirely sure how to bring this out other than suggesting that the core of the problem is, perhaps, in the notion of the bridging principle itself, and a clash that arises with extended cognition on the basis of the parity principle, phenomenal aspects of consciousness and problematic causal principles that start to develop once we begin to individuate cognition on the basis of effects.

side-step the problem of introspection and the need for a theory of consciousness. The benefit of this step is that he can simply map the roles of language onto the neurological data, when it becomes available, and thus produce a telescoping of man in psychology and the behavioural sciences. This eventually leads to a' rival image' which produces the trilemma at the end of that paper[498]. But while this works for 'higher' cognitive states Sellars realizes he's going to run into a problem with sensations. This problem with sensations is part of what leads to the 'faculty-dualism' of *Empiricism and the Philosophy of Mind* between sensations, which are represented in the 'residue' of a descriptive statement in looks talk, and the fact stating roles which are affirmed or denied in the subjunctive conditional structure of 'looks talk' against a background of social and standard conditions. Brandom, of course, characterizes this distinction as 'Sentience' and 'Sapience'.

Now my argument, to be clear, is that he's developed a far more sophisticated version of the illusion in Ryle that treats language as thinking, but is missing the essential element inside of thinking that consciousness would give. But Sellars is not oblivious to this point. He's being highly pragmatic about it.

Sellars writes

> The point is an important one, for if the concept of a thought is the concept of an inner state analogous to speech, this leaves open the possibility that the inner state conceived in terms of this analogy is *in its qualitative character* a neurophysiological process[499].

He can thus, get around this problem of introspection by ignoring the intrinsic qualities of thoughts except as neurological processes, and focus on the patterns, and their sentence-like structures. Indeed by

---

[498] Sellars, *Philosophy and the Scientific Image of Man*, 1991

[499] Sellars, *Philosophy and the Scientific Image of Man*, 1991. Pg 34

treating thoughts as analogous to speech, more specifically sentences, or sentence like structures, he can avoid other problems like the causal problem of William James, which I look at in the next sub-section which trades on the strength of the normative power in phenomenological arguments at the end of the section just prior[500]. The benefit of Sellars' project is that he can map the neurological data on to the roles identified in the language without raising the dilemma about the causal chain of impact on the individuation of the emotions by either objects and events or neurological agents. He can do this because he can avoid a theory of consciousness.

Indeed, this move altogether dispenses with the introspective properties of thoughts, and the problem of individuating them, and it is where the lingua-sensory faculty-dualism in *Empiricism and the Philosophy of Mind* properly originates as we discussed earlier in the paper[501].

Sellars writes

> Thus our concept of 'what thoughts are' might . . . be abstract in the sense that it does not concern itself with the *intrinsic* character of thoughts, *save as items which can occur in patterns of relationships which are analogous to the way in which sentences are related to one another and* to the contexts in which they are used[502].

Brandom, in his commentary to *Empiricism and the Philosophy of Mind* represents this faculty-dualism with four levels with the bottom containing sentence structures linked by inferences[503] what Sellars calls in his other papers concepts[504] the top containing sensations, while the

---

[500]See '8.1 Occurrences; Moods, States, Acts and Feelings' in this paper.
[501]See 'Chapter Seven : Motives, Moods and Reasons' in this paper.
[502]Sellars, *Philosophy and the Scientific Image of Man*, 1991.. Pg 34 Italics are his.
[503]Brandom, Sellars, *Empiricism & the Philosophy of Mind,* 1997. Pg 126
[504]Sellars, *Philosophy and the Scientific Image of Man*, 1991. Pp 8-9 Sellars defines 'conceptual'

middle houses the point in the argument where theoretical entities become visible via the machinery of the residuum and dispositions for the space of reasons to open up. This is the machinery of the Sentience and Sapience distinction and what ultimately 'plugs' discourse into the space of reasons[505].

On the other hand one might make an appeal to an introspective or theitic act, as Sartre does in his 'Anger is not ultrajoy' argument to differentiate between two neurological states[506]. Take the case of scepticism or disbelief which triggers a reaction from the insula cortex. The insula cortex is involved in one of those higher cognitive states Descartes thought could not be found to have a corresponding state in the neurology like wishing, choosing or believing[507]. The insula cortex is an area that sits on the in-fold between the frontal and temporal lobes. Through research activity in this area has also been shown to be also associated with people's reactions to wholesomeness and disgust[508]. This suggests to neuroscientists, by neural proximity and the similarity it displays in areas and types of activity, that disbelief, scepticism and the feeling one is being lied to are neurologically similar to the feeling of disgust[509].

Now also note this last claim, and the similarity it shares with Sartre's original argument, that is, let us assume as we did to simplify the argument, that anger and joy have the same physiological symptoms, namely 'quicker respiratory rhythm, slight augmentation of muscular tone, increase of biochemical exchanges' but that the two, anger and joy, may still be differentiated on the basis of a theitic argument, that is, by an appeal to introspective scrutiny. That is on a par with a neuroscientific claim that if two symptoms of neural activity,

---

thinking as that belonging to a framework in which thoughts can be measured in terms of correctness, relevance and evidence. This is, of course, a more sophisticated version of Carnap's framework. In Sellars, *Truth and Correspondence*, 1991, he develops the concept-framework structure out of a critique of Carnap, see pg 203, where he defines 'concepts' as distinguishable by relations of inference between them. Not all language roles are conceptual. For example 'helas' and 'alas' do not express concepts pg 203.

[505]Indeed it is this very complexity in Wilfrid Sellars thinking which is what separates his treatment from earlier positivist strains, and thus gives him a more sophisticated framework, which in turn allows him to critique Carnap in the papers published between and around *Empiricism and the Philosophy of Mind* See Sellars, *Philosophy and the Scientific Image of Man*, 1991.

[506]Sartre, *Sketch for a Theory of the Emotions*, 2004. Pg 23

[507]Sellars, *Philosophy and the Scientific Image of Man*, 1991.

[508]Carter, *Mapping the Mind*, 2010 Pg 170 – 172

[509]Carter, *Mapping the Mind*, 2010. Pg 171

that is, say, the symptoms, made up from the neurochemistry and activity in a specific part of the brain is identified in two different claims, one for whether the subject believes a person is being deceitful, and the other for whether something is disgusting or unwholesome, are both identical, it is then, still, left open to a third claim, either introspective scrutiny or a linguistic behaviourist argument, whether the two are in fact the same thing. That is, there is a third source which can differentiate between one's disbelief, and one's disgust, and whether the two can be co-identified, even if, the neuroscience gives us the same 'symptoms'.

Now, of course, this sketch undermines some of the complexities of the limbic system, and neuro-chemical structure of the brain for the purposes of brevity. It may well turn out that joy can be associated with the stimulation of dopamine in the temporal system, triggering a response in the ventromedial cortex which has been labelled the brain's 'reward circuit' or the medial forebrain bundle, which, of course, is made up of the nucleus accumbens, ventral tegemental area to the septum, amygadala and prefrontal cortext. But that claim[510] once the neuroscience has had its say has to be made on the basis of a normative source and the two on offer for the purposes of this paper are linguistic behaviours available from the direct third person, and the direct first person with indirect claims in each, comprising the area we might refer back to as the old function of a normative argument but made in the domain of the ordinary language user. We abandoned this indirect domain on pain of admitting a contradiction. What we have left to individuate neurophysiologistic information by is the behaviour of linguistic utterances, and structures and appeals to phenomenological style arguments.

My argument is simply that the third person direct claim is a weaker claim than the first person direct claim. The reason why is that an argument that utilizes an appeal to the first person direct claim can access areas of introspection that contain non-linguistic knowledge. That is, it is not hindered by the need to encode information in the structures of a communicable medium. In fact, communicating insights and information from this source can be quite hard. There is often no easy or generic way to communicate these insights. To be sure we can sometimes communicate them, but the linguistic structures we use call

[510] See Carter, *Mapping the Mind*, 2010. Pp 56-57, 64-66.

upon something else not available at the level of linguistic behavioural descriptions like Ryle attempts to use. This is like the 'Witness/Reader argument' where we need complex exposition involving figurative devices, comparative analogies, stories and comparisons to convey the phenomenological insight. In other ways there are times when we can see the linguistic side and the phenomenological side together. This is like the 'Remember-when, Remember-how' configuration. Sometimes we can grasp the insight in linguistic behavioural analysis of the forms first and then think about it without words or linguistic structures to see what is not obvious in the linguistic behavior. This is like the argument where Ryle asks us to 'imagine lightening so bright it hurts your eyes' or trying to anticipate your next thought before one has it. Thus the first person direct claim does two things. First it advocates a theory of consciousness to take an account of these elements that point to a domain above linguistic behavioural arguments or linguistic knowledge but it does so on condition of a theory of consciousness and introspection by implication. That is, we need a theory of consciousness to account for the level where linguistic behavioural arguments break down like we saw with the 'flash-bangs' and like I pointed out above with the phenomenological insight style arguments. Secondly, it undermines the authority of a third person direct claim, namely, a Linguistic Behaviourist argument, because it does something that a third person direct claim does not do, it enables us to grasp this introspective insight directly from our own perspective.

8.3 Notes From Sartre Towards A Theory of Consciousness

As I have been at pains to point out the problem with a Linguistic Behavioural analysis of the emotions is that it either assumes non-linguistic knowledge in the audience and thus contains occult strains of phenomenology, or it commits a category mistake by treating language structures and behaviours as though they were the emotions themselves. The final insight will be to show how talk about the emotions is not the same thing as the emotions. Rather it is a linguistic codification for something happening at a non-linguistic level. To lead up to this insight, now, I want to draw out some notes from Sartre's theory of consciousness. Specifically, I want to draw attention to what lay at the base of the linguistic behavorist category mistake. In talking about a 'flash of anger' or a 'glow of pride' he falls subject to an illusion caused by categorization of the emotion into its linguistic structure. That is, he misses structurally, what is important about the 'flash of anger' which is that it is a 'flash of anger' about something. Thus in order to talk about a 'flash of anger' as a 'flash of anger' qua flash of anger from or about something, he needs at least some structural elements that directly implicate a theory of consciousness in an account of the emotions. As such I'm going to import these structural elements from Sartre's model to start building an analytic model[511]. This will lead up to the final insight on offer in this paper.

Sartre points out

> A few simple observations will suffice, and it is remarkable that the psychologists of emotion have never thought of making them. It is obvious indeed that the man who is frightened, is frightened of something. Even if it is a case of one of those indefinite anxieties that one feels in the dark, in a sinister and deserted alley, etc, it is still of certain aspects of the night, or of the world that one is afraid[512].

Sartre's argument, first of all, to be clear, is that an emotion, like fear, like joy, like anger, is always of, or about something. Fear is a fear of something[513]. This occurs at the unreflective level in the moment. In the same way that we don't see ourselves looking out the window of the bus

---

[511]David Chalmers is perhaps the best place to begin with this. My apprehension is that while *The Conscious Mind* is a brilliant piece of analyses and first class philosophy, and deserves a place in the philosophical classics of last century, it is ambiguous about language. I sympathize with Chalmers point in *How Can We Construct a Science of Consciousness*, 2004, on the difficulty of training subjects for exploration of introspective analysis for experimental purposes, and sometimes it takes a rare or talented mind to grasp a particular insight or discover it, but I argue, this position isn't that different to mathematics, as well as Jewish and Eastern forms of meditation. There is however traces of an introspective tradition in *The Conscious Mind* that could be developed into an 'Analytic Phenomenology' along similar lines to some of the insights in this paper. See Pp 175-76 for the three orders of judgment, and pp 236-233 for his point on registrations.

[512]Sartre, *Sketch for a Theory of the Emotions*, 2004. Pp 34-35

[513]See Sousa, *Emotions*, 2003, Section 3. 'Emotions and Intentional objects', de Sousa actually picks up on this point.

at the woman in the car, or hear ourselves hearing a piece of music, most emotions are immersed in the world at the unreflective level. That is, of course, they are positional towards an object. Anger is about something. Irritation is from something. Emotions follow the same patterns as consciousness, that is, they occur in direct relation to an object or situation on which consciousness can fixate.

It is only later when we reflect on our anger or irritation in a theitic act, that we are taken in by this illusive aspect of perceiving emotional reactions as states, that is, we think of our prior selves, as being in an angry 'state'.

Sartre writes

> For the majority of psychologists everything happens as though the consciousness of emotions were primarily a reflective consciousness, that is, as if the primary form of emotion, as a fact of consciousness, were its appearance to us as a modification of our psychic being, - or to use ordinary language, its being grasped first of all, as a state of mind. And certainly, it is always possible to become aware of emotion as a fact of consciousness, as when we say; I am angry, I am afraid, etc. But the fear does not begin as consciousness of being afraid, any more than the perception of this book is of perceiving it. The emotional consciousness is at first non-reflective, and upon that plane it cannot be consciousness itself, except in the non-positional mode[514].

---

[514]Sartre, *Sketch for a Theory of the Emotions,* 2004. Pg 34 by non-positional, here, Sartre means consciousness is non-positional towards itself. When it becomes positional, it becomes a theitic act. Sartre thinks Descartes was wrong in the 'cogito'. To get the 'cogito' to actually work in positing existence as a phenomenological property, one needs to reflect. The mistake was to think that the theitic act can happen in the moment in order to perform a cogito. Thinking that the theitic act happens in the moment leads to the mistaken view of thinking the cogito can work as a present act of thought grasping itself simultaneously. There are rare states of anxiety where this occurs and it is these Sartre originally develops his model of consciousness from Psychasthenia for. These are like Kierkegaard's fear of his own ability to sin, or the gambler who vowed to stop gambling, when he comes near the gambling tables, or the woman who each time her husband leaves their Parisian

The non-self-positional mode, or unreflective level of consciousness which is where emotions start, when reflected on or brought before consciousness in a theitic act becomes the 'me' of the emotions and this forms the other side of the Egoto which apperception of the I forms the initial part. He comes to this side of the emotions, the 'me', in a refutation of the French Moralists and Psychologists who were looking for a material presence to explain actions, and in particular Rochefoucauld[515]. The me is what happens later when we think back to how something made us feel. But *in the moment* the emotion arises as a way of apprehending the world.

Sartre writes in *The Transcendence of the Ego.*

> But we must, before we go any further, rid ourselves of a purely psychological theory that affirms, for psychological reasons, the material presence of the me in all our consciousnesses. This is the theory of the amour-propre put forward by the French moralists. In their view, the love of self – and consequently the me – is hidden in all feelings, in a thousand different disguises. In a very general way, the me, by virtue of this love that it bears to itself, is seen as desiring for itself all the objects that it desires. The essential structure of each of my acts would then be a reference to myself[516]
.

---

apartment she will run to the window and start hailing men like a prostitute. See ;4.2.2 Freedom, consciousness and Psychasthenia' in this paper. These are anxious moments of flight from the sudden realization of freedom and not existential affirmation of existence. Rather than 'I think therefore I am', Sartre thinks the Cogito should be reflective of a past consciousness, such as 'I think therefore I was'. The past consciousness comes before the present one as the object of consciousness to prove to consciousness that it existed in a prior state. Only when a prior consciousness is brought before a present one does the ego of the cogito become visible to consciousness.

[515]Sartre, Transcendence of the Ego, 2004. Pg 17 - 21

[516]Sartre, Transcendence of the Ego, 2004. Pg 17. See also Solomon, Robert. "Emotions in Phenomenology and Existentialism." In *A Companion to Phenomenology and Existentialism*, edited

This, is rather like the common stock in trade '101 Introductory Ethics Course' where they trot out the Ayn Rand expositions, and introduce the meta-ethical positions on the Objectivist argument, which runs, more or less, towards a claim that all actions, even seemingly benevolent, voluntarily altruistic and unselfish ones are themselves derived on some secret or overt selfish basis[517]. The typical claim is, of course, that the person who does a good deed, does so because they want the feeling of having done a good deed. On this view the good deed doer has a bit of a hidden endeavour behind the seemingly selfless actions. The French Moralists, on this view, were looking at the 'me' as a material force, which moves the person to act, in that they can only think of themselves at the kernel of their actions and this is a compelling force[518]. Sartre, however, doesn't think it quite happens like that in the real world. On the unreflective level, there is no 'me' to consult or consider, because consciousness has not taken itself as an object. There is no reference back to the self as a 'me'.

Sartre writes

> I feel pity for Peter and I come to his aid. For my consciousness, one thing alone exists at that moment: Peter-having-to-be-aided. This quality of 'having-to-be-aided' is to be found in Peter. It acts on me like a force. Aristotle had already said as much: it is the desirable that moves the desirer[519].

---

by Hubert L. Dreyfus and Mark A. Wrathall.
http://onlinelibrary.wiley.com/doi/10.1002/9780470996508.ch21/: Wiley Online, 2007.

[517] I.e. see, for instance, Rachels, James. *The Moral Elements of Moral Philosophy*. 5th ed. Boston: McGraw Hill, 2007. Pp 70-71

[518]Rand, Ayn. *The Virtues of Selfishness*. New York: New American Library, 2007. Pp 24-32. See also Branden, Nathaniel. "5. Isn't Everyone Selfish." In *The Virtues of Selfishness*, edited by Ayn Rand, 66-70. New York: New American Library, 2007.

[519]Sartre, Transcendence of the Ego, 2004. Pg 18

In Sartre's picture we see Peter in pain, and either because we feel pity for Peter or we don't like pain, we act on it, but we don't stop to think at the reflective level of 'this will benefit me'. There's no 'me' on this level as a presence to our thoughts there is only Peter and his pain. Even if we do it out of selfish reasons like to 'shut him up'. No, Sartre thinks it is the dislike of seeing Peter in pain at the unreflective level that moves us. The 'I' and the 'me' does not appear as part of this process, rather what consciousness apprehends of itself are the qualities it projects in the world. The pity or the pain appears on Peter's face and that is what consciousness grasps and in turn this is what Sartre thinks moves us.

Indeed we might put this case a different way. When we have before us a hamburger, it is the deliciousness of the hamburger which moves us to eat it. We don't stop and conspire 'oh I will eat this hamburger in order to enjoy it' or 'the taste of this burger will be of benefit to me'– we might only think that if we were on the reflective level, and already performing a theitic act, but rather, *in the moment*, there is only the hamburger before us and the deliciousness of the hamburger, as it appears to us, like the pity for Peter that moves us. The deliciousness is of the kind that moves us to eat it. The deliciousness is a reflection of the self in the world as a quality but it is not consciousness reflecting back on itself but rather consciousness as reflected in to the qualities of the world. This leads to Sartre's main criticism of the French moralists, which is, of course that

> (I)t is thus a waste of time to place behind the unreflected consciousness of pity an unpleasant state that will then be viewed as the profound cause of the act of pity. If this consciousness of displeasure does not turn back on itself in order to posit itself by itself as an unpleasant state, we will remain, indefinitely in the impersonal and unreflected domain.[520]

---

[520]Sartre, *Transcendence of the Ego*, 2004. Pg 19

In short, the *me* can only appear in the reflected state. The private intimacy of an emotion occurs at a later state, during reflection. *In the moment* we see the raw emotional charge of an event, such as Peter in distress, at the impersonal level. It appears in the world, and not at the state of the me, observing it, not upon reflection.

Sartre, later, uses the phrases 'tete la gifles' in *Being and Nothingness*[521] for the way an annoying person appears in the world. For Sartre 'tete la gifles' means appearing in the world with 'a head for slaps'. 'Tete la gifles' is the quality of the person's appearance as one and the same act of apprehension in grasping them and the annoyingness that appears before consciousness and in the person's visage. It is there upon their countenance as a real and genuine quality of the world. Their head appears to us from the very midst of the world, from the horizon of the world, amongst the objects of the world and its appearance is one that is crying out for slaps. The annoyingness and head for slaps isn't something extra or additional that we grasp about the person after the fact, or that we include or conclude from further analysis, rather, it is in the very essence of the way they appear to us. 'Tete la gifles' thus describes this strange aspect of our emotional lives.

At the unreflective level, in the moment, there is something aggravating about the person who makes us angry, something distressing in the person who makes us upset. At the unreflective level it is something about *that* person. It is, indeed *they*, who are upsetting. It is *they*, who are aggravating. *They*, who make us angry! Later, when we look back, of course, we say things like 'it was the anger that made me say it' and view our prior selves as being in a 'state' but in the unreflective moment that an emotion organizes the world for us, it is not 'us' who are angry, no, indeed not. It is *they* who make us angry. It is there in their very mode of appearing to us.

So take for instance the case in Original Ryle's moodology of an agitation which occurs as a clash between an inclination, or an 'Original Rylean' motive, as opposed to one of Wolff's third personal descriptions, or White's overall set of motives which we began the investigation of the string of arguments in this part of the chapter with. Say an Original Rylean impediment, for instance the inclination to buy an ice cream

[521]Sartre, *Being and Nothingness*, 2003. Pg 186

from the store, or the romantic motive of a date, and getting in one's car, putting the key in the ignition and finding the 'blasted' thing won't start.

At that moment, the whole world leaps into a little spiral of agitation, in which ice cream tubs, blind dates, hypotheses about battery problems, shopping centres and parking lots all begin to rotate around the hand holding the key in the ignition barrel of the car that won't start. The object of irritation is, to be sure, the impediment. The impediment of course is the ignition barrel which won't start the car, and potentially, the starter motor and the head-lights which might have been left on. We might also with a flash, recall, then raise some hopes related to the jumper-leads in the boot and start thinking about where we might boost a charge from. Then our hopes fade as we find ourselves remembering, almost with that very same thought, that we had lent them to our neighbour who never returned them and that he was still up the coast. Perhaps, *in the moment,* we might even have not very nice things to say about him.

But in that first primal moment of agitation one is not aware of one's self as being irritated at the barrel. The 'I' and the 'me' have not yet appeared because we are not at the reflective level. Rather, one is aware of the barrel with the key in it not turning over the engine. This is followed, perhaps, by the accusation that either the car, cosmos or Gods are in a league of jest against one. One might even proclaim 'aw you're joking', followed by several more tugs, a sound like one imagines a mechanical cow might make upon being kicked, and the sudden feel of the dashboard, sharp, hard and hot against one's palm as one slaps it.

Later, of course, we might think back, and *reflect* on the moment, and admit we shouldn't have said the things we did. We might reflect that it wasn't the ignition barrel's fault, and probably not the neighbours.It wasn't a 'bastard of a thing' and neither was he, or indeed any of the other names we used. That is it was purely a mechanical fault. We'll admit, the barrel, of course, had no 'will' of its own. It was not in league with the rest of the car, the gods, or the cosmos, and that our irritation was the guiding cause for our misguided anger and accusations of such. But *in the moment* the world leaps up and around the ignition barrel. *In the moment* it is the barrel and then by extension the car which is the cause of our irritation. The irritation is a way of seeing the factual impediment; the ignition barrel and by extension the car. It is not linguistic or context free *anger* like the linguistic behavioural analysis of the things we said in the moment might suggest. Rather it is an apprehension of an obstacle in the world full of aggravation and irritation.

Similarly, supposing we got the thing started, and took a journey down to the shopping centre, and were just about to park when someone darted in, in front of us and stole our spot then gave us the unfriendly finger.

We wouldn't be in an abstract state of anger, caused by the secretion of hormonal changes in the limbic system, and their sudden discharge through the ventromedial cortex, into the anterior-cingulate cortex, causing us to suddenly find ourselves floating context free in a sudden neurological 'state' as an emotional angry man. No. Rather we are angry at the man with his finger up at us and his car parked in our spot. Earlier we were angry at the ignition barrel, the head lights and the next door neighbour. The 'surge of anger' is a 'surge of anger' at *something*. This is the insight lurking behind White's argument about Ryle's category mistake about motives and things that can serve as motives. What he had in fact picked up on was a structural element of consciousness that had become confused with the way we talk about motives.

Indeed the illusion of 'emotional states' governing 'thrills of anticipation' and 'glows of pride', I argue, is in fact a phenomena that occurs or arises after the event and the attempt at trying to explain the emotions. This last process however I argue is a linguistic codification of an internal moment of introspection and reflection on a prior consciousness. In the ego complex the emotional data of the past, in a prior consciousness, is grasped by the reflecting consciousness of this moment as its own object. *In the moment* the person or object tied up in the situation is the target of our anger. The anger is a way of apprehending them. Later when we talk about them the anger becomes a separate issue. This later abstraction I argue arises from linguistic codification of the emotion in explanation as part of the act of reflection, and this is supported by the fact the reasons we give change. In the moment we are angry with the person and what they have done. If asked we might say 'I'm yelling at him because he took my car spot and gave me the unfriendly finger.' Later we say things like 'Oh, I was yelling *because I was angry.*' That is *in the moment* the object of anger is illuminated by consciousness of it, which is why we swear or curse at them, or project the emotional content in their direction or warn them that they are making us angry. *In the moment* it's something about the situation, ignition barrel or man with his finger up that we don't like. *Later* we tend to say we acted that way *because* we were angry.

The illusion comes from the narrative structure of past statements which induce us to think of emotions as abstract states rather than simply manifestations of our consciousness of the events. 'I got so angry'. But in the moment we aren't 'getting so angry' rather, something is upsetting us and the anger is tied directly into our consciousness indeed it is a manifestation of our consciousness of the thing we are angry at. The anger is not a statement, or a bit of linguistic behaviour or a proposition. These are some of the after effects of the anger itself. This is where White, following Ryle, went wrong when he began looking around for the logical characteristics of motives. Once this is understood we can see how auto-phenomenological arguments can be used for what White was looking for; a way of individuating the sorts of things that can serve in place of a logical class of motives developed from ordinary language usage. The pieces are all in place. We can now see how Ryle's framework was in the main dysfunctional since it relies on a confused sense of the first and third personal positions and a log keeper role. From Brandom we saw that reasons as linguistic utterances lead back to non-linguistic knowledge. From our discussion of Ayer we saw that beliefs do not need linguistic structures to be true or false. From our discussion of Brochard, Brandom and Kant we saw that we can apprehend the world without an 'I' attached to propositional statements. From Wolff we saw that tendency dispositions dissolve between the first and third personal perspectives into phenomenological style arguments and mere linguistic descriptions while from Sartre's theory of consciousness 'flash-bangs' or 'feelings proper' are ways of apprehending objects and events.

Thus, I argue, firstly, that a theory of the emotions implicates a theory of consciousness to individuate the emotions, which is to say, that the identity of the emotions rests upon a direct first person claim which is a stronger source thus than linguistic behaviourism and neurophysiology once the latter two are stripped of all their assumptions. Secondly, this is a more powerful normative force than a direct third person claim, because the force of a third person claim cannot distinguish an object at the level of a descriptive claim because the person and not what they are conscious of is the subject of a descriptive third person claim. This, combined with a first person linguistic codification *after the fact* creates the illusion of 'flash-bangs' or 'feelings-proper' in Ryle's original terminology. A flash of anger is a flash of anger about something. It has an object and a context. But taken out of that context it loses its object. Third person linguistic claims are linguistic uses which create the illusion of 'states'. That illusion contributes, with the illusion that arises from the linguistic codification of the emotion during reflection from the first personal perspective to the illusion of an emotional linguistic structure. That is, the two together, the first person report as a linguistic codification of something that happened at a non-linguistic level, and the third person linguistic description, end in taking the bit of linguistic utterance to be the emotion itself. This results in ascribing the properties of linguistic usage to non-linguistic knowledge. This, of course, is the indirect domain I pointed out earlier with Robert Wolff's linguistic behavioural analysis of Ryle's dispositions, which as we saw, were really third person holistic descriptions. The confusion which created the 'agitational calculus' was generated from an analogical structure. The analogical structure arose from tying first person experience with the behaviour of third person descriptions. Here we see the same process, tying linguistic codification from a direct first person domain of experience, to third person descriptions as an analogical structure in a shared domain. The shared domain was treated by Ryle as a normative source; Ordinary Language. The problem with the shared domain is that it admits contradictions. This is why we had to abandon the implicit ordinary language position in the philosophy of mind, because to accept it was to accept the possibility of rival sources with contradictory claims. I hope that the skeleton key I have been at pains to bring out has now become visible, and we can see where the cracks begin to systematically appear in Ryle's account. The problem arises from the indirect domain because it can inherit assumptions created by confusions and concealed phenomenological appeals in analogical structures. We mistake the first person codification *after the fact* and the third person descriptive use, as the 'flash of anger'. This is why Ryle's feelings-proper don't have objects or events to fixate on, and why it seems he can get away with a theory of mind without a theory of consciousness. The description happens either from the third person, or from the first person as a linguistic codification of first personal non-linguistic information put into a third personal medium. Treating linguistic codification and third person descriptions as though they were the emotion is the true category mistake.

Chapter Nine : The causal argument and the problem of consciousness.

Now the causal argument has two sides. I shall go into both below. But first it is important to note my position on this. Once we have climbed out of an implicit ordinary language argument we inherit the problems that the implicit position did away with and one of the problems that re-emerges is the causal problem of consciousness, only here, it re-emerges in a new sophisticated form. This will be the terminal problem of the thesis and it is one I do not have a solution for. However, in these last pages I will sketch the problem out and illuminate it with the thought experiment of the 'Roid Rager In the Void'. Thereafter I will draw some threads together and offer a final conclusion tying all of the pieces together.

In his *Sketch for a Theory of the Emotions* Sartre assumed that by building a phenomenology of the emotions out of what we shall call the 'fixation model' he could avoid the errors of the 'Intellectualist' and 'Periperic' theories of the emotions. The strategy he employs is by avoiding 'states' altogether and he can concentrate on the fixation by consciousness in the form of an emotion on an object. I argue, however, he has merely shifted the problem. Moreover a closer reading of William James, as Russell points out, shifts James closer to panpsychism[522]. But let us, for the purposes of saving space, and ease of exposition, start with Sartre's reading, digress momentarily with Russell, then have a closer look at what this new problem looks like.

Indeed, as Sartre formulates the problem of the Periperic school, the Periperic school being the position on which Sartre reads

---

[522]See Russell later this chapter. Also see Chalmers, David J. "The Metaphysics of Consciousness." In *The Character of Consciousness*, Pp 103 - 205. New York: Oxford, 2010. for where I'm drawing my definition of Panpsychism from. James is possibly closer to panprotopsychism, based on the patterns between protophenomenal qualities.

James, posits that emotions are the projections of physiological phenomena into mental phenomena. The second school on this problem we might call the 'Intellectualist theory' as Sartre does, which argues the reverse[523], which is that someone feels sad so they cry. The 'Intellectualist theory' according to this view, holds that physiologistic symptoms are produced by mental 'states'. The way Sartre puts these distinctions between mental and physiological phenomena makes it sound like there are two radically different substances, one mental and in the mind and one physiological and in the body. I do not think this was James' actual position. I should like to slightly correct this on a Russell reading because I think Russell's is the correct and quickest way without engaging William James at length.

Sartre writes

> William James distinguishes in emotions two groups of phenomena; a group of physiological phenomena and a group of psychological phenomena which we shall call, as he does, the state of consciousness. The essence of his thesis is that the states of consciousness called joy, anger and so forth are nothing but the consciousness of physiological manifestations – or if you will, their projections into consciousness[524].

Now rather than, at length, go into James' paper *Does Consciousness Exist,* or *A World of Pure Experience* and what I think is a panpsychistic position, or Type-F Monism in Chalmers taxonomy[525],

---

[523]Of course, Sartre's 'Intellectualist Theory' is not to be confused with the 'Intellectualist Legend' which is one formulation of Ryle's Regress, nor the 'Intellectualist school' which unlike Forodrian Neo-Cartesians as defined in LOT2, try to reduce 'knowledge how' into a species of 'knowledge that', that is, knowledge 'that' is foundational to knowledge 'how', both strains of which we looked at in '3.4 Linguistic Behaviourism, Ordinary Language arguments and the Normative Value of Science'. Sartre's 'Intellectualist theory' is a classification, for the division of mental phenomena and physiological symptoms which holds that mental 'states' are prior to the physiological symptoms. Sartre thinks James Periperic and the Intellectualist strains are wrong because they leave out the perception of objects and consciousness. Both the 'Periperic' and Sartreian classification of the 'Intellectualist' theory rely on a division between substance and matter that I think is problematic to a reading of James as Russell points out.

[524] Sartre, *Sketch for a Theory of the Emotions,* 2002. Pg 16

[525] See Chalmers taxonomy in Chalmers, *The Metaphysics of Consciousness,* 2010.

I'll just present a 'Russellian reading' and, for this paper, leave it at that.

Now as Bertrand Russell pointed out, James didn't think the world was built up of two radically different substances or phenomena, as Sartre seems to think James did, that project themselves onto each other. But rather, Russell thinks James thought of them as two radically different patterns of relations.

Russell writes

> James view is that the raw material out of which the world is built up is not of two sorts, one matter and the other mind, but that it is arranged in different patterns by its inter-relations, and that some arrangements may be called mental, while others may be called physical[526].

So the 'Peripheric Theory' on a slightly corrected and clarified reading we might say claims that emotions, or mental phenomena are one particular collection of inter-relations of patterns into consciousness by physiological causes. That is if "a man is afraid because he runs away" then the state of fear is one way we talk about the inter-relations of patterns projected into consciousness by the action of the man running away and we call this physical. The other arrangement of inter-related patterns where the person is afraid, and running away, we might call mental. This is slightly different to the Sartre reading, and I think this is the right one.

Thus on a corrected Sartre reading we might say the woman's sadness is a mental pattern created because she is crying, which is to say, the act of crying is a set of interconnected causal patterns that projects the state of sadness and this constitutes her consciousness.

Thus Sartre argues

> If, 'I agree with the advocates of the peripheric theory (that a mother is sad

---

[526] Russell, Bertrand. *The Analysis of Mind*. London: The Muirhead Library of Philosophy, 1951.Pg 23

because she weeps) I shall limit myself to
the reverse order of the factors[527].

The reverse order of factors is the Intellectualist Theory, that is, a
mother weeps because she is sad, being sad, the emotional state of
consciousness, or the pattern of inter-relations on Russell's reading,
would, of course, be what makes a mother weep.

Sartre writes

> If I am a supporter of the Intellectualist
> Theory, for example, I shall set up a
> constant and irreversible succession
> between the interior state of consciousness
> considered as antecedent and the
> physiological disturbances considered as
> consequences[528].

Now, as with Sellars sentence structures analogous to speech, the
appeal of Ryle's analysis of emotion into linguistic behavioural traits, is
that it breaks emotional terms like the 'flash of anger' or the 'agitation
of the car that won't start' into episodic forms in which events take
place, with specific verb structures and behaviours we can talk about
like an English Grammarian. This is appealing because it solves causal
paradoxes, like the one above which arises in trying to give an account
of emotions. But Ryle's solution is faulty and further it also boxes
consciousness out, and as such precludes the possibility of discovering
any properties or laws of consciousness, like the one above, where we
argued that the properties belonging to motives need to be individuated
by the sorts of things that can feature as motives if we are to avoid a
category mistake. Further, we argued that the properties of motives like
anger, anticipation, grief, sadness, while possessing the trait that
anger, for instance, is anger at something, anticipation is anticipation
for something, grief is grief from someone, sadness is sadness over
something. These are best individuated by an appeal to the first person

[527]Sartre, *Sketch for a Theory of the Emotions,* 2004. Pg 14
[528]Sartre, *Sketch for a Theory of the Emotions,* 2004. Pg 6

direct experience of consciousness, rather than on the basis of the former's normative appeal[529].

The more pressing issue is that with the sort of move we made on the basis of White's re-categorization, we also, inadvertently, re-introduced the causal problem of the emotions into consciousness, since now, we are using phenomenological data to individuate between the sorts of things that we can class as mental phenomena.

One might argue, for example, that the chill of fear one feels while sitting in front of the fire while being the general unknown is a fear of something unknown out there in the darkness as indeed Sartre does above. If one appeals to first person direct experience to individuate the 'chill' of fear, then one also opens the doorway up for questions about what causes that fear. Similarly is it the unknown which causes the fear or is it neurological agents working on consciousness? The door now swings both ways. In siding with a phenomenological account we re-introduce causation into the mix. As soon as we begin to use consciousness to individuate neurological agents or physiological data we open up a causal Pandora 's Box and inside we find causal problems reminiscent of the flag-pole example and of the same species that plagued Hemple's deductive nomological model.

There are exotic fears like anguish which we've already discussed and which don't fixate on objects or specific situations, but rather, reflect an awareness of one's self in inner states of turmoil or indecision. Other stranger species exist. One might be afraid of what happens after death as not knowing what happens next like the Irishman in Sartre's short story *The Wall* and which Sartre discusses at the end of *The Imaginary*[530]. Following along the same lines, the dread that something is going to happen is a dread of not knowing what that something is. The more abstract the fear, the more difficult it becomes

---

[529]Indeed, in Chalmers case, as we will see, it's arguable on the taxonomy I develop in this paper for the first and third person, and direct, indirect appeals that a move like Sellars or Ryle's by Chalmers, is simply mapping third person data onto third person data, which isn't what he wants to do to find a Neural Correlate of Consciousness. He wants to maps third person onto first person. This, I argue, is what lay at the base of his debates with Dennett.

[530]Sartre. *The Imaginary,* 2004. Pg 187. 'an attempt to conceive death or nothingness of existence directly is by nature doomed to fail', see his short story *The Wall*, where Sartre gives an example of this in the Irishman's dialogue. Sartre, Jean Paul. "The Wall." In *The Wall and Other Stories*, 1 - 17. New York: New Directions Books, 1948.

to instantiate the predicate of that fear with an object class or category to make the claim that it is a *something*.

Returning by a side route to the causal problem, consider, now, the following two thought experiments. The first is "The 'Roid Rager in the Void" and the second is "the man having a 'Prozac sunset'". I want to consider these two cases, particularly the former in some depth, because they push the limitations on this tangible fixation model for the emotions, which I've picked up from Sartre[531]. Now it follows on from this theory that if a man is happy, he's not floating in a general state of 'happy' he's 'happy' about something, and the same can be said, in this model, for the case of the man who is sad. The man who is sad is because it is raining, or his football team lost, or he lost his job. A man who is angry, is angry about something, a man who is in love is in love with someone. Emotion points to objects and events, and objects and events are emotionally *'irradiated[532]'* in consciousness. Now taking the second case first imagine the case where a man takes Prozac, or a similar antidepressant[533]and now finds himself suddenly in a pleasant, perhaps, euphoric mood, standing on a hill, smiling and nodding quietly to himself, viewing a sun set he's seen a thousand times before, but has never stopped to notice it. Did the sun set cause his mood, his fixation on it, or was it the increased oxytocin levels in his brain chemistry?

It is hard to say. He is conscious and feels eurphoric about the sunset. It forms a locale and focus for smiling and nodding. You could say the sun set is at the centre of his smug inner state of euphoria, but did the sunset, or did the presence of neurological agents and stimulants in his brain chemistry produce this euphoric state in his consciousness? This challenges the tangible fixation model, because it suggests that neurological agents may have an effect on consciousness, but it also doesn't reduce consciousness down to brain chemistry, he still needs consciousness for the neurological agents to produce euphoria when he stares at the sun. Let's push this case a little by introducing the problem of the 'Roid Rager in the Void'.

Let's say, as the steroids simulate raised levels of testosterone in the 'Roid Rager's system, and flush it with increased symptoms his

---

[531]Sartre, *Sketch for a Theory of the Emotions*, 2004.
[532]Ryle, *Concept of Mind*, 1983. See earlier footnotes.
[533] An oxytocin neurotransmitter stimulant for the purposes of the argument

anger increases and as it increases it locks on to something be it simple, profound, or silly, like say the blue shirt of his physical trainer. If there's no physical trainer when he experiences this sudden burst, then let's say, he fixates on the annoying blue shade of the vase with the flower in it, or if he is not in the house then he locks on to the car in front of him, or the set of traffic lights that have been red for far too long, or even God forbid, the little old lady driving around in people's blind spots.

Now there is an interesting question to be asked here.

What happens to the 'Roid Rager if we cast him into a void? That is, we pump the 'Roid Rager up on steroids, and drop him into an objectless void, and watch him float there for awhile with no objects for him to fixate on?

What will the 'Roid Rager do, do you think?

Obviously, sooner or later, he will apprehend that he is in a void perhaps blink and suddenly rage at it.

Now suppose as he floats there we had a nozzle to turn down the luminosity of the void and another one whereby we could increase the amount of 'roids in his system and for the purposes of the argument we do, and we look on in at the 'Roid Rager, floating there, tumbling through space in a red faced flail of punches and yells as the void vanishes around him. What is his anger fixated consciousness going to do?

What bothers me, and what I don't have an answer for in this thesis, is once you remove all of the objects of consciousness that contribute to an 'external fixation model for an emotion' whether that person will turn their emotion on itself and it will become something else with a new structure. I don't know whether that emotion will somehow turn on them and gain a substructure comparable with melancholy, nostalgic drift, with things like Sarteian anguish and like in the fear of the alcoholic or gambler it becomes redirected at the self and gains a structure like Kierkegaard's fear of his own capacity to sin, or whether objects are essential to these emotions that mimic and follow structured patterns of consciousness. Indeed there is here the question of whether an object apprehended by an act of consciousness as not consciousness, is that philosophically prior and fundamental to the possibility for experiencing an emotion?

But we can learn something, from this question if we examine it closely and the way the three central philosophers could deal with it, to help us align our terms.

Ryle, himself, lacks the resources to answer this question because he hasn't developed an explicit theory of consciousness other than the one I have given him from his inconsistencies and here drawn out in the exposition of this paper. We might ask the 'Roid Rager for one of Ryle's special status reports, and see if he's 'alive to what he is doing' but this relates to a complication I'll mention in a moment.

Sartre would echo Bergson on this and say the example is impossible. It is an abstraction that could never happen and so practically is an irrelevant problem. He'd probably be right, since it is hard to formulate the central terms like whether the void can appear as a 'thing' before consciousness as not consciousness or whether a void properly would just be consciousness by itself like in mystic states[534]occ upying itself as its own boundary conditions since there'd be nothing outside of consciousness and thus a phenomenological void proper has no object. Perhaps, such a void would be more like the problem the Irishman faces in Sartre's short story trying to imagine what comes next after they shoot him. What is death to the Irishman in *The Wall?* He simply can't imagine it.

Consciousness inside the void without an object to fixate on might simply cease.. Such a ceasing of consciousness could take either of two forms both of them dissolving consciousness in different, but ultimately the same final way which is removing the condition that Sartre thinks allows consciousness to exist in the world. The prior we might imagine with a bit of onomatopoeia to accompany it like 'zzzzip' or 'pfft' in which consciousness simply vanishes and we don't know why. The latter, I suspect, would be by what Sartre calls 'fascination' which he draws out of an analysis of Rousseau. Fascination occurs when the negation of the self to the object in Sartre's framework doesn't exist and thus the form

---

[534]Sartre, *Transcendence of the Ego*, 2004. Pg 37 Bergson, *Time and Free Will, an Essay on the Immediate Data of Consciousness*, 2001. Pg 63.

of the negation could be neither internal nor external[535]. Consciousness simply melts into the world. That is normally consciousness as consciousness of an object at the non-theitic level is a negation of the world as background, to the object foregrounded, which of course, is an internal negation of consciousness to the external and then the object is an external negation of the object as an external negation to the background, while a quality is an internal negation of the self to the object. Since there is neither a background, nor an object, consciousness cannot exist, since it is contingent on being conscious of something and since the anger mimics this consciousness in its way of organizing the world it too would vanish.

Indeed questions about the void arise like whether, if it be a proper void, it eliminates memory, the imagination, the positional attitudes of the imagination and conceivability. These questions introduce new problems that make it even harder to formulate the problem in Sartrean terms. Would the 'Roid Rager have memories, for instance, and given those would he turn his anger on those memories and fossick up some past event, say, a memory of his mother over perhaps some spinach issue that occurred decades before? On the other hand, if the 'Roid Rager in the void lacks memories how is he to process or understand his sensations of 'rage'. Indeed could he explain them to himself in a way he might understand it without precedents, or even the acquisition of a language? This last point on language is the complication I mentioned a moment ago with Ryle. This, of course, pushes us in trying to conceptualize the problem in the direction of Donaldson's Swampman argument and I think this line of thought offers an attractive insight in that it shows us some broad brush stroke limits of the conceivability for an emotional context.

Sellars I think would be on the right track to solving it, in that, the experiences of the 'Roid Rager in the void would be private. Private in the sense that we can make inferences to thoughts, standing on the outside of the void, with our clipboards, watching him tumble through space in a spiral of kicks, flailing punches and profanity and posit theoretical entities for what's going on in the 'Roid Rager's mind. These thoughts, the earlier Sellars would treat as analogous with language

---

[535]Sartre, *Being and Nothingness*, 2003. Pg 200- 201

ignoring the sorts of intrinsic qualities that Sartre is interested in. But the latter Sellars of *Empiricism and the Philosophy of Mind* might also point out while this is an observational language only the 'Roid Rager in the void could give us the report. But if we watched closely enough one of us might make the leap between an observation language, and a report language. It may begin one day when he or she bangs their foot on the way to the void, or when their car won't start, or one of the other void watchers steals their parking spot. This in turn might allow us to begin to attribute internal experiences to yells and flailing fists.

Indeed it is to this very possibility of developing a report language I now turn.

Chapter Ten : Conclusion, discussion of the implications of a return to Pre-Fregeian Psychologism and final objections.

Now I want to develop some of the above into a fruitful set of features that can be used to forward the central thesis of this paper, offer a unique insight, and finally bring this chapter to a close.

Consider the case of the man who stole our car spot and gave us the finger. We could, on the one hand, have pity on him for not knowing any better, wish him well, offer up a benediction to the Lord to watch over him, but be fairly certain that if he doesn't change his ways, sooner or later something will probably happen to his car. We could, alternatively think 'you rat bastard, I'll show you' park our car around the corner and put on our little cowboy hat, to go and give his car a seeing too personally.

The former, of course, is the appropriate response, but nonetheless notice here that the way we react to the situation is tied into the way we interpret it. If we have pity on him, for not knowing any better, and realize his arrogance must make things very hard for him, and wish him well, that's a different response to 'you rat bastard, I'm going to show you'. It involves a different way of seeing the situation at the level of the flash of anger, and the recognition, or act of consciousness that is tied into that anger as being angry at the man, as a way of seeing him take our car spot and give us the finger.

But following on from this the question of the meaning and how to interpret the act has two levels to it. There is the moment when we are seeing the act and interpreting it, literally, him, at the moment he is pulling into our spot, and giving us the finger and the anger, that occurs at the moment of the event, in interpreting it, and following that, the benediction on his behalf or the cowboy hat and cricket bat. Both of these contain the 'flash of anger'. Both may even have the same type of neurophysiological data with activity in the same parts of the brain, but the way the person deals with the situation can occur in two different ways.

Moreover, there is also the point that what they feel *in the moment* is different to what happens *later* when they reflect back and talk about it. Thinking back on the event and interpreting one's actions is a different sort of case to being there *in the moment*.

This is important, because the way we feel about something in the moment, and the way we describe it later, may not be the same thing. *In the moment* we might wind down the window and yell at the man who took our parking spot "You utter bastard. That was my spot. I was waiting for it. You utter, utter, utter bastard" and accompany such a statement with any number of explicatives, rude gestures, epithets, insults and exclamatory statements it is, perhaps, not polite to publish in a philosophical paper. Later, we might, reflecting back, say, 'no, he wasn't really a bastard, it was the anger that made me say those things.' – If such is the case, and I should argue it is, then indeed, there is a gap between how we perceive the world, and how we might describe ourselves seeing the world. Indeed it would seem to suggest that language is like a code we pound thought into.

Further in this earlier state, this *'in the moment'* if there is a non-linguistic element, something that later we may be able to describe linguistically, but *in the moment* contains something non-linguistic about it, and that the linguistic descriptions or rationalization depends upon for its meaning, then I should argue this constitutes the central thesis of this paper and the insight I have been at labour to bring about, that non-linguistic knowledge is prior to knowledge of linguistic meaning. That's not to say that language plays no part in linguistic meaning, or that it doesn't shape our perception of our emotions, how we handle, describe or display them. All it says is that non-linguistic meaning is prior to and fundamental in relation to knowledge of linguistic meaning in certain contexts where we talk about the mind. The relation part is the bit that covers the linguistic shaping of our emotion.

This position, for which I am arguing, has significant ramifications across neuroscience and philosophy of mind, if it can be upheld. For instance, it brings into question David J. Chalmers' search for a formalism to record linguistic statements or write first person information down in, in order to collect data in order to construct a 'Neural Correlate of Consciousness'[536].

For instance Chalmers argues

> The third-person data relevant to the science of consciousness include both behavioural and neural data. The availability of behavioural data is reasonably straightforward: one is constrained only by the ingenuity of the experimenter and the limitations of experimental contexts. In practice, researchers have accumulated a rich body of behavioural data relevant to consciousness. In contrast, the availability of neural data is much more constrained by technological and ethical limitations[537].

---

[536]See Chalmers notes on *"Conference for the "Neural Correlates of Consciousness"* given by The Association for the Scientific Study of Consciousness, 1998, Chalmers takes his definition of a 'Neural Correlate for Consciousness' from the 1998 conference. See Chalmers, *What is a Neural Correlate of Consciousness*, Pg 60, 61, in *The Character of Consciousness*, Pp 103 - 205. New York: Oxford, 2010. where it was defined, as 'a specific system in the brain whose activity correlates directly with states of conscious experience', and David Chalmers takes to mean, I take it, as a bit of tautological tongue-in-cheek, as 'A neural system N is an NCC if the state of N correlates directly with states of consciousness[536]' and which he builds into a definition for a set of co-varying properties, that will allow him to map a link between first person subjective data and third person objective systems. See also Pp 103 to 207, "The Metaphysics of Consciousness." In *The Character of Consciousness*, Pp 103 - 205. New York: Oxford, 2010, for behavioral research, along with the neurological data gathered by PET, fMIR, EEG, MEG, and other 'doo-hickey' forms of technology, along with anatomy, physiology, measurable changes in brain chemistry, and so forth, constitute what he takes to be third person data, since such data can be gathered inter-subjectively.
[537]Chalmers, *How Can We Construct a Science of Consciousness*, 2004. Pg 47

350

First person data, for Chalmers, is what he calls and discusses as the 'hard problem' of consciousness[538]. What he is looking for is a method of collecting data, which he thinks, can only be got from descriptive statements collected from subjects. What he needs is a 'formalism' that offers a general form, structure, or criteria that such statements might come in for the purposes of collecting data. What I have argued here suggests that attempts at standardizing such a 'formalism' for collecting linguistic data may be problematic. Reporting an experience, like a 'flash of anger' may change the very nature of the mental phenomena and create problems like those I have laid out in this paper which arise from the linguistic treatment. The illusion that a 'flash of anger' is context and object free, or that its neurological correlate, mapped by activity in the limbic system, is merely a neurological state caused by brain chemistry with no relation to awareness of the event or object in consciousness like we saw with the man locking his car door, then turning around, giving us the finger, and walking off. There is a causal relationship between the event, our anger and neurological changes in the limbic system. Here, to ignore consciousness, is to lose sight of an important set of factors bottoming out in a causal relation. The illusion created by linguistic behaviourism and neuroscience costs us dearly. To fall into this illusion, to go forward without our eyes open to these factors, could cost us years of wasted research and effort, producing countless amounts of inaccurate and problematic data and set us back in any number of related ways from developing an account of the nature of mind.

David Chalmers identifies the lack of a general acceptable formalism, as one of the major problems in constructing a science of consciousness.

> (An) obstacle is posed by the absence of
> general formalism with which first-person
> data can be expressed. Formalisms are
> important for two purposes. First they are
> needed for data gathering: it is not enough

---

[538]Chalmers, *How Can We Construct a Science of Consciousness*, 2004. Pp 47 - 100

to simply know what one is experiencing:
one must also write it down. Second they
are needed for theory construction: to
formulate principles that connect first-
person data with third-person data, we need
to present the data in a way such that such
principles can exploit[539].

The problem is, of course, if what I have argued is correct, which I argue
it is, then linguistic descriptions of consciousness, in the case above,
emotional content, may not be strictly indicative of a person's 'state' of
consciousness. Writing it down or giving it as a special status report
may in fact change or contaminate the data. This is because, if what I
have argued is correct, treating linguistic behaviours and ways we talk
about aspects of mind, consciousness and mental phenomena, as the
phenomena itself, produces inconsistencies and pitfalls, like the missing
causal relation I was at pains to draw out in the example of the man
who gave us the unfriendly finger and made us angry. Here
consciousness and its fixation upon that unfriendly finger played a
pivotal causal role in altering our brain chemistry. This causal role that
came from apprehending the man and his unfriendly finger was
obscured by the nature of the linguistic analysis of Ryle's feelings
proper, and the illusion created the linguistic treatment of 'flash-bangs'
along with the neurophysiologistic side of the story. The illusion arose
because our reporting the event, the narrative structures we use later
on, may be different to the structure of consciousness in the moment.
The danger involved in developing 'formalisms' is that we may develop
the very same faults, flaws and illusions that Ryle's Linguistic
Behaviourism unwittingly lead us into.

Indeed, if what I have argued is correct, and indeed I argue it is,
emotions are not 'states' at all, nor are they reductive to linguistic
expressions, rather, emotions are involved in our ways of seeing the
world. The illusion of a prior 'state' only arises when people reflect on a
prior act of consciousness. On this modified view of Sartre's that I am
presenting linguistic descriptions of emotions are a deceptive

---

[539]Chalmers, *How Can We Construct a Science of Consciousness*, 2004. Pp 51 -52

codification of a prior moment of consciousness from the first personal point of view or they are third person descriptions which feature the person as the object. The melding of these two, the linguistic codification from the first person and the third person description, creates the Linguistic Behaviourist illusion of the 'flash-bang'. The domain it occurs within is the one I pointed out in the section on Wolff, it is the indirect domain, where assumptions are carried over from the direct first and third person positions creating analogical structures. On this view, an attempt at making the phenomena linguistic, indeed, talking about emotions or writing them down, transforming them into some sort of 'formalism', perhaps any sort of formalism, may change the very nature of the phenomena.

We might describe the point like this. In the moment that the star struck poet first sees her, he is afflicted by all sorts of phenomena. Later, reflecting back, under a tree, while chewing his pen, he searches for the words to describe that phenomena, that very moment he was struck, when 'all that's best of dark and bright' he felt 'met in her aspect and her eyes' such that he was 'mellowed to that tender light' which of course 'heaven to gaudy day denies'[540]. It is in such a 'state of enchantment' that it seems he found himself bereft of his senses. However if the central driving thrust of this thesis is correct that's not the phenomena of the moment. Rather, that's a codification of consciousness reflecting back on itself, as theitic consciousness, as a reflective consciousness, as consciousness taking itself as its own subject for analysis. The words our poet is using, these are not the phenomena of consciousness as it happens in the moment when he sees her but rather they are like a code, he uses later, analysing this earlier state, in order to describe to us what he felt at the time.

Likewise later when we reflect on the prior state and begin to describe how the man who took our car spot at the supermarket made us feel that is, entirely, a different sort of thing to how we felt *in the moment* when he gave us the finger. The long and short of it is that linguistic statements, words, bits of language, descriptive or otherwise, are not, strictly speaking, the emotional meanings to which, we assume they ascribe themselves. We don't think 'sudden, uncontrollable, RAGE!' like Bruce Banner does in the comics, moments before turning

---

[540] Lord Byron, of course.

into the Hulk. The word 'rage' is something that enters our vocabulary when we think back on the situation or the scenario that happened in the moment which made us angry. It's a piece of 'code' that we appeal to from a reflective state of mind in order to transmit something about how we view our earlier consciousness.

That's not to say that consciousness is somehow unconscious, or a form of unconsciousness because it is not linguistically expressed *in the moment* or propositionally affirmed, verbally or privately. On the contrary, in our earlier state, we were angry, precisely because we found ourselves angry at something. The anger, although not explicitly linguistic is a manifestation of consciousness. We are of course angry about something and in order to be angry about something we must indeed be conscious of it. The illusion of the 'state' is something that enters our vocabulary when we think back and reflect. That is to say, moreover, descriptions of our emotional lives are not our emotional lives. They are, rather indeed, like a code, or a codification of our emotional lives as presented to consciousness when that consciousness takes a prior consciousness as its object. Indeed it is this theory of a code, or a 'non-linguistic' theory of the emotions and reflective consciousness that sees emotional talk as merely a cipher we use to talk about our emotional lives. Such a code or cipher begins where the I or me appears in a narrative structure as part of a reflective act and the I, or the me, is indicative of a prior consciousness brought before the present consciousness for analysis and codification. It is this codification that constitutes the central movement of this thesis for a move towards a reconsideration of a Pre-Fregeian psychologism.

It opens up analytic philosophy on any number of fronts and possibilities for investigating these structures and developing new methodologies. In what ways is a psychologistic theory of meaning beneficial and significant for de-codification? Is de-codification possible? Are my en-codifications different on a phenomenological level to your de-codifications? Is it conceivable that they are different from my perspective? Might we ever prove this? Can codification happen and progress with each party not knowing the meaning or how to unpack the codification of the other organism's codifications? We're not talking the whole gambit here, at this peak of the thesis, only about emotional contents and insights into the ways in which we draw upon them and express them. Even so these are the types of 'difficult problems' an analytical psychologism will have to face.

As controversial as this position is, it is the one I am arguing for, that is, I am arguing we need to return to a view

> that what a speaker knows is a kind of code. Concepts are coded into words and thoughts which are compounded out of concepts, into sentences, whose structure mirrors, by and large, the complexity of the thoughts[541].

I am arguing that

> Communication is, thus essentially like the use of a telephone: the speaker codes his thoughts in a transmissible medium, which is then decoded by the hearer[542].

Even though

> The whole analytical school of philosophy is founded on the rejection of this conception, first clearly repudiated by Frege. The conception of language as a code requires that we ascribe concepts and thoughts to people independently of their knowledge of language; and one strand of objection is that, for any but the simplest concepts, we cannot explain what it is to grasp them

---

[541]Dummett, *What Do I Know When I Know a Language?*, 1993
[542]Dummett, *What Do I Know When I Know a Language?*, 1993

independently of the ability to express them
in language[543].

Indeed I'm not moving mountains merely pebbles though it might seem
that way on first glance. All I'm arguing, indeed the position I'm
advocating, is that at least some language is a mere code, at least, for
describing some of what goes on in consciousness and specifically that
which we codify as the 'emotions'. We use it when reflecting on and
analysing a prior state of consciousness. It's a unique code, we utilize, in
theitic acts, that is, when consciousness takes itself as its own object,
and we try to 'transmit' the information of this analysis to another
person. Language is the form of the code that we attempt that
transmission in. The meaning of that language depends upon non-
linguistic knowledge.

Indeed the evidence is that there are some emotions which we
struggle and simply can't find the words to codify. The upshot is, of
course, that if we can't describe them within a language then they can
occur independently of language. If we struggle to linguistically codify
them then it implies that we can grasp them independently of
expressing them in a language. Moreover we saw that some 'wants' and
'needs' cannot be justified or explained within a language but refer us
beyond the game of giving and asking for reasons. This came about
because reasons as linguistic utterances in the game of giving and
asking for them are founded on non-linguistic knowledge, knowledge
one must arrive with and posses to play the game. We also saw that
attempts at avoiding the game could be broken down into special moves
in the game that terminated in non-linguistic knowledge or dispositions
which were in Ryle a confusion of first and third personal perspectives.
The third personal perspective contained mere linguistic descriptions
with no 'power' or 'force' behind them, and that these contained
'sympathetic' or occult phenomenological arguments, while the first
personal perspective contained strictly phenomenological style
arguments. We saw once we had this insight, that inheritance
conditions usually given in adverbs were either descriptions from a
third person point of view, or linguistic codification of something that
was happening on a non-linguistic level. This allowed us to understand

---

[543]Dummett, *What Do I Know When I Know a Language?*, 1993

White's insight, while simultaneously noting what was wrong with it. White attempted to distinguish motives by the ways people spoke about them, but lacked a theory to individuate the types of things that could be motives. What White had picked up on was a structural element of consciousness. What he was looking for, however, was an indirect assumption with elements inherited from the direct first and third person. Motives qua motives are internal reasons which refer us back to non-linguistic sources. Motives qua motives as individuating things that can serve as motives by their logical characteristics are non-linguistic pre-codifications.I argued that what White needed was an auto-phenomenological argument, and now that we have it, it in turn allows us to bring light on the mind to individuate elements of the mental. That insight also gave us applications in neuroscience like we saw with Sellars position and Ledoux's research into Post Traumatic Stress Syndrome. This insight arises once we see the mistake of Linguistic Behaviourism which is to take the behaviour of the linguistic codification in the ways we talk about elements of the mind like emotions, for those elements themselves.

That is to say, to offer a thumbnail sketch of the threads in this thesis, I have provided arguments that suggest, what we might refer to as our emotional 'states' are in fact, manifestations of patterns inside consciousness that imply they are not 'states' at all, but that discussion of 'states' is an illusion that arises as something that happens in the process of the codification of emotions into words. Once grasped this insight leads us to the view that emotions are forms of consciousness, and embedded in our ways of seeing the world. The immediate upshot of this, is, of course, an argument for a return to Pre-Fregeian psychologism, but it does not bring us back to psychologism empty handed. Already within this paper alone we have a wealth of phenomenological insight. Moreover, we now have a phenomenological method, the auto-phenomenological argument.

If you recall back to the start of the paper I outlined three theses that I would argue. These were of course (I) The Short Narrow, or 'Sharp Thesis' which was simply that an appeal to phenomenology was a stronger basis for an argument than hetero-phenomenological sources for individuation on the mind. The two hetero-phenomenological sources we looked at were Linguistic Behaviourism and Neurophysiology. Then we had (II) The Strong Narrow Thesis which claimed that emotions are irreducible to linguistic behaviours. (III) The General Broad Thesis which puts it in the same area, roughly, as David Chalmers in the call for a theory of consciousness. For this, I argued that there are mental states, acts and operations that are irreducible to Cognitivism or Functionalism without a theory of consciousness, ergo, we need a theory of consciousness. We can now see how all of the pieces of the paper fit together.

The Short Narrow Thesis undermines an Ordinary Language cum Neuroscientific position with its claim that non-linguistic sources are fundamental for establishing meaning in some mental phenomena. First it undermines an implicit meaning of language position in the philosophy of mind on psychologistic grounds, this in turn undermines an implicit Ordinary Language cum Neuroscientific position if they are taken together in a feet-of-clay style argument. This was supported by (A) the claim that hetero-phenomenological sources usually contain hidden assumptions concealed in analogical structures as part of a first and third person indirect domain. The problem with using this domain is it allowed for contradictory claims to arise. These hidden assumptions and concealed analogical claims may arise like when neuroscientists first start to experiment with finding activity in the brain related to some linguistic conception or label like 'disgust'. The neuroscientist experimenting with disgust may come up with a series of experiments utilizing disgusting objects drawn from his own experience, or case studies. What these actually are, are analogical claims referring back to himself or to subjects of a case study. The problem with these, as we've seen, is there is room for contradictions to arise between third and first personal positions if the defining structures are left vague.

There are further problems with using analogical structures inherited from the indirect domain of codified ordinary language usage without reference back to first or third personal perspective. This is like the case of 'flash-bangs'. Straight hetero-phenomenological linguistic analysis fails to take into account causal relations because the linguistic codification is a different sort of thing to the conscious apprehension of the event or the object the emotion is related to.

(B) When a hetero-phenomenological claim individuates mental phenomena it is still open to a third claim, an auto-phenomenological claim whether that is indeed the same phenomena. This is like Sartre's argument that anger is not ultra-joy. Thus, I argue that an appeal to phenomenology is a stronger appeal than either of the two hetero-phenomenological sources.

(C) There are just some insights into the nature of mind that we simply can't get from either neuroscience, linguistic-behaviourism, or the two taken together. This is like the 'Witness/Reader argument' or the problem with trying to anticipate your next thought before you have it. In time we might come to be able to distinguish these in ordinary language by adopting or developing behavioural traits as part of a linguistic community, but these will be mere linguistic wrinkles on the surface. We might learn how to put quotation marks around see, to distinguish between see, as in perceptual vision, and "see" like when exercising the mind's eye, but these differences are merely indicative of something that is occurring on a non-linguistic level. The linguistic behaviour and devices that we may employ only point towards the non-linguistic knowledge. In this sense they are a codification. The linguistic behaviour is merely a surface ripple on the pond that obscures the nature of mind. The challenge of Psychologism is to tell us what lays under those ripples in that pond.

The 'Sharp' Thesis in turn supports the Strong Narrow Thesis. Emotions are irreducible to linguistic behaviours. They are irreducible, firstly, because there is a causal relationship between the person experiencing the emotions and the emotions themselves. I argue this implies a theory of consciousness, which of course brings us to the third thesis, the Broad Narrow Thesis. We can not reduce emotions to talk about the emotions without giving an account of consciousness. This is evident in the missing set of causal relations between the brain undergoing sudden chemical changes in the limbic system, associating these changes with the linguistic codification of the term 'rage' and the man with his car parked in our parking spot giving us the finger. Secondly, as I have drawn out with Ryle's 'Agitational Calculus' and the development of the flash-bang strain above, the way we talk about the emotions are not the emotions themselves.

I consider the strongest argument of the thesis the 'flash-bang' strain but there is more here on offer from Psychologism. Indeed what the strong narrow thesis, in effect, sets up is the argument that some mental phenomena are irreducible to either linguistic behaviours, brain chemistry or both. Since phenomenological insight gives us a stronger argument because there are non-linguistic domains it can access, but which, we can't think about in linguistic terms, like the difference between the percept of an event and an imaged act of consciousness in the 'Witness/Reader argument' or the claim that there is knowledge that it is necessary to arrive with and possess in order to play Brandom's game of giving and asking for reasons, then it follows that some mental phenomena are simply irreducible to linguistic behaviours and utterances. This in turn supports and advances my position on psychologism against the implicit ordinary language position on mind. This brings us to lastly, the third thesis, which puts it in the general category of Chalmers with the need for a theory of consciousness, except, that here, we arrive with insights to contribute to the search for an adequate theory. These are insights that tell us that linguistic codification is a different sort of thing to some forms of mental phenomena. We can see this with Brandom's game of giving and asking for reasons, Hoffe's three levels between Original Synthesis in Kant, and the Apperception of the I, and with practical thought experiments that arise from some of the inconsistencies in Ryle like the difference between analysing the 'Remember when/ remember how' configuration as a bit of Linguistic Behavioural analysis, or the phenomenal exercise of thinking back to what Reg got up to last Christmas and comparing that with backing down the drive, or trying to recall how to solve second degree equations using the Quadratic formula.

But this also leaves us with unsolved problems. Once we recognize that consciousness individuates the emotions with the 'identity argument', we still have the 'causal argument'. The benefit and draw back of a return to Psychologism, is firstly, it frees us from an illusion that left us blind to the causal powers and sets of relations between consciousness and other mental processes like emotion. Secondly, it reintroduces causation along with all of its problems. A return to psychologism is both a blessing and a curse. Nonetheless, I argue, it is the right move to make. I have briefly outlined the problematic side of this blessing with the thought experiment of the 'Roid Rager in the Void. While this is a figurative device, it doesn't quite bring out what is at the heart of the problem because it presents the 'Roid Rager from a third person perspective. The problem I am trying to get at is the complex relationship between consciousness and brain chemistry. If you think back to the man who gave us the unfriendly finger, then you think of the physical trainer with his annoying blue shirt, then there is a causal question that arises. Solving this problem would let us draw several different lines of thought about how to approach the causal model, and the problem that arises between neurophysiological stimulants and consciousness and what this means for the nature of mind.

Related to this we have arguments about the mind that may draw on elements outside of the authority of auto-phenomenology, like for instance, states when the person is not conscious but there is neurological evidence of brain activity, like when a patient is heavily under hypnosis or in some state of sleep, or similar situations where the person can be observed to have learned something or does something, that can be catalogued from a third person point of view, but doesn't have a conscious memory of this learning process or the action. It seems these contravene my position on auto-phenomenology. This would be to misread my argument. My argument is that auto-phenomenology is a stronger source for individuating on the nature of the mind because emotions are irreducible to linguistic behaviours. This, of course, is the Strong Narrow Thesis of the paper. Since non-linguistic knowledge is necessary for meaning in talk about the emotions then language can't be the whole story about the mind. Since language isn't the whole story then the emotions are not reducible to their linguistic counterparts. This supports the Short Sharp Thesis: an appeal to auto-phenomenology is a stronger basis for an argument, but it is not the only basis for an argument. The reason why it is a stronger source has three interrelated sides. Firstly it allows us to gain insight into phenomena that is not available at the linguistic behavioural level as I've defined the linguistic behavioural level in this paper. Secondly it concentrates on the phenomena itself and not the linguistic codifications and thus thirdly, avoids distortions and category mistakes that may arise from taking the linguistic behaviour of talk about the phenomena for the phenomena itself. Lastly in relation to the broad narrow thesis, this point about non-conscious states like sleep or deep forms of hypnosis like clinically induced somnambulism, supports it. A theory of consciousness, a good one, will tell us when we can use auto-phenomenological arguments to individuate elements of the mind. A good theory of consciousness will give us the conditions of the authority for individuation by consciousness and its processes.

Appendices

Appendix A

Counterfactual logic, Linguistic Behaviourism and Ordinary Language arguments.

I am going to spend a few brief paragraphs exploring a clash between Ordinary Language and Modal Realism that arises from Linguistic Behaviorist style analysis of the English subjunctive conditional and counterfactual logic which David Lewis points out in his work *Counterfactuals*. The clash is interesting since, as Lewis points out, the normative values in the case of the English subjunctive is that of mistaken usage of the language, and its appeal likewise is to 'what it makes sense to say' in that language. The normative force behind Lewis' own counterfactuals is in his Modal Logic and as he points out, ultimately rests with his conditions for truth values[544].

---

[544] I'm using Lewis as a comparative example because the clash he presents between the English subjunctive and his theory of modal logic is a genuine rivalry over how we might read an argument and the criteria we might use to judge the argument. One part of the rivalry is mistaken linguistic usage and the other is relevant to determining the validity or invalidity of the argument. There are also problems with how we might take the term 'conversational implicature' and Jackson's 'conventional implicature'. Lewis' reading of the English subjunctive follows more closely with Jackson's term, since it relies on a specific structure rather than a set of principles. See footnotes below.

David Lewis writes

> Granted, the counterfactual constructions of English do
> carry some sort of presupposition that the antecedent is
> false. It is some sort of mistake to use them unless the
> speaker does take the antecedent to be false, and some sort
> of mishap to use them when the speaker wrongly takes the
> antecedent to be false. But there is no reason to suppose
> that every sort of presupposition failure must produce
> automatic falsity or truth-value gap. Some or all sorts of
> presupposition, and in particular the presupposition that the
> antecedent of a counterfactual is false, may be mere matters
> of conversational implicature[545], without any effect on truth
> conditions. Though it is difficult to find out the truth
> conditions of counterfactuals with true antecedents, since
> they would be asserted only by mistake[546].

This clash between Ordinary Language and a system of logic based on the conditions for truth
values that Lewis provides, that is, the clash of normative force between ordinary language usage

---

[545] There is a problem with this term although the technical specification of its usage is anachronistic and relates to developments after the publication of Lewis' work. See Grice, Paul. "Logic and Conversation." In *Studies in the Way of Words*, Pp 22-40. Harvard: Harvard University Press, 1991. 'Conversational implicature' contains a a set of principles which we might call an 'etiquette', for convenience in this footnote, and a logic, such that something may be true on logical grounds, but still violates one of the principles of an etiquette in conversation. For instance, suppose our etiquette contains a number of principles like 'be informative', 'be truthful' 'be brief', 'be helpful', and so on, and suppose also that I knew that the Paramatta Eels had won the rugby. If you asked me who won the game and I replied 'it was either the Bulldogs or the Eels', what I say may be *logically* true, but it may also violate one of the principles, in this case to 'be helpful' and to 'be informative'. Lewis' formulation of the English subjunctive conditional follows more closely with Frank Jackson's reading, since it contains an implicit assertion that the antecedent is false. See Jackson, Frank. "On Assertion and Indicative Conditionals." *The Philosophical Review* 88, no. 4 (1976): Pp 565-589. Also Grice, Paul. "Indicative Conditionals." In *Studies in the Way of Words*, Pp 58-87. Harvard: Harvard University Press, 1991. And Bennett, Jonathen. *A Philosophical Guide to Conditionals*. Oxford: Oxford University Press, 2006. Pp 22-24, and also pp 31-33 for the Ramsey test.
[546] Lewis, David. *Counterfactuals*. Malden Blackwell, 2001.pg 3

and logical values is similar to the clash between linguistic behaviorism and phenomenological style arguments that I'm going to be bringing out later in this paper in a dispute that arises between Ryle and Robert Wolff[547] on the nature of dispositional statements.

As such Lewis' work affords us a valuable insight into how these types of clashes can have a normative appeal. For Lewis' system, once understood, has its own force of appeal, while ordinary language usage has another. In Lewis' example, such a claim might arise in an argument where someone is using the subjunctive conditional to illustrate a counterfactual case, and we should have to decide whether to use Lewis system or to use the ordinary language form inherited by common English usage. Since the two forms make contradictory claims about what the counterfactual might be, either claim may shape the argument in a different way. By spending a few paragraphs illustrating both the common English form and Lewis' special form I can draw out the full force of the normative claim and provide a comparative case of what a clash between rival forces of appeal is like. For when we understand Lewis' Counterfactual Logic, we genuinely feel compelled towards the theory on logical grounds, yet when we return to ordinary language usage of the English subjunctive conditional, we do feel that that cases of use where the antecedent isn't taken to be false are some sort of mistake.

Secondly I want to distinguish between a formal logical use of the subjunctive conditional, the ordinary language usage and a special use I'm going to use later for Wilfrid Sellars that focuses on the 'residuum' in descriptive and fact stating roles. A careful analysis of Sellars argument shows that it actually anticipates and guards against a 'New-look psychology' criticism like Jerry Fodor makes[548]. Getting this framework in place is important for precise and shrewd philosophical practice.

According to Lewis theory

---

[547]See '7.3 Dispositions, Phenomenology and Law-Like Statements' in 'Chapter Seven : Motives, Moods and Reasons' in this paper.
[548]See 'Chapter Seven : Motives, Moods and Reasons' in this paper. Fodor's criticism isn't actually damaging to Sellars provided one conducts a close reading of the subjunctive conditional structure of the logic hidden in Sellars analysis of looks talk, but it is problematic to the development of a report language and the implicit language position on the mind.

$$\Box\,(\varphi \supset \psi)$$

is enough to guarantee the truth of

$$(\varphi \;\Box\!\!\rightarrow \psi)$$

Which is to say if $\Box\,(\varphi \supset \psi)$ is true, then $(\varphi \;\Box\!\!\rightarrow \psi)$ is also true.

That is

$$((\Box\,(\varphi \supset \psi)) \supset (\varphi \;\Box\!\!\rightarrow \psi))$$

However $(\varphi \;\Box\!\!\rightarrow \psi)$ can be true while $\Box\,(\varphi \supset \psi)$ can be false. The reason is the limit assumption[549] which claims that if there are infinitely many spheres around i[550] and i is the actual world[551], then any non-empty set of spheres has a smallest member. As such each sphere has a set of worlds and the smallest sphere is the first sphere out from i where the antecedent is true of some world and $(\varphi \supset \psi)$ holds at every world in that sphere.

---

[549]Lewis, *Counterfactuals*, 2001 Pg 19
[550]$ is centred on i, see Lewis, *Counterfactuals*, 2001 pg 14-15. If each world, i, is as similar to itself as any other world to it, then *i* should, itself, belong to every non-empty sphere around it.
[551] See Lewis, *Counterfactuals*, 2001 Pg 13 – 19. $ is nested, closed under unions and closed under non-empty intersections.

Later he does away with spheres for a system of comparative similarity[552]and comparative possibility [553] and utilizes a selection function[554] but we can illustrate the point he is after in reguard to the clash between the normative force behind a theory of logic and standard English use of the subjunctive conditional without going much beyond his system of spheres ($).

Thus in the case where $(\Phi \supset \psi)$ is true of every world in the innermost antecedent permitting sphere, Lewis argues that $(\Phi \,\Box\!\!\rightarrow \psi)$ is true, but if an outer sphere has a Φ-permitting world where ψ is also not true then $\Box(\Phi \supset \psi)$ would not be true because the outer sphere contains a world or worlds where Φ is true and ψ is not. This would make the necessary conditional false, since we have an incidence of a true antecedent and a false consequent. However $(\Phi \,\Box\!\!\rightarrow \psi)$ would still be true since $(\Phi. \psi)$ is true of every world in the smallest Φ-sphere.

Interestingly enough in Lewis' system a necessary consequent guarantees the truth value of the conditional. $\Box\psi$ for instance guarantees the truth of $(\Phi \,\Box\!\!\rightarrow \psi)$. The reason is that the would-modal operator has a vacuous and non-vacuous form of truth. In a set of spheres where ψ is necessary, i.e. $\Box\psi$ and Φ is entertainable at some antecedent permitting sphere, $(\Phi \supset \psi)$ would be true in the smallest sphere that Φ appeared in since the consequent ψ is necessarily true. So according to Lewis $(\Phi \,\Box\!\!\rightarrow \psi)$ would be true and would look like Diagram One

---

[552]Lewis, *Counterfactuals*, 2001. Pg 48
[553]Lewis, *Counterfactuals*, 2001. Pg 52
[554]Lewis, *Counterfactuals*, 2001. Pg 61

Diagram One

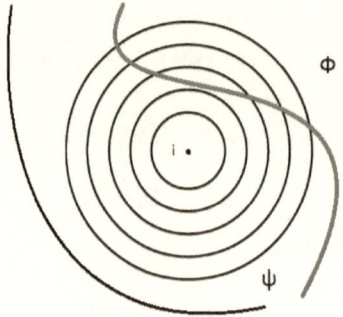

Diagram two is vacuous, because the antecedent occurs outside the system of spheres. I will give the formal conditions for the would-modal operator that Lewis uses in a moment, but it is interesting to see the logical relationship between these two forms.

Diagram Two

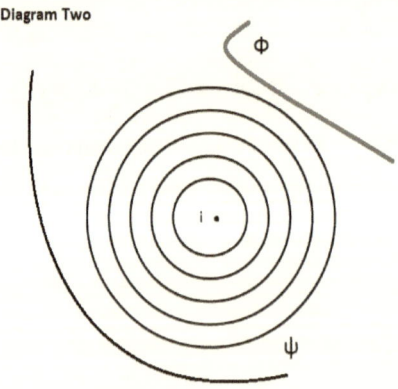

We can now see why $\Box (\phi \supset \psi)$ and $(\phi \; \Box\!\!\rightarrow \psi)$ are not equivalent. A situation where $(\phi \; \Box\!\!\rightarrow \psi)$ was true while $\Box (\phi \supset \psi)$ was false would look like diagram three according to Lewis system of spheres. Here $(\phi \supset \psi)$ holds for an inner sphere but is false in an outer sphere. I have included arrows to show where this is so.

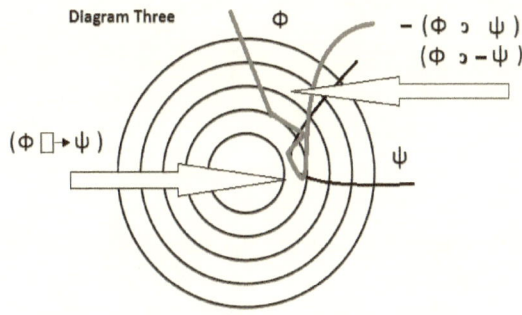

Diagram Three

The truth of $(\phi \; \Box\!\!\rightarrow \psi)$ is not enough to guarantee the truth of $\Box (\phi \supset \psi)$. However if $\Box (\phi \supset \psi)$ is true then $(\phi \; \Box\!\!\rightarrow \psi)$ must be true since the modal operator guarantees the truth of the inner-most antecedent permitting sphere as well as the outer-most. We can see how this logical relationship of the conditional works between the 'de dicto' necessary modal conditional and Lewis' counterfactual would-modal operator if we apply modus tollens and arrive at a world where the conditional does not hold.

Diagram Four

1) $((\Box(\Phi \supset \psi)) \supset (\Phi \Box\!\!\rightarrow \psi))$

2) $\;\;\vert\;\; -(\Phi \Box\!\!\rightarrow \psi)$ Assume for a demonstration

3) $\;\;\vert\;\; -\Box(\Phi \supset \psi)$ Modus Tollens 1, 2

4) $\;\;\vert\;\; \Diamond -(\Phi \supset \psi)$ RS, Reverse 3

5) $\;\;\vert\;\; w\;\; -(\Phi \supset \psi)$ DD Drop Diamond 4

6) $\;\;\vert\;\; w\;\; \Phi$ Negative Conditional 5

7) $\;\;\vert\;\; w\;\; -\psi$ Negative Conditional 5

[555]

Which gives us $(\Phi. -\psi)$, this is the inner most world where Φ and not ψ obtain, and also the world that violates $\Box(\Phi \supset \psi)$ since the antecedent Φ is true while the consequent ψ is not. In Lewis system this would look like diagram five since this is the innermost sphere with a world where Φ is true and ψ is not.

---

[555]This differs from Lewis' work and his quantified translation key in Lewis, David. "Counterpart Theory and Quantified Modal Logic." In *Philosophy of Logic: An Anthology*, edited by Dale Jacquette. Oxford: Blackwell, 2002. Pp 292-293. There seems to be a problem with Lewis claim that 'nothing is in anything except a world' since it seems he's dodging the problem that for a world to be something it needs to be in another world. Here, however is not the place to bring it out. The logic I've suggested, however, seems elementary and I argue, is implied by his claim regarding de dicto necessity and the 'would' counterfactual logic operator.

Diagram Five

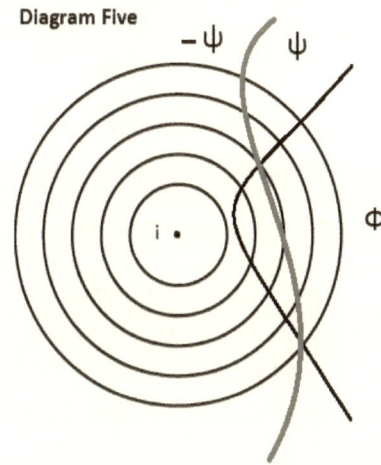

This is because $(\Phi. - \psi)$, does not meet the conditions for his would-modal counterfactual operator, which he gives on page 16.

Lewis writes

$(\Phi \,\square\!\!\rightarrow\, \psi)$ is true at a world i (according to a system of spheres $) if and only if either

(1) No Φ-world belongs to any sphere S in $ or
(2) Some sphere S in $, does contain at least one Φ-world, and
$(\Phi \,\supset\, \psi)$ holds at every world in S.[556]

We can see that $(\Phi. - \psi)$ breaks the first formal condition of the would-modal operator since Φ appears in the system in some sphere in ($). It also breaks the second since the smallest

---

[556]Lewis, *Counterfactuals*, 2001. Pg 16

antecedent permitting sphere contains at least one world where Φ is true and ψ is not, in that same world, and thus $-(\Phi \; \Box\!\!\rightarrow \psi)$.

If we return to Lewis' formal conditions, the two cases where $(\Phi \; \Box\!\!\rightarrow \psi)$ would be true are firstly the non vacuous case and secondly the vacuous case. The conditions for truth in the non-vacuous case are thus $(\Phi \; \Box\!\!\rightarrow \psi)$ and $-(\Phi \; \Box\!\!\rightarrow -\psi)$. The conditions for truth in the vacuous case are $(\Phi \; \Box\!\!\rightarrow -\psi)$ and $(\Phi \; \Box\!\!\rightarrow \psi)$. The vacuous case looks particularly disturbing to a logician since it seems to imply a contradiction, however, as Lewis assures us, since the antecedent never appears in the system, a contradiction cannot arise. As such Lewis can treat both conditions of the vacuous case as true even though were an antecedent to arise he would have a contradiction in the smallest antecedent sphere.

In general he treats impossible antecedents, like say a self contradictory antecedent, as vacuous cases where the antecedent can't appear in the system. Later on though, he introduces a strong modal operator to replace these. This is the 'double arrow' would-modal operator, which obtain from a different set of conditions[557].

Lewis writes

$(\Phi \; \Box\!\!\Rightarrow \psi)$ is true at a world I (according to a system of spheres $) if and only if there is some S in $, such that S contains at least one Φ-world, and $(\Phi \; \supset \; \psi)$ holds at every world in S[558].

[557]Lewis, *Counterfactuals*, 2001. Pg 24
[558]Lewis, *Counterfactuals*, 2001. Pg 24

Then on page 26 he defines the would-modal operator in terms of the stronger 'double arrow' would-modal operator.

$$(\phi \ \Box\!\!\rightarrow \psi) \ \overset{df}{=} \ (\phi \Box\!\!\Rightarrow \psi) \supset (\phi \Box\!\!\Rightarrow \psi)$$

According to Lewis' formal conditions for the would-counterfactual in a case where $-(\phi \ \Box\!\!\rightarrow \psi)$ and $(\phi \ \Box\!\!\rightarrow -\psi)$ we would say the opposite of the counterfactual was true. This was the case that we arrived at from the steps in Diagram Four, the case at which the nearest possible world, or the smallest antecedent permitting sphere contained a world or worlds where the antecedent was true and the consequent was false.

In a case where $-(\phi \ \Box\!\!\rightarrow \psi)$ and $-(\phi \ \Box\!\!\rightarrow -\psi)$ obtain we would say the opposite would be false. The reason for this last case is that $(\phi \supset \psi)$ is true in some antecedent worlds in the smallest antecedent permitting sphere, while $(\phi \supset \psi)$ is false at other antecedent worlds in the smallest antecedent permitting sphere. This violates condition (2) of Lewis' formal truth conditions for the would-modal operator. For while it is true of the first proposition of condition (2) "some sphere (S) in \$ does contain one ϕ-world" it is not true of the conjugation, that $(\phi \supset \psi)$ holds at every world in that sphere (S).

We can understand this relationship better if we look at how Lewis would express these conditions using the might-modal operator. He defines both the might-modal operator and the would-modal operator in terms of each other[559].

---

[559]Lewis, *Counterfactuals*, 2001. Pg Page 2

$$(\Phi \,\square\!\!\rightarrow\, \psi\,) \stackrel{df}{=} \,-(\Phi \,\diamondsuit\!\!\rightarrow\, -\psi\,)$$

$$(\Phi \,\diamondsuit\!\!\rightarrow\, \psi\,) \stackrel{df}{=} \,-(\Phi \,\diamondsuit\!\!\rightarrow\, -\psi\,)$$

And defines the truth conditions for $(\Phi \,\diamondsuit\!\!\rightarrow\, \psi\,)$ itself on page 21.

Lewis writes

$(\Phi \,\diamondsuit\!\!\rightarrow\, \psi\,)$ is true at a world (according to a system of spheres $) if and only if both

(1) Some Φ-world belongs to some sphere S in $, and

(2) Every sphere S in $, that contains at least one Φ-world contains at least one world where $(\Phi \,.\, \psi)$ holds.[560]

Notice here that unlike the would-modal operator both conditions must be fulfilled. Also unlike the would-modal operator there is no vacuous case. The significance of the conditions for the would-modal vacuous case, if given for the might-modal operator would render the operator false. That is if the conditions for truth in the vacuous case are both true, that is if

$(\Phi \,\square\!\!\rightarrow\, -\psi\,)$ and $(\Phi \,\square\!\!\rightarrow\, \psi\,)$ are true, because Φ doesn't appear in the system then the might-modal counterfactual operator would be simultaneously false $-(\Phi \,\diamondsuit\!\!\rightarrow\, \psi\,)$.

Like the vacuous case the 'opposite true' produces a false might-modal counterfactual $-(\Phi \,\diamondsuit\!\!\rightarrow\, \psi\,)$.

---

[560]Lewis, *Counterfactuals*, 2001. Pg 21

Like the 'vacuous' and 'opposite true' the 'opposite false' produces

$-(\Phi \diamondsuit\!\!\rightarrow \psi)$ but unlike the 'vacuous' and 'opposite true' it also produces

$(\Phi \diamondsuit\!\!\rightarrow \psi)$. The reason for this is that in the smallest antecedent sphere ψ holds for some Φ-worlds, while at other Φ-worlds ψ is false.

These relationships are important, because at i, if we take i to be the actual world,

according to Lewis $(\Phi \,\square\!\!\rightarrow \psi)$ reduces down to $(\Phi \supset \psi)$ if i is an antecedent

permitting world[561]. That is, if Φ is true at i, the conditions for $(\Phi \,\square\!\!\rightarrow \psi)$ become

$(\Phi \supset \psi)$.

So if $(\Phi \,\square\!\!\rightarrow \psi)$ is true and Φ is true at i, we should be able to go

$$\frac{\begin{array}{c}(\Phi \,\square\!\!\rightarrow \psi)\\ \Phi\end{array}}{\therefore \psi}$$

And since Lewis is claiming

$$\frac{(\psi \,\square\!\!\rightarrow \chi)}{\therefore (\psi \supset \chi)}$$

And we know ψ is true from the prior argument, we should be able to go

$$\frac{\begin{array}{c}(\psi \,\square\!\!\rightarrow \chi)\\ \psi\end{array}}{\therefore \chi}$$

---

[561]Lewis, *Counterfactuals*, 2001. Pg 27

Thus if $(\Phi \;\Box\!\!\!\to\; \psi)$ is true at i, and Φ is true at i we should be able to make the moves in diagram six.

Which explains why transitivity works in the propositional calculus but once we step outside of it transitivity breaks down. Consider the following diagram, diagram seven.

Diagram Six

$$\Phi = A \qquad (\Phi \;\Box\!\!\!\to\; \psi)$$
$$\psi = B \qquad \underline{(\psi \;\Box\!\!\!\to\; X)}$$
$$X = C \qquad (\Phi \;\Box\!\!\!\to\; X)$$

---

1) $(A \; c \; B)$
2) $(B \; c \; C)$
   $\therefore (A \; c \; C)$

3) ⎡ Asm - $(A \; c \; C)$
4) ⎜ A   NC 3
5) ⎜ -C   NC 3
6) ⎜ -B   MT 1, 6
7) ⎜ -A   MT 1, 6
8) ⎣ $(A \; . - A)$ CD 3, 7

   $\therefore (A \; . \; C)$

Diagram Seven

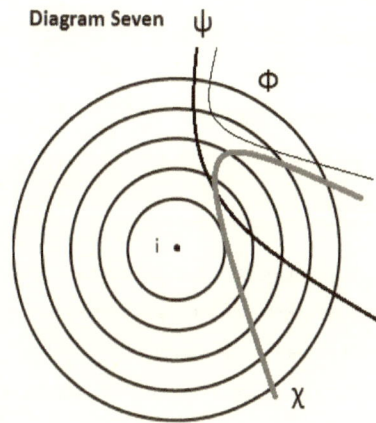

As you can see $(\Phi \ \Box\!\!\rightarrow \psi)$, and $(\psi \ \Box\!\!\rightarrow \chi)$, but $-(\Phi \ \Box\!\!\rightarrow \chi)$. Transitivity breaks down in counterfactual logic. The same happens to contraposition[562].

When Φ is true at the actual world and ψ is not then we get $-(\Phi \ \Box\!\!\rightarrow \psi)$.

Similarly when Φ is true and ψ is true, Lewis argues the counterfactual is true

$$\frac{(\Phi \ . \ \psi)}{\therefore (\Phi \ \Box\!\!\rightarrow \psi)}$$

I find this troubling because Φ and ψ might only be coincidently true. However, here is not the place to go into it. The point I wish to draw out, is that once we understand Lewis' theory on logical grounds standard material implication is a counterfactual that is true at i. This leads Lewis to argue

---

[562]Lewis, *Counterfactuals*, 2001. Pg 35

According to the truth conditions I have given, a counterfactual with true antecedents is true if and only if the consequent is true. This is so both for 'would' and 'might' counterfactuals. . . In short: counterfactuals with true antecedents reduce to material conditionals[563].

In this way the English subjunctive conditional may have a linguistic behavioral analysis, such as Lewis makes where the conditional must be false or it is mistaken, even though the truth values may be, and in fact there are conditions where subjunctive conditionals presented in counterfactual logic with a false antecedent may be true or false. Whether we pick the common English subjunctive conditional or we use Lewis quantification over worlds where propositions might or would be true has an effect on how we might read an argument that uses counterfactual cases. A premise that might otherwise be correct or true on Goodman's reading[564] or standard use of the English subjunctive conditional, may turn out to be false on Lewis' and this in turn will effect both the soundness and the validity of a particular argument.

This is a genuine clash and we should have to decide one way or another on some basis in a particular instance. The clash that occurs between the normative force in Lewis' theory, once one understands it, and correct ordinary language use of the English subjunctive conditional is a rivalry. It is a rivalry between truth values and the mistaken use of a language. More than this

---

[563] Lewis, *Counterfactuals*, 2001. Pg 26

[564]Goodman, Nelson. "The Problem of Counterfactual Conditionals." *Journal of Philosophy* 44, no. 5 (1947): 113-28. See also Sellars, Wilfrid. "Counterfactuals, Dispositions and the Causal Modalities " *Minnesota Studies in the Philosophy of Science* II, (1958): Pp 225-308. See also Plantinga, Alvin. *The Nature of Necessity*. Oxford: Claredon Press, 1978. Pp 174-180., Stalnaker, Robert. "A Theory of Conditionals." In *Studies in Logical Theory*, edited by N. Rescher. Oxford Blackwell, 1968. 98-112., Robert Stalnaker, Richard Thomason. "A Semantic Analysis of Conditional Logic." *Theoria* 36, no. 1 (1970): Pp 23-42. And Lewis own discussion of Stalnaker's theory, Lewis, *Counterfactuals*, 2001. Pp 77 – 83. Also Chalmers, David. "The Tyranny of the Subjunctive." http://consc.net/papers/tyranny.html, 1998. Presented at Princeton, for Chalmers analysis and 2D semantic framework, etcetera, for some other leading views.

once we understand Lewis' theory we can genuinely 'feel' this rivalry between his counterfactuals and our standard use of the English subjunctive conditional. Familiar and accepted usage pulls us one way, while Lewis' conditions for counterfactuals pulls us the other.

It is a rivalry between which we use to read an argument using. One reading would make the argument sensitive to truth values and a critique of invalidity, the other appeals to language intuitions and can claim the argument is mistaken.

Understanding the way that a normative force can affect us in reading an argument is important because I suggest in the paper a similar rivalry exists between linguistic behaviorist arguments and phenomenology, and that this rivalry between two different forms of analysis is obscured and what lays behind Ryle's Ordinary Language arguments. This rivalry inside of Ryle's Ordinary Language arguments is a rivalry between the two strains of argument in Ryle that I have been at pains to bring out through the paper. The strains are of course the Linguistic Behaviourist strain, which analyzes the behaviour of language and the 'occult' phenomenological strain that makes its claim on the basis of introspective scrutiny. If we include Lewis' argument then there can be seen three types of normative force that can draw us in different ways, a linguistic behavioural force like we get from Ryle, a phenomenological reflective force like we see in Sartre and a logical force like that we see in Lewis. However determining the relationship between these forces is a paper for another occasion.

Appendix B

Thesis Synopsis with detailed chapter summary.

Aim

This paper sets out to study the sources of normative appeal in the language of claims and arguments made about the nature of mind. It draws heavily on an analysis of Gilbert Ryle's arguments in *The Concept of Mind* in order to reveal a central flaw in anti-psychologistic theories of the mind. This paper argues for a return to a Pre-Fregeian Psychologism based on three central theses. The first thesis argues that auto-phenomenomenology, as elucidated in the paper, is a stronger source of appeal in arguments that involve individuation on the mind. The second thesis argues that emotions are irreducible to linguistic behaviors and that some pre-linguistic knowledge is necessary for establishing the grounds for various claims in what Robert Brandom calls 'the game of giving and asking for reasons'. Both of these theses undermine one of two sides in the anti-psychologistic position. The argument that emotions are irreducible to linguistic behavours is the capstone of a series of arguments that undermine an implicit language theory of meaning like that evinced by either Gilbert Ryle or Michael Dummet. The argument that pre-linguistic knowledge is necessarily prior to and fundamental for playing the game of giving and asking for reasons undermines the anti-psychologistic foundations in an explicit theory of meaning like that offered by Robert Brandom or Alfred Jules Ayer. The third thesis argues that since there are mental states, operations and acts that are irreducible to linguistic accounts, the anti-psychologistic attempt to search for meaning within language is heavily flawed, since both require non-linguistic meaning, one by showing that non-linguistic foundations are fundamental in establishing meaning within the game of giving and asking for reasons, and the other for individuating between different types of emotions and arguments that make claims about emotions since emotions cannot be reduced to linguistic behavours, nor can linguistic behavours describe or individuate between them. This paper argues, thus for the third major thesis. It argues that what we need prior to a theory of linguistic meaning is a theory about the nature and structure of consciousness. This paper argues that we need to return to a Pre-Fregeian psychologism in our search for a theory of mind.This is because the paper argues that a theory of consciousness undermines the linguistic foundations of Twentieth Century anti-psychologism that has dominated analytic philosophy of mind since the work of Gottlob Frege.

Method

The central method of the paper is to draw upon distinctions designed to reveal hidden sources of appeal within Ryle's arguments along with elements drawn from the philosophy of Jean Paul Sartre. There are strong historical grounds presented in the paper for arguing that Sartre's philosophy was instrumental in influencing a shift in Ryle's philosophy. This shift in Ryle's philosophy creates a key set of inconsistencies in his classic work *The Concept of Mind*. By examining these key inconsistencies we can learn to identify theories about the mind that commit a critical error and avoid this same mistake. Towards these ends the paper employs a wider distinction between theories of meaning in talk about the mind that distinguishes an anti-psychologistic strain of argument, this distinction involves an implicit and explicit distinction in approaches to a theory of meaning. The paper argues both attempts are mistaken as they depend upon a shared common error. The anti-psychologistic strain of argument takes it that a theory of linguistic meaning is prior to and fundamental for a theory about the mind. It does this for any number of reasons, such as psychologists can not explain or talk about their theories of mind without utilizing a language and a language draws upon the implicit knowledge of meaning the user has when he uses it. At the core of these arguments, Michael Dummett idenitifies the founding tenant of anti-psychologism as the rejection of the view that our words are mere codes for thoughts, and that we can not imagine what our thoughts might be like without language, hence fundamental to the search for a theory of mind is the search for a theory of meaning. This paper rejects this view. This paper argues that linguistic descriptions of emotions, utterences like 'a stab of anger'. 'a throb of regret' and 'a glow of pride' are mere descriptions of emotional phenomena that occur after the fact and that these are part of an elaborate illusion that leads many philosophers to see language as fundamental to mind. This paper also explores two other insidious forms of this illusion in the linguistic content of reasons or motives, and beliefs, using Alfred Jules Ayer and Robert Brandom as examples of the same elaborate illusion where language is taken for the thing that the language is describing, except in this case the illusion takes the form of a complex doctrine of judgments for Robert Brandom and propositions asserted on the grounds of a semantic theory of truth for Alfed Jules Ayer.

There are three stages to the methodology of the paper. The first stage divides Ryle's arguments by their appeal into three different types. The second stage of the paper dissolves the normative authority of one type of argument, the Ordinary Language argument, as merely a set of assumptions based on the other two. Dissolving the authority of the Ordinary Language argument allows us to see through an illusion that arises on the surface of Ryle's philosophy. Grasping this insight allows us to access a deeper insight into theories about meaning in the philosophy of mind. This is because Ryle uses Ordinary Language arguments overtly as a normative source to justify two other types of arguments he draws upon. By showing an instance where treating the source of an appeal in an Ordinary Language argument as an authoritative source would lead to a contradiction between claims made by the other two types of argument, we can recognize that underneath 'Ordinary Language arguments' are several different types of appeal and that rather than these types of appeal being variations on the same type of argument, these can be rival sources that support contradictory claims. The third stage of the paper then reorganizes both types of argument left over from the breakdown of the third type of argument, the Ordinary Language argument, then pits these two other rival arguments against each other in order to argue which of the two is the stronger source for making claims that individuate on the nature of mind.

The synopses will now turn to the first stage before turning to exposition on analysis into the three types of argument in Ryle. For the first stage the paper develops its taxonomy of Ryle's arguments based on the types of appeal the argument draws upon and classifies the argument according to that source of appeal. As already noted by this synopses there are three distinct types of argument in Ryle. Recognizing these arguments allows us to break them up and analyze the source of normative appeal inside of them. The first type of argument we will look at are ones that depend upon a linguistic behavioral distinction. These are arguments that depend upon differences in linguistic behaviors to make a claim about the mind. The second type are arguments that depend upon the appeal to what an 'Ordinary language speaker' knows about a language. These are appeals to what it *makes sense to say* in a language. The third type of argument is hidden. These are arguments that involve phenomenological appeals to introspective elements of consciousness but often don't acknowledge the form of the appeal or its source. It is difficult to detect these in Ryle and a confusion often arises because Ryle himself was polemical about consciousness as a doctrine within philosophy. He thinks the term 'consciousness' arose as part of a technical vocabulary that was picked up out of Protestant Reformation theology before being developed into various historical philosophical doctrines. It will thus be necessary to extract a phenomenological method as a way of recognizing this third type of argument and separating Ryle's polemical views on consciousness and his arguments against it from his actual use of consciousness in the form of various acts he asks us to perform for the sake of making claims about the mind.

Turning now to the first type of argument, the Linguistic Behavioral argument, the thread for distinguishing these begins in the chapter *The Difference Between Linguistic Behaviorism and Logical Behaviorism* and is based on the inability of Proffesor Weitz's propositional model to capture what it is about language Ryle is specifically interested in. Professor Weitz offers three types of model propositional sentence he argues can capture the essence of what he calls 'Ryle's Logical Behaviorism'. However none of the three models offered by Weitz capture the specific features of Ryle's interest in language at the level of interest that Ryle entertains. The reason for this is Weitz's model is an explicit reading of Ryle, rather than an implicit reading. Weitz tries to capture Ryle's model at the level of whole propositions or model sentences rather than at the level of relationships between groups of verbs, nouns and adjectives creating various families of dispositions, episodes and heed verbs like Ryle does. Ryle is interested in the behavior of words. In particular, Weitz makes an error of assuming that all dispositions follow a uniform function. The paper points out that Weitz cannot distinguish between one family of dispositions that are based on a family of verbs, adjectives and nouns that account for skills, abilities and capacities, and another family of dispositions that Ryle groups together and which account for motives, beliefs, tendencies and inclinations. Weitz's model fails to make the distinction between the two families of dispositions because his model cannot observe linguistic behaviors at a levelthat can recognize the relationship between the different type of epithet applied to people who fake or lie about motives, beliefs and tendencies and those who fake or lie about abilities, skills and capacities. Those who lie in regards to the family of capacity dispositions, Ryle argues, we call cranks, charlatans and frauds. Those who lie about their beliefs, motives and inclinations, Ryle argues, we call hypocrites. To observe the difference you have to observe the relationship between the sets of verbs and linguistic structures that make up claims about motives, beliefs and inclinations, against those that make up the family of dispositions relevant to capacities, abilities and skills, and the different types of epithet applied to people who fake one of either, while also recognizing the sets of relationships that hold both families together and discerning the difference between them. Weitz's model also fails on any number of fronts. It cannot capture the knowledge/belief distinction made on the behavior of 'how' and 'that' clauses, as well as adverbs of manner like carefully, recklessly, and heed verbs which refer to skills, abilities, tendencies and capacities and not, motives, beliefs, tendencies or habits.

In 'Linguistic Behaviorism' Episodes and dispositions. What exemplifies the Linguistic Behaviorist's claims? this paper continues to build on the linguistic behavioral phylum of Ryle's arguments by looking into what exemplifies a 'linguistic behaviorist' claim. Here the paper examines the distinction between episodes and dispositions to see what exactly Ryle bases that distinction on and discovers he bases it specifically on an analysis into the behavior of words. The paper notes that Ryle bases the distinction between episodes and dispositions on the fact that verbs like 'know, believe, aspire and posses' do not behave in the manner of verbs like 'run, wake up and tingle'. One set of verbs refers to a person's temperament, motives, outlook or nature while the other set refers to specific instances in time.

In Capacity and Tendency Dispositions. How do you know? The Primacy of Concern and Knowledge How and Making an Ordinary Language claim the paper follows up on the 'capacities' and 'tendencies' distinction and eliminates 'capacity dispositions' including the groupings of 'abilities' and 'skills' from the central thread of the paper as inapplicable to the anti-psychologisitc schools the paper is attacking so as not to build a strawman. It notes various Neo-Ryleans who lend themselves towards building this strawman but also notes that Ryle himself supplies the arguments necessary for excluding capacity dispositions from an anti-psychologistic thesis.

It is also here in these two chapters, *Capacity and Tendency Dispositions. How do you know? The Primacy of Concern* and *Knowledge How and Making an Ordinary Language claim* we conclude our phylum of Ryle's Linguistic Behaviorist style argument. These are arguments that make their distinction on specific behavioral grounds of linguistic structures, utterances or what French linguistics call a parole. Ryle usually makes these claims about the mind in the grammatical currency of groups of verbs, nouns, adverbs and adjectives based on their behavior and various relationships between different sets at a level of analysis that goes beyond propositional models or whole sentences. Part of what makes Ryle an implicit language theorist is he makes the distinctions in the words themselves without abstracting to any attempt at an explicit statement of meaning. We will return to this point later in the synopses when we explore the second type of argument Ryle uses. The Ordinary Language claim.

In *Linguistic Behaviorist Claims, the Occult stream of consciousness. Propensities, Occurrences and the Agitational Calculus* the paper finishes building its analytical phyla of Ryle's Linguistic Behavioral cartography with a look at the Occurrences. The Occurrences are important because these contain feelings proper. 'Feelings proper', which the paper gives the title 'flash-bang', so as not to lose them in the hills and dales of Ryle's Linguistic Behavioral Cartography, will be important in our final account because they offer us a key and critical insight into the problem underlying linguistic theories about the mind, but which won't make sense until we build the steps that lead up to that insight and we are ready to understand it.

Feelings proper or 'flash-bangs' and tendency dispositions, which are made up of beliefs, motives and inclinations will be the focus of the core of our attack on Ryle's Linguistic Behaviorism, since we have excluded the 'capacity dispositions' which includes skills, abilitiesand capacities. The 'flash-bangs' or feelings proper of the 'Occurrences' are a strain of linguistic behaviors defined by Ryle as utterances like 'a glow of pain' or 'a stab of anger'. Here I will draw attention to the fact that Ryle has trouble distinguishing between a 'glow of pride' and a 'glow of warmth' on purely Linguistic Behavoural grounds. This will be the first step of three that will give us our final insight and form the backbone of the argument for a return to a Pre-Fregian Psychologism. But we need a number of stages and steps in place to reach that insight.

The Occurrences will eventuate in one half of the second central thesis of the paper and the claim that emotions are irreducible and can not be individuated by linguistic behaviors. The other half of the second half of the second central thesis will arise from the tendency dispositions. Both provide support for the argument the thesis puts forward that we need a return to a Pre-Fregeian psychologism in the philosophy of mind.

The second type of argument Ryle uses is a Linguistic Behavioral argument. The thread for this argument also starts in *The Difference Between Linguistic*

Behaviorism and Logical Behaviorism and like the Linguistic Behavioral argument, the phylum for this type of Rylean argument draws upon the deficiencies in attempts by prior philosophers to capture the illusive forms of appeal that characterize Ryle's arguments on the mind. In this case David J. Chalmers account of Ryle's philosophy as a codification of the Freudian-Behaviorist orthodoxy in to an argument that our mental concepts can be analyzed in terms of (i) dispositions to behave in certain ways and (ii) associated behaviors, fails to capture the normative grounds that Ryle makes those claims on. Ryle was first and foremost an Ordinary Language philosopher and Chalmers account leaves out the normative basis of Ryle's Ordinary Language argument. What Chalmers leaves out of his account of Ryle's 'Logical Behaviorism" is that Ryle's philosophy is heavily characterized by an appeal to the authority of Ordinary Language and what it *makes sense to say* within a language.

Ryle justifies his arguments in *The Concept of Mind* with the view that people already know how to talk about the mind and the role that a philosopher of mind need be is one of taking that knowledge and correcting its logical geography. Confusion arises, Ryle argues, when highly technical theoretical vocabularies like those of the sciences, Reformation theology, philosophy, the laws of motion, and so on, are applied to areas people already know how to talk about using everyday language. The shift from a vocabulary of everyday language about the mind to a technical vocabulary creates confusions between the two when one is applied to the other. In *Linguistic Behaviourism: Episodes and dispositions. What exemplifies the Linguistic Behaviourist Claims?* I build on Ryle's insight by drawing on Michael Dummett's paper *What do I know when I know a language?* Both Ryle and Dummett are implicit language philosophers. By that they think that meaning can only be talked about implicitly inside of a language. Dummett is arguing against one particular type of theory about meaning which is forwarded by people like Robert Brandom and Alfred Jules Ayer and of which, Frege's is the earliest view. This position is the explicit language position which takes it that a theory of meaning can be stated explicitly. Dummett thinks that any explicit attempt at formulating a theory to explain meaning is going to invoke necessarily unavoidable implicit knowledge in stating and communicating that theory.

Ryle argues along similar lines tp Dummett in regards to psychology. Ryle argues that if a psychologist were to discover a new psychological theory about either the mind or meaning in language, the psychologist is going to be in the curious position that he can only communicate his new theory through the medium of a language he already knows.

Both Ryle and Dummett occupy the implicit side of an implicit and explicit distinction. On the other side we have Brandom, who offers an argument based on judgments and propositions and Ayer who has a theory about language that involves semantic claims about truth and a criteria stating explicitly what the meaning for that truth is, whether it involves an appeal to endorsement in a belief or action like in emotivism or propositionally structured beliefs based on arguments that the ability to be mistaken only arises with language because the truth or falsity of a statement or proposition is only possible once a language is on the scene. Brandom is post-Sellarian, Ayer is pre-Sellarian. The four, Ryle and Dummett as implicit language theorists and Brandom and Ayer as explicit language theorists are all anti-psychologistic. This is because all four are united by a common argument. This is important enough to place the entire quote in the synopsis.

> Philosophers before Frege assumed. . . that what a speaker knows is a kind of code. Concepts are coded into words and thoughts which are compounded out of concepts, into sentences, whose structure mirrors, by and large, the complexity of the thoughts. We need language, on this view, only because we happen to lack the faculty, that is, of the direct transmission of thoughts. Communication is, thus essentially like the use of a telephone: the speaker codes his thoughts in a transmissible medium, which is then decoded by the hearer. The whole analytical school of philosophy is founded on the rejection of this conception, first clearly repudiated by Frege. The conception of language as a code requires that we ascribe concepts and thoughts to people independently of their knowledge of language; and one strand of objection is that, for any but the simplest concepts, we cannot explain what it is to grasp them independently of the ability to express them in language[565].

This is the anti-psychologistic position underlying all four philosophers. The anti-psychologistic position argues that language is fundamental to and essential for thought, thus a theory of meaning is prior and fundamental to a theory of mind. The paper argues this is not the case. Language is something that happens to thought. In some cases language offers us a descriptive vocabulary for thought processes and experiences but it is not these thought processes, it is more like a code we use to describe these processes since we lack the faculty of direct transmission. The paper is going to argue that one needs a return to a Pre-Fregeian psychologism to understand the mind. It is only after we have a theory of mind that we can work towards a complete theory of meaning. The final insight offered by this paper will argue that language is a code we use to communicate thoughts. It will do this through phenomenological steps that lead up to that insight.

The difference between the Linguistic Behavoural and the implicit language theorist's position, and the explicit language theorist's position on a theory of meaning in talk about the mind is that the explicit language theorist's position

[565]Dummett, *What Do I Know When I Know a Language?*, 1993

utilizes a theory about meaning that requires explicit statements about what meaning is. In this case Fregeian meaning, of which Brandom and Ayer form the explicit branch, argues that a word can only have meaning in the context the word can or does appear in, and most commonly represented in the Pre-Sellarian strain with propositional forms or theories about meaning that involve semantic claims about truth in statements, in this case represented by Alfred Jules Ayer's position on statements about beliefs and grounds for proving theories. Brandom similarly follows a semi-propositional structure which he calls judgments. He views judgments as the minimum unit one is able to think and affirm. Implicit theories like Ryle and Dummett, and Sellars to a certain extent, are less interested in theories of meaning that rely on sentences or propositional structures, and more interested in exploring relationships between meaning and words. All four use Ordinary Language arguments, or arguments that appeal to what *it makes sense to say* in a particular language.

In *Linguistic Behaviorism, Ordinary Language arguments and the Normative Value of Science*, the paper differentiates between Ordinary Language arguments and Ordinary Language statements. Ordinary Language statements are arguments about why ordinary language is a good source for telling us things about the mind. Ordinary Language arguments, however, are arguments that appeal to what it makes sense to say in a language. They are arguments where the terms '*it makes sense to say*' or '*it doesn't make sense to say*' enters into an argument to affirm or deny a claim about the mind. The two are different. An Ordinary Language claim or argument is an appeal to a user's knowledge of a language as a normative source in order to make some claim about what it makes sense to say in a language. An Ordinary Language *statement* is a claim as to why the normative value of an Ordinary Language argument or claim is a better source than some other normative source like an appeal to a meta-theory of logical syntax about possible worlds or the finding from a particular branch of science. In *Ordinary Language, the Manifest Image and the Philosophy of Mind*, this paper illustrates the difference between Ordinary Language arguments and Ordinary Language statements with a debate that arises between Wilfrid Sellars work in *Philosophy and the Scientific Image of Man* and Ryle's work *Dilemmas*. The debate centers over whether 'Ordinary Language' or 'Science' is a better source for telling us things about the mind. The core of this debate is a contention over whether science is the flowering of processes continuous with ordinary language, which Sellars holds, or whether the vocabularies of science are too theoretical and abstract, and merely contaminates ordinary language discourse about the mind leading to language confusions and philosophical puzzles, which of course is the view Ryle holds. The paper does not try to resolve the argument over 'Ordinary Language' and 'Science' here. The reason for including this debate is important to what will become a distinction between autophenomenological sources and heterophenomenological sources later in the methodology of the paper.

To recap on the two types of argument we have developed phyla for thus far we have the Ordinary Language argument that draws its appeal on claims that involve the terms 'it makes sense to say' or 'it does not make sense to say' to affirm or deny a claim. In the chapters *Ordinary Language Arguments, Analysis of Ordinary Language Claims and their ability to affirm or negate on a reputed source,* and *Exemplar: Ghostographic analysis of Ryle's philosophical practice* this paper offers several examples of these taken from Ryle. For instance in *The Concept of Mind* Ryle lodges an argument against Augustine's Volitions on the basis that '*it doesn't make sense to say* it took two, or three volitions to get out of bed', or '*it doesn't make sense to* talk about one's daily activities in this way'. I point out in this chapter that while Linguistic Behavioral arguments map onto Ordinary Language arguments, there are often Ordinary Language arguments that don't map onto Linguistic Behavioral arguments. That is to say there are arguments that Ryle makes that involve '*it makes sense to say*' type of claims, but which don't have Linguistic Behavioral distinctions in the form of specific structures or relationships between sets of verbs, nouns, adjectives, epithets and so on attached to them. Drawing attention to this opens the way for exploration of the third type of argument. The Phenomenological argument.

This leads us to the third type of argument which is the phenomenological argument. The foundation for this argument begins in the historical thesis at the start of the paper. The historical theses is made up of the chapters *The Two Accounts of the Imagination, The Mysterious Philosopher Hypothesis* and *Sartre's Irreal.* The historical thesis begins with and an examination in to the difference between Ryle's account of the imagination in *The Concept of Mind,* and *On Thinking,* and inconsistencies that run through Ryle's account of the imagination and the rest of his account in *The Concept of Mind.* The historical thesis draws evidence froma transcript from an interview with Bryan Magee, along with a confession in a little known article where Ryle admits he uses Sartre's argument against Hume in his account of the imagination, to posit that the inconsistencies in *The Concept of Mind* arise from little acknowledged influences on Ryle's work by Jean-Paul Sartre.

The paper posits that prior to this contamination of Ryle's philosophy and the influence of Sartre's philosophy on *The Concept of Mind,* Ryle's original concept of mind was that in thinking one is talking to one's self. Thought can thus be treated like language according to this view. Since people already know how to talk about the mind all one needed to do was to correct the geography of every day terminology on the mind to untangle the sorts of mirages and confusions that arise from imposing scientific, theological and philosophical vocabularies on to ordinary language. The paper then posits that Sartre's influence caused numerous problems for Ryle's idea that thought is simply the act of speaking to one's self in a number of ways which resulted in a number of inconsistencies in *The Concept of Mind* starting with Sartre's arguments against Hume which Ryle latter admitted were the very same arguments that he used in his work on the imagination in *The Concept of Mind.* This historical thesis introduces us to the phenomenological arguments of *The Concept of Mind.*

This third strain of argument, the 'phenomenological strain' of argument in Ryle's *The Concept of Mind* are an occult form of argument. They are an 'occult' form of argument because the appeal they make is often hidden and not openly acknowledged by Ryle. The overall goal of this part of the paper is to return to Sartre and sift out from his phenomenology a methodology for distinguishing these types of arguments and drawing attention to the source of appeal they make. Part of what obfuscates these arguments is the fact that Ryle has a polemic against consciousness that begins with a diachronic attack on the meaning of the term 'consciousness' and includes arguments against using the term 'introspection' and 'reflection' which are based on the view that these arise as philosophical doctrines from confusions and puzzles that are introduced by applying the technical vocabularies of early modern science, theories of optics and Protestant Reformation theology to what we know about the mind in our ordinary language talk. Ryle utilizes various 'log keeper' and internal monologue explanations to account for what we take to be 'reflection', 'consciousness' and 'introspection' which he calls retrospection and are essentially linguistically natured internalizations of our ability to report on these processes.

Before turning to Sartre, the paper first lays out Ryle's linguistic behavioral concept of mind and his attack on consciousness in order to show that Ryle does not have the resources to account for his own use of acts of consciousness and introspection methodologically in one particular strain of arguments that arise from *The Concept of Mind*. These arguments of course are the phenomenological strain of arguments. His rejection of consciousness, introspection and reflection, over and against his use of introspective acts that rely upon distinguishing different types of consciousness for making specific arguments are at the core of the inconsistencies in *The Concept of Mind*. For many of the things he overtly denies about consciousness, he himself does or gets the audience to do in making various arguments.

In the chapter *Ryle's use of the terms 'consciousness' and 'introspection'* the paper lays out all of the ways in which Ryle limits the uoe of the terms 'consciousness' and 'introspection'. All of these limitations arise from the way people use the terms 'consciousness' and 'introspection' in everyday conversation. For instance Ryle allows consciousness in the sense that one can be conscious that something in the room is different, or that they've lost feeling from the knees down, or in the sense that they're self conscious and concerned how others might view them. He makes these sorts of distinctions in *The Concept of Mind* and the paper exhaustively lists them.Then in *The Species of Mindologue* the paper chronicles the various types of 'thought speak' Ryle identifies along with the foundations for his theory that thinking is like talking and there are different varieties.

In *Ryle's Diachronic attack on Consciousness* the paper lays out Ryle's theory that consciousness is merely jargon left over from the Reformation that philosophers have picked up on and expanded.

All of these are compared and illustrated by the time the paper concludes the argument it started in *Ryle's use of the terms 'consciousness' and 'introspection'*, which was of course that Ryle does not have enough resources to account for consciousness and acts of conscious introspection in his phenomenological strain of arguments. The argument that Ryle does not have the resources for his phenomenological arguments in his everyday language philosophy concludes in the chapter entitled *Phenomenology* by showing that Ryle's linguistic behavioral account of the mind along with the different species of language and the log keeper account he offers cannot account for his use of introspection and consciousness in the acts and forms of his phenomenology. One example, his 'Reader/witness' argument, serves to illustrate this point.

In the 'Reader/witness argument, Ryle asks us to compare the account of the reader of a race with a witness of a race to show that the reader can imagine the race from any angle he likes while the reader is limited to the view from which he watched the race. The argument depends upon a perception and visualization difference that only becomes apparent once the audience has attempted the exercise to see the point. The point cannot be broken down into linguistic behaviors, nor does it depend upon the normative value of what it *'makes sense to say'* within a language. Another phenomenological argument might serve to illustrate this point further. In *The Concept of Mind* Ryle points out that one can not anticipate one's next thought before one has it. It is not until one actually tries the exercise that one sees the problem. These are arguments that cannot be made on the Linguistic Behavioral or Ordinary Language resources that Ryle offers for his theory of mind. They cannot be broken down into log keeper accounts or explained in terms of different types or species of internal language, nor can they be explained away in Linguistic Behavioral terms or by an appeal to Ordinary Language as a normative basis for such claims. These are phenomenological arguments that depend, for their strength, upon some process of phenomenological exercise. One must try the argument out before one can see the point of it.

In order to develop a methodology for isolating these phenomenological arguments and revealing the source of appeal they make this paper carries out an exposition of Sartre's philosophy. The paper begins with a short literary review in order to correct a number of mistakes in the Sartre literature. Most important is a reading of Sartre's process ontology against the background of Hegel's work on logic. In *Did Sartre read Hegel's Logic and Why is this Important?* The paper argues, pace Professor Spade, that Sartre did read Hegel's work on logic and argues this on the basis that Sartre said that he did. It backs Sartre's claim that he had read Hegel's work on logic by showing that, in fact, there are numerous references back to Hegel's work on logic throughout *Being and Nothingness* along with an analysis of various stages of the logic.

The relationship between Hegel's logic and Sartre's *Being and Nothingness* is important to understand Sartre's process ontology. Sartre's process ontology is important for this paper because it reveals how Sartre viewed consciousness as an 'ego creating force'. The 'ego' is contained in the reflective consciousness and contains what Sartre calls the 'cortege of the psyche' and appears in linguistic descriptions of emotions and actions such as 'I was angry' or 'she hurt me'. The ego appears before a present consciousness reflecting on a prior consciousness, which is positional towards an object, situation or circumstance in a way that is described in far greater detail inside *Being and Nothingness* and elucidated in the paper. However what is important is that the ego is accessible to conscious thought and can be used in an argument. The ego appears with a linguistic marker in statements or descriptions about how objects or situations make us feel, or actions carried out for different reasons, and is referred back to in linguistic statements describing elements of the 'cortege of the psyche' with an 'I' or a 'me'. A correct reading of Sartre against a background of his references to Hegel's logic and the translator's notes, published separately after the original translation, allows us to identify and access the 'cortege of the psyche' which contains information about these prior acts and the way consciousness perceives or views them. It is the 'cortege of the psyche' that Ryle engages in many of his phenomenological arguments. A reading of Hegel's logic and Sartre's engagement with it over the course of *Being and Nothingness* is critical for understanding why Sartre views consciousness as an ego creating force and for unlocking puzzling statements about the mind.

Sartre argues that consciousness arises in the world oriented towards an object which becomes its ontological foundation for its own negation and thus the grounds of existence. This is a process that Sartre refers to as producing the 'in-itself of facticity'. This process is critical to understanding how Sartre uses thetic consciousness to uncover the 'cortege of the psyche' in arguing about consciousness.

Sartre demonstrates how, while reflecting upon the object of a prior consciousness the present reflecting consciousness can bring a prior act of consciousness before the present one in order to view the 'cortège of the psyche' and make arguments based upon it. The act of a reflective consciousness on a reflecting consciousness for an argument is clarified in a technical sense and simplified by the paper into 'introspection' while the initial positional act of consciousness that refers us to either an object or an object of a prior consciousness without engaging the 'cortege of the psyche' is referred to in the paper in the technical sense as 'consciousness'.

Identifying the presence and nature of this structure inside of an argument, that is, reflection on a prior act of consciousness, what Sartre calls 'thetic' consciousness, is essential for identifying the third type of argument and why Ryle undermines his own polemic. The act of bringing consciousness before itself as an object of scrutiny to view the 'cortege of the psyche' for an act of introspection in order to make a claim about the mind, reveals to us the hidden phenomenological appeal in many of Ryle's occult arguments. When we step into the argument and compare prior acts of consciousness like when an argument asks us to compare the time we watched a race and compare those instances with times we read about the race or listened to it on the radio in order to see the difference between perception and visualization, being of course, as a witness we are limited to one view up in the stadium while as a reader we can envisage the entire event from whatever angle we like, in accordance with limitations imposed on us by the account we read, we are engaging with a thetic act inside of an argument and engaging in a form of phenomenology. Ryle does this, of course, when he takes up Sartre's arguments against Hume and argues that 'impressions' and 'ideas' are fundamentally a different sort of thing based on reflective acts like getting us to compare our experience of lightening or light so bright it hurts our eyes with an attempt to imagine light so bright it hurts our eyes. These are phenomenological arguments that take consciousness as their object and require us to examine our consciousness in a thetic way to see the point of them.

This gives us our central three forms of recognizing, analyzing and organizing the different types of appeal in arguments that Ryle makes about the mind. To recap, the first type of argument the synopses looked at in depth, examined in the paper, was the Linguistic Behavioral argument which depended upon examining linguistic behaviors and identifying structures at the level of relationships between verbs, nouns, adverbs, adjectives, epithets and so on, in order to make claims about the mind. The second type we looked at was the Ordinary Language argument, which drew its appeal from what *'it makes sense to say'* or *'doesn't make sense to say'* within a language or an equivalent claim. The third type of argument we looked at in this synopsis was the Phenomenological argument which utilizes a structure such that consciousness takes itself as its own object in order to make a claim about the mind. Here, we drew upon the methodology of Jean-Paul Sartre, which we have strong historical grounds for believing influenced Ryle. It was this influence, it was argued in the historical thesis, created the inconsistencies that riddle Ryle's *The Concept of Mind.* From Sartre we developed a method for recognizing and clarifying thetic arguments. This recognition process will be essential later when we reach Professor Wolff's criticism of Ryle, and will pave the way for the transition between the second and third part of the methodology.

In the next part of the synopsis I will describe the middle part of the methodology, which is to challenge and breakdown the normative authority of Ordinary Language as a unified source for justifying arguments or claims. It is important to do this in order to challenge the implicit anti-psychologistic strain behind Ryle and elucidated by Dummett, because Ryle thinks and treats both phenomenological style arguments and Linguistic Behavioral arguments as drawing their authority from what the ordinary language user knows. That is to say, Ryle thinks that both phenomenological arguments and Linguistic Behavioural arguments draw their authority from Ordinary Language arguments. By breaking down Ordinary Language arguments and showing that various claims that an implicit language philosopher would take to be grounded in what the ordinary language user knows about the use of a language are in fact rival claims, we lay the grounds to compare those rival sources of appeal and decide which of them is the stronger source. This forms the central shape of the argument of the thesis, but to understand how this shape works it is necessary to understand the way the machinery inside of that framework operates. One key piece of that machinery involves a two sided 'feet-of-clay' argument, which reveals the flaws in explicit and implicit arguments by critiquing Brandom and Ryle. The critique of Ryle will end in an exploded model of Ryle's implicit linguistic behavioral concept of mind, involving a systematic break down of dispositions, occurrences and propensities. The other side of the 'feet-of-clay' argument in the paper arises over settling an argument in the way analytic and continental philosophers read Kant. The Brandom side of the feet-of-clay argument will compare Brandom's reading of Kant against Sartre's. It is to this 'feet-of-clay' argument the paper will turn, before moving into the breakdown of Ordinary Language arguments as a unified source of appeal to ordinary language theorists, before finishing off with the third part of this synopsis, which will be a comparison of those rival sources which will lead us to our three central theses and the argument for a return to a Pre-Fregeian psychologism.

The Sartre-Brandom work starts in *Sartre, Phenomenology and the Apperception of the I,* and forms one side of the 'feet-of-clay' argument. This has its origins in the way the continental and analytic traditions read Kant. Explaining this work in the synopsis is important for understanding the detail in the structure of the overall argument the paper presents because the Brandom-Sartre argument forms one side of the two sided 'feet-of-clay' argument. On one side of the 'feet-of-clay' argument is an argument against explicit language theorists in the philosophy of mind. This side will argue that we do not walk around perceiving the world in the form of propositions, representations, judgments or statements. On the other side of the 'feet-of-clay' argument we have an argument against an implicit language theorist, in this case an exploded model of Ryle's concept of mind, systematically showing where all of the flaws are. The reason why the paper uses the specific instances of Brandom and Ryle is that it follows Sellars example of using actual philosophers to systematically expose flaws in styles of theoretical reasoning and offer insights into these classes of errors.

On Ryle's side of the 'feet-of-clay' argument, by the end of the paper we will have systematically exploded Ryle's tendency dispositions which are made up of 'inclinations', 'motives' and 'beliefs'. 'Alongside this are Ryle's Occurrences in the form of the feelings proper, or flash-bangs as the paper calls them. The tendency dispositions and capacity dispositions taken together make up both sides of 'dispositions' in Ryle's linguistic behavioral concept of mind. The capacity dispositions have already been removed early in the paper as irrelevant and offering us little more than a straw man in the way of a good philosophical investigation and not of any core significance to the implicit language thesis as it relates to philosophy of mind. Ryle himself provides arguments to this effect which place examples of unrolling a skein of wool outside of a linguistic theory of meaning or an implicit theory about the mind relevant to the mind as viewed by a language theorist since the mind is not occupied in linguistic activity. There are examples of 'knowledge-how' relevant to an implicit theory of meaning and we discussed these briefly with Dummett and the paradoxical problems they present to stating an explicit theory of meaning. As already noted the paper picks up several Neo-Ryleans using explicit propositional and implicit language models to try and capture the know-how and knowledge-that debates and dismisses them on the basis of Ryle's argument.

The Brandom-Ayer explicit side of the 'feet-of-clay' argument starts with the Brandom-Sartre analytic/continental disagreement over how to read Kant's apperception of the I. The argument starts in the chapter titled *Thetic Consciousness and the Apperception of the I at the reflective level.* The continental tradition, here represented by Sartre, approaches Kant more in terms of guided analysis of experience and investigation into the possibility for the form that experience might take. The analytic tradition approaches Kant and the appecrception of the I heavily influenced by semantic insight from Frege about the way propositions and words work, which of course Brandom freely admits.

Sartre follows the phenomenological maxim introduced by Brentano, and argues that consciousness, in order to be conscious, must be conscious of something. Thus consciousness can be positional towards an object like an apple. Sartre argues that Kant argues it is a de jure condition for the possibility of perceiving an apple that one can think "I saw an apple" but the I does not exist as a de-facto reality always present in our thoughts. This is because the "I" or the "me" are both parts of a linguistic marker for Sartre's "ego" and only appear on reflection into a prior act of consciousness. This act of reflection in which the 'I' or the 'me' appear as linguistic markers for the ego is a form of thetic consciousness. Thetic consciousness is when consciousness takes itself as its own object. Sartre argues that it is a condition of consciousness, that is being conscious of something, that a later act of consciousness can reflect back on it at which point the "I" appears. As already noted the reflected ego linguistically marked by the "I" contains what Sartre refers to as the 'cortège of the psyche'. These are aspects or elements of the prior consciousness the present one can apprehend on reflection like how the apple makes us feel. If the apple made us hungry then the memory of that hunger will appear as part of the 'cortege of the psyche' upon reflection into that prior act of consciousness in which we first perceived the apple and it made us hungry. The insights from analysing and comparing information gleaned from the 'cortege of the psyche' is what Ryle draws upon in many of his phenomenological arguments.

Brandom on the other hand argues that the minimum unit of thought someone is capable of is the proposition, since this is the minimum unit that one can affirm. He reads Kant's apperception of the I as the first level of thought in the "*I think* such and such is the case.' Brandom argues that the I is a 'de facto' inhabitant of consciousness that can either affirm or deny a judgment.

Rather than engaging the debate between Sartre and Brandom over Kant's apperception of the I on scholarly grounds and arguing whether Kant thought that the I is present de facto, or necessary as a de jure condition, the paper turns to noted Kant scholar Otfried Hoffe who argues that the key to understanding the apperception of the I in Kant is to understand the process as part of a larger structure which has three levels. On the most fundamental level is the process that Hoffe refers to as 'Original Synthesis'. This is the level where a unity of 'intuition' and 'concept' provide basic notions like weight. The second level of Hoffe's schematic is the point at which judgment appears such as 'the body is heavy'. It is only at the third level of Hoffe's schematic that the I appears in reflection on a judgment. Here we have the 'I think. . . '. If Brandom is right, on philosophical grounds, then we occupy thought on the third level and an I is always present in our thoughts. If

Sartre is right then we occupy it at the second level and can apprehend an object like a tree or a bird without an I appearing in our thoughts.

The paper argues that Sartre is right. This is important because there is a key insight that will start to become apparent in the next section of the synopses, when we move into the break-up of ordinary language arguments. In effect what I will do is move linguistic behaviorism and ordinary language arguments in to one of two sides of a distinction between first personal and third personal views. The paper will argue that there is a difference between thought and linguistic expression of a thought. This will be based on an insight the paper is leading up to that language as a description of thought processes is something fundamentally different to those processes. The argument once fully understood will demonstrate that language is something much like a code in which the speaker ciphers his thoughts into. This position once elucidated will of course undermine the fundamental tenant in Dummett's argument against psychologism and the root on which anti-psychologism is based as defined by Dummett. This argument will come to fruition in the conclusion to the occurrences thread with the argument against flash-bangs.

This synopsis will now move into the middle part of the methodology where it undermines and breaks down the authority of Ordinary Language arguments and develops two other forms of argument from the original three explained in the first part of the methodology. The synopses will then move into the final part of the paper before concluding followed by a brief discussion.

The middle part of the three part methodology begins in the chapter titled *Dispositions, Phenomenology and Law-Like Statements* and is based around a critique of Ryle by Professor Wolff. The critique is based on arguing that Ryle hypostatizes tendency and inclination dispositions into metaphysical forces compelling a person to do something. For instance, Wolff notes that Ryle argues that two contrary inclinations, two contrary tendencies, or two contrary motives, or any one against a factual impediment will produce an agitation. A man who is inclined towards patriotism and has a tendency to be cowardly will find himself in an agitated state when in a situation where he must fight for his country and there is danger present. The tendency and inclination dispositions, Ryle argues, act like forces impelling him to do two different things and this in turn makes him agitated. Wolff's criticism is that Ryle is imbuing 'dispositions' with a metaphysical force. Rather than possessing this force, Wolff notes the linguistic behavior of these dispositions in our ordinary everyday language is holistic and descriptive. Wolff argues pace Ryle that words do not act like a force compelling us in contrary ways creating an agitation. When we describe a man as patriotic and cowardly, we are describing a man who shies away from danger except in the case when he needs to fight for his country. These are merely descriptions of somebody we use in ordinary language without the sort of impelling force that Wolff criticizes Ryle of predicating them with. The paper agrees with Robert Wolff's point.

But the paper also returns to the case of the agitated man, and draws an example from Plato's Republic which involves a story told by Leontius, which is used by Plato as a part of an argument for the tripartite theory of the soul. The story involves a man who did not want to look at corpses that had been piled up and an inner compulsion that drove him to it. Everyone is familiar with these sorts of cases of self conflict. We know we should not eat the chocolate bar, it is bad for us, but some part of us argues that we should. The paper picks up this thread of the 'agitation calculus' and separates it from the holistic nature of the dispositions in Wolff's linguistic behavioral analysis.

If we return to the taxonomy of arguments developed at the start of the synopsis we can see what is going on here. Robert Wolff is in fact making a Linguistic Behavioral argument about the way that dispositions work holistically to describe people from the third person view. "Bob is a shy fellow, but he will fight given the right cause." This is a holistic description of what Bob is like given by someone from the third person point of view. The conflict that Plato points out, and Ryle is arguing for, between contrary motives, inclinations and tendencies, that same conflict we feel between eating a chocolate bar and knowing it isn't good for us, is a phenomenological argument that is best understood from introspective analysis from the first personal view. Where the two cross over, Ryle's phenomenological argument about agitated self-conflict, and Wolff's linguistic behavioural analysis of dispositions, is inside ordinary language. The way they cross over is an indirect domain where assumptions are inherited from the first and third personal points of view. This indirect domain occurs when we see a person acting a certain way, such as an agitated state, and assume that they are having an experience, that is, a clash between two inclinations, motives or tendencies, or one of either with a factual impediment. When we do this for the sake of an Ordinary Language argument about *what it makes sense to say* in a language, that is we argue that ordinary language is an authoritative source of appeal for either Ryle or Wolff, we are doing so on assumptions inherited from either position. What we are left with, are contrary claims indorsed by the same source, 'ordinary language'. The upshot is a contradiction on what dispositions entail.

The reason why the contradiction appears is because the two sources; the phenomenological insight of the first personal point of view and the linguistic behavioral appeal of third personal point of view, are actually rival sources of normative appeal that share assumptions in common. These assumptions are what lead to the illusion holding up ordinary language as a single unified normative source of appeal and is one version of an illusion that anti-psychologism is based on,. The illusion is that we can unlock the mind with a semantic theory about language since the mind thinks in terms of words. There is a problem with this the paper will bring out. It starts with realizing that'Ordinary Language' isn't a unified normative source of appeal, it's a field of mixed assumptions from two rival sources, and contains the indirect first personal point of view and the indirect third personal point of view. These mixed assumptions create analogical structures between the first person direct and first person indirect, and the third person direct and third person indirect. The first person direct contains phenomenological insights and the third person indirect contains observations about someone else from the first person view but is limited in this way; if we were at a library and someone was clicking a pen and that stopped us from concentrating and created irritation then we are viewing that person from analogical structures from the third person indirect. The third person indirect contains information about others and how their actions, motives and behaviours impact on us. The first person indirect, likewise, contains information we apply to others from the first personal direct view in analogical structures. This is like when we try to assign motives or reasons to why the person is trying to click the pen.

The paper argues if we collapse the indirect third personal point of view into the direct first personal point of view, and collapse the indirect first personal point of view into the direct third personal point of view, and admit that the first and third personal points of view are irreducible to each other, then the contradiction disappears. This in effect costs us Ordinary Language arguments as a normative source of appeal in arguments about the mind.

Let us recap the stages so far to see where this puts us. What we have done so far is to develop a taxonomy of Ryle's arguments based on the sources of appeal in those arguments. This gave us Linguistic Behavioral arguments which drew their appeal based on the behavior of words in utterances, pieces of language, what the French semioticians refer to as a 'parole' and in Ryle usually contain distinctions made at the level of structures arising from verbs, nouns, adjectives, epithets, adverbs and so on. The next type of argument we derived from the inadequacies of David Chalmers account of Logical Behaviourism. This was the Ordinary Language argument which based its appeal on what it *makes sense to say* in a language. It seemingly drew upon some body of knowledge that allowed speakers to do things like differentiate between a 'glow of warmth' and a 'glow of pride' without linguistic behavioural distinctions to uphold such a difference. The third type of argument is based on phenomenological insight. These we discovered with Sartre were arguments where consciousness takes itself as an object to peer into the 'cortege of the psyche' for the purposes of an argument. At the start of the paper I pointed out that while Linguistic Behavioral arguments depended upon Ordinary Language arguments since they draw their linguistic behavioural distinctions on what it makes sense to say within a language. However some Ordinary Language arguments were not Linguistic Behavioral arguments. Some of the ones that weren't Linguistic Behavioral arguments but made an appeal to Ordinary Language for their meaning turned out to be phenomenological arguments. The 'flash-bang' or 'feelings proper' strain of Ryle's Occurrences were particularly important for that, and as I pointed out, the ability to differentiate between a 'glow of warmth' and 'glow of pride' would be a thread running through this paper and to keep an eye on it. Since both Linguistic Behavioral arguments and Phenomenological arguments shared Ordinary Language claims in common, it made sense for Ryle to treat these two as the same type of argument, both justified by our common knowledge of language. The source of normative appeal for this phyla was revealed by two insights, the first argues the implicit language theorist position and here Ryle makes the point that psychologists would not know how to communicate their theories without language. The second was Ryle's argument people already know how to talk about the mind. However, as we just discovered, we found phenomenological insight at the back of one area of 'ordinary language' and linguistic behaviors at the back of another. These give us rival sources of appeal rather than a unified one.

This lead us to dissolve Ordinary Language arguments because to continue to use them as a normative source for all our arguments would be to admit a normative source that is capable of producing and upholding a contradiction. The domain of assumptions carried over from direct phenomenological insight, like in the Ryle-Plato case of self conflict, and the holistic nature of a linguistic description of someone's dispositions from the third person like that offered by Wolff divided our original arguments into four new types. These four types were the first person direct, third person direct, first person indirect, third person indirect. The paper collapses the first personal direct perspective into the third personal indirect view, and the first personal indirect view into the direct third personal perspective. This gives us two new perspectives for the types of appeal in either arguments or claims about the mind. One, the third personal direct and first personal indirect is hetero-phenomenology. Two, the first personal direct perspective with the third personal indirect view is auto-phenomenology.

It is in this way the paper moves from the taxonomy of the three types of argument in Ryle which were explained at the start of the paper and form the first part of the methodology, these were (1) Linguistic Behavioral analysis, (2) Ordinary Language claims and (3) Phenomenological arguments into two sources of appeal. The two sources of appeal were (1) Auto-phenomenology and (2) Hetero-phenomenology. This transformation of the original taxonomy into two new sources of appeal completes the middle part of the methodology. The third part of the methodology builds an argument as to which of these two sources is the strongest. The building of that argument involves the completion of a number of threads started earlier in the paper including both sides of the 'feet-of-clay' argument and the completion of Ryle's Occurrences that started with 'flash-bangs'.

In *Motives, Moods and Reasons* the paper picks up on the Brandom-Ayer explicit view of language and meaning in relation to mind, and focuses on Brandom's account of reasons. *Motives, Moods and Reasons*, contains lots of caveats and fine line distinctions. For instance, it defines Brandom's position against Sellars in relation to Gilbert Ryle's original linguistic structures. When Sellars uses Konstatierung statements, the paper argues he does so as capacity dispositions. The move Sellars discuses, from an observation language to a report language within the space of reasons is one based on knowledge-how and as such it falls outside of the original guidelines set earlier in the paper. Brandom however moves dispositions over to tendencies because he focuses specifically on motives, not on a descriptive language keyed into language dispositions. The paper also points out some of the key differences between what Brandom takes to be a piece of reasoning and what Ryle does.

In *Motives, Moods and Reasons* the paper presents five cases from the philospher's own experience and one presented from Sartre's which challenge

Brandom's reason-explanation structure and what Brandom calls 'the game of giving and asking for reasons'. The strategy here is to argue that pre-linguistic knowledge is necessarily prior to and fundamental for playing the game of giving and asking for reasons, because the person must arrive at the game with a special sort of knowledge, that cannot be established in the game in order to play it. Brandom argues that we can explain intention as practical reasoning using beliefs and desires. Beliefs must be utterances that can be taken as true or false since they must be believable to the people that hold them, while desires are linguistically stateable goals that can be used as steps in inferences from beliefs to getting what they want in order to account for intention.

It is best to focus on one of the cases presented to show how these cases work. Imagine you are an inner city bouncer at a relatively popular night spot and the past three weekends you have noticed a woman who arrives, talks with two different men, and shortly after the men she has spoken to erupt into a violent brawl that must be broken up. Each weekend there are different men involved but the same scenario unfolds. Why might this woman be creating this scenario? What are her reasons?

One might posit something psychological such as an Electra complex and the desire to have sex with her father leads her each week to create a scenario where two total strangers end up fighting each other. Similarly one might posit that the contingencies of natural evolution lead her to create scenarios where men must fight in order for her to discover a strong mate. Both of these types of explanations are flawed if we return to the first and second personal points of view. The paper argues as a condition of understanding what she is actually thinking the reason given must be one that can actually figure as part of a process of practical inferences in the girl's own reasoning deriving from beliefs and desires to practical inferences to getting what she wants. In both cases it is a difficult hypothesis to argue that the girl reasons to herself that she has an unresolved issue with her father that leads her that night to go out and cause two unknown strangers to fight in a bar. The absurdity arises because there is something fundamentally wrong with these types of psychological theory. These are third personal direct views that by their nature exclude the types of things people think about and muse on in their practical reasoning.

One might likewise posit that the girl does it from a sense of dissatisfaction in her life, boredom, a need for amusement or reassurance. Any one of these classes of reasons might do and we might from them select the last. The need for reassurance. If we are then to play Brandom's game of giving and asking for reasons, we might then ask, why does the girl need reassurance? Here, once more we are prohibited from positing third personal direct views such as childhood trauma or an Electra complex. The reason must be one that she might recognize and that would figure in her own reasoning processes. Upon further examination and several more caveats the paper arrives at the conclusion that the need for reassurance, or the need for a need for reassurance itself points to something non-linguistic, a special type of knowledge that cannot be derived from or established within Brandom's game of 'giving and asking for reasons' but must pre-exist in order for the game to be played. This is knowledge one must posses in order to arrive and play the game.

In *Motives that are neither Dispositions nor Propensities* the paper concludes the exploded model for Ryle's Linguistic Behavioral cartography of the mind by drawing on A. R. White's paper and the claim that motives are not propensities nor are they dispositions to do some logical work and draw attention to the problem of individuation on what can serve as a motive once we accept White's criticism. White has uncovered part of the logical structure of consciousness, but he is missing an important piece of the puzzle. That piece will come from the Occurrences and the sequence of arguments that started with individuation between 'a glow of warmth' and 'a glow of pride' and Ryle's inability to individuate either from the other on the basis of linguistic behaviors alone. We will return to this below.

So to recap. We blocked off the capacity family of dispositions early in the paper as irrelevant to the implicit language side of the anti-psychologistic argument about meaning and offering us little more than a strawman and poor fare for philosophical musings. We eliminated Ryle's beliefs with Ayer and Brandom and siding with Sartre that we can perceive objects with consciousness at a level below reflection without the I as a necessary de facto inhabitant for affirmation. We took apart motives with Brandom in the game of giving and asking for reasons and demonstrated that one must arrive with special linguistic knowledge in order to play the game. We pulled apart tendencies and inclinations with the Robert Wolff debate over meta-physical substances and found that they lead back to either a linguistic behavioral argument about holistic descriptions from the third person or phenomenological insight from the direct first person and an indirect domain of inherited assumptions in between. Since motives, tendencies and inclinations taken together are the propensities, and we've eliminated all three of them as the impelling sources of behaviour since they are third person holistic desctiptions, we've eliminated the propensities, and since we've eliminated the propensities we've also eliminated agitations from the third person direct and first person indirect heterophenomenological side of the debate. This thus gives us our feet of clay argument and what I will explain as (c:i) in a moment. We have also eliminated the explicit side of the feet-of-clay argument, since we've shown flaws and faults in Brandom's account of beliefs and motives. We do not apprehend the world in sets of propositions or judgements we silently affirm to ourselves. Reasons and beliefs are a third person phenomena. What the Brandom-Hoffe-Kant demonstrates is that beliefs as linguistic statements are not beliefs we hold at the perceptual level. Linguistic statements of beliefs about the world are something that happen to our perception. We developed this argument form Hoffe's analysis, Sartre's phenomenology and Brandom's philsophy. Beliefs we looked at with Sartre, Hoffe and the apperception of the I, while motives we took apart with the game of giving and asking for reasons to show the game rests upon pre-supposed non-linguistic foundations that could not be set up in the game. This is because if we push reasons far enough we discover that there is a point where they are no longer linguistic

This is important because this side of the 'feet-of-clay' argument will become the proposition C. C is the proposition that some motives and reasons are unutterable.

IF we relate this back to a linguistic behaviourist map of the mind all that is left now in Ryle's original cartography are the Occurrences and in particular 'feelings-proper' or 'flash-bangs' as I labelled them, and the final argument which shows that autophenomenological sources are foundational in our talk about the mind.

Now we will move into the third and final part of the paper.

The final part of the paper consists of two central arguments derived from the thread of the paper overall. The structure of the arguments is explained in the chapter *Psychasthenia, Dispositions and the Meta-Concern*. Both arguments argue that autophenomenology is a better source for individuation on claims about the nature of mind. This is the first of the three original theses presented at the start of the paper and explained in the 'aim' at the start of the synopsis. One argument argues this position from the first personal direct point of view, the other argues it from the third personal direct point of view. The argument is based on a number of cases where we ask a person for an explanation and they can not give one. There are two cited in the paper. One is an old lady who drives around in people's blindspots and when confronted about it at a set of lights gets angry and drives off. Another is taken from one of Sartre's case studies into the work of the psychologist Pierre Janet and a patient who would cry each time she came to see the psychologist.

The argument involves the disjunction either (A) that the person cannot give a reason or (B) the person does something in order not to give a reason, like crying or driving off. The difference between A and B is B involves a story about consciousness, A posits a dispositional account. (A) is broken down into two further disjunctions. (a:ii) posits that there is something unutterable about these people's actions that prevents them from making their actions, motives and reasons linguistic. This returns us to the limits of Brandom's game of giving and asking for reasons which we already discussed. (a:ii) is also compatible with B, the person might do something in order not to give a reason because it is unutterable. (a:i) However argues that the person's dispositions prevent them from doing so, they are either (c:i) linguistically dispositional in the way Ryle thinks the tendencies and inclinations operate in order to create the compelling forces that produce the agitations or (c:ii) an account that approximates the way most people read William James, while noting several caveats and points of scholarship. (c:ii) offers us an explanation in terms of mechanistic dispositions. An account of human behaviour based on mechanistic dispositions is the view people are physiologically wired up that way. The person is happy because they smile. The person is afraid because they are running away. They run away because they are pre-disposed by their dispositions to run away in a given circumstance. Both (c:i) and (c:ii) try to give an account from the third person view that leaves out consciousness or the possibility of consciousness. Ryle does it with language, a mechanistic explanation uses different types of dispositions.

We have already removed the other type of third person explanation earlier with our stipulations against psychology and the girl who starts fights at the bar on a Friday night. The stipulation we introduced was that the thought must be capable of serving as part of the person's reasoning processes. Here we are bringing the argument to a close by arguing against the remaining possibilities, one, the person has internal linguistic dispositions like the forces producing Ryle's agitations, two the person is wired up in the presence of a certain observable chemical, a certain smile, a certain action to behave in a certain way and that gives us an account of the mind.

As noted, (c:i) has already been eliminated by the Ryle-Wolff discussion on dispositions. Either the disposition is given from a third personal point of view and merely descriptive like Wolff argued, or there is a phenomenological story involving consciousness. This phenomenal story involving consciousness ties in directly to the second thesis of the paper and will be concluded with the occurrences flash-bang argument. This occurrences flash-bang argument will show that language is merely a code for thought and this leaves us with psychologism as the only viable option.

(c:ii) argues that the reason the person does this is mechanistic and there is no consciousness present to account for this. Here the paper uses a distinction based on a lie detector test. A lie detector test can only tell us 'when' someone is lying it can not tell us why the person is lying. It is not enough to merely know when someone is lying by the light flashing on, for an account of why the person is lying, we must have an account of the reason they are lying. Likewise, it is not merely enough to know when someone is experiencing emotions from the presence of trace changes in their brain chemistry to explain the mind, we must also have an account of those emotions themselves. Further there is a problem with trying to individuate emotions merely on physiologistic symptoms. Sartre points out that joy displays more intense physiological symptoms of rage. The person experiencing joy displays many of the same physiological symptoms as the person experiencing anger, for instance fixated pupils, increased heartbeat, faster breathing. But joy is not a form of anger. The two are individuated and authenticated by a difference experienced in consciousness. If we accept this position put forward by Sartre and the distinctions between first and third, direct and indirect views, then it becomes clear that many of our claims, like those of neuroscience or psychology, that come from third and first person indirect views use analogical structures that depend upon consciousness and insights derived from the first personal direct view to make their argument. When the neuroscientist makes a claim about 'anger' or 'joy' those claims are *indirectly* linked to his own experiences of what anger and joy are. The neuroscientist gets his concepts about anger and joy, initially, from his own direct first personal perspective. Thus, the paper argues, heterophenomenological sources are fundamentally based on autophenomenological sources.

So if we put the case of the lady who cries when she sees the psychologist or the little old lady who travels around in people's blind spots and drives off when we ask her why, then the argument looks like this

(a): The person can't give a reason
B: The person does what they do in order not to give a reason.

(a) breaks down into two further premises as we discussed above

(a:ii) There is something unutterable that stops the person giving a reason
(a:i) The person is impelled or prevented to because of their dispositions

A moment of reflection will show that (a:ii) is compatible with B, because the person drives off or cries or performs some task in order to not give a reason. This in turn implicates a theory of consciousness and establishes grounds for our third thesis in the paper, that is we need a theory of consciousness. Since (a:ii) is compatible with both A and B we will call it C. We will call (a) minus (a:ii), A. This gives us the first line of an argument with this shape

[(A . C) v (B . C)]

Since either side of this premise is a conjunction divided by a disjunction, in order to make the argument we need to disprove A or B and conclude with the proposition that is not negated. Our strategy in the paper is to go for not A. Since the conjunction of a false with a true, produces a false, and we have established grounds for the truth of C with the otherside of the feet-of-clay argument and the nee to arrive with special pre-linguistic knowledge to play Brandom's game of giving and asking for reasons, all we need to do is disprove A. So the basic strategy we are going to adopt and the shape of the overall argument looks like this

1)	[(A . C) v (B . C)]		(a:i) = A
2)	- A		(b) = B
	(B . C)		(a:ii) = C
3)	-(B . C)	Ass	
4)	(A . C)	SD 3, 1	
5)	A	AD 4	
6)	(A . –A)	CD 2, 4	
	(B . C) RAA Ass		

A (a:i) is made up of two types of disposition. (c:i) which is an account of the mind like that offered by Ryle's Linguistic Behavioural tendency dispositions, and which we will call D, and a mechanistic dispositional account (c:ii) which we will call E. A is true if and only if either D or E is true.

Now C clashes with D, since if a linguistic behavioural account of dispositions can account for the person's behaviour, then there is something 'utterable' about the person's behaviour. Since C is the claim there is something unutterable about the person's dispositions they both can't be right. (C > –D), and of course, (D > –C). But this does not automatically make (A . C) false, giving us one side of the disjunction, [(A . C) v (B . C)] because A is made up of (D v E). So even though (C > - D) and the elements of A are D and E, E can still be true thus proving A because A is made up of (D v E). D can be false while E is true making A true. (A > (D v E)). True hook true is a true if E is true. This is because true disjunct false is true.

Indeed we have shown D, (c:i), false by the Ryle-Wolff discussion, since linguistic dispositions are either (1) holistic and descriptive with no actual impelling force, which is to say that Ryle's dispositions are an analogical construct made up from the third person direct and first person indirect position, and are thus a heterohenomenological source of appeal, or (2) part of an account that depends upon a story about consciousness and part of Ryle-Plato's self conflict. If they are (2) then that account ultimately bottoms out in an anti anti-psychologistic argument that concludes with Ryle's Occurrences argument which will show that language is merely a code for thought and thought is something fundamentally non-linguistic. Also note we removed Brandom style reason-explanations as pointing towards non-linguistic knowledge necessary and foundational to playing the game of giving and asking for reasons. This supports C, the argument that there is often something unutterable about reasons, which is compatible with B, forming one possible side of our disjunction, (A . C] v {B , C}], and affirming (B . C). Notice how tightly the logic fits together?

The paper argues as well as (c:i) E being false, (c:ii), D is also false. The mechanistic dispositional account of the mind is flawed from the first person direct view because consciousness is involved in the way we experience and produce emotions, the argument for this I will introduce in a moment with the 'flash-bang' account. The mechanistic account is flawed from the third personal view because phenomenological insight is a stronger source for individuation on the mind than third person observation. This is established by Sartre's argument that anger is not ultra-joy.

Sartre's argument is that anger is not ultra joy even though joy and anger both have similar or the same symptoms. This argument also establishes the authority of autophenomenology as a source of appeal since it places the authority of a direct first person source over that of a third person source. The reason why it does this is Sartre's argument rests upon phenomenological insight into what anger and joy are, in order to argue that third person observations about the similarity, or exact physiological likeness of two states, do not establish the identity of the emotions. Physiological accounts can be over-ruled by an appeal into phenomenological insight. The 'cortege of the psyche" is a stronger source to appeal to in arguments about the nature of mind.

If we agree with Sartre's argument that anger is not ultra-joy then we must also agree with the authority of a phenomenological insight as a claim about the nature of mind. This also supports our first thesis introduced at the start of the paper. Sartre's 'anger and ultra joy' argument undermines the third-person direct, first person indirect view, the heterophenomenoogical source of appeal, and leaves us with the first person direct, third person indirect view, which of course is the autophenomenological source of appeal.

Here at last we arrive at the conclusion of the flash-bang strain that started in Ryle's Occurrences armed ready for the final insight of the paper. This will argue against the mechanistic account of the first person point of view in the way described with the lie detector test. That is it will show the problem with an account of the person's behaviour by a mechanistic view and argue - E. The link between D and E is they both attempted to give an account of a person's actions without an account of consciousness. Since we have disproved both halves of the disjunction we have shown A to be false.

However it is important to note a version of the mechanistic account survives in the problem of the "'Roid Rager in the Void" which takes account of the third thesis, that is, the need for a theory of consciousness. We will return to this is a brief discussion at the end.

Thus armed we are at last ready to conclude the Occurrences thread with our account of 'flash-bangs' or as Ryle calls them 'feelings proper'. The paper argues that a 'glow of pride' or a 'flash of anger' isn't just a flash or a glow, but rather it is a flash or a glow about something. A 'flash of anger' is a flash of anger at the man who stole our car spot at the shopping centre. In the moment we are angry with the man, it is his behaviour which causes us to yell and stick our rude fingers up. It is his behaviour that makes us angry. Later when we reflect back and try to describe our actions the anger enters into our descriptions, which occur after the fact as reflective consciousness apprehends the earlier consciousness, and produce the illusion of anger acting like a physical force. We say things like 'it was the anger that made me do it' or 'that's not true, it was the anger that made me say it'. But in the moment we are angry with the person. Our conscious apprehension of the event or situation is part of the anger and what produces it. This is the illusion that an anti-psychologistic theory of mind falls into. Once we understand that the sudden surge of anger we feel in the moment is different from the description we give it later, we can see that the mistake the anti-psychologistic philosophers made was to mistake thoughts for the words we use to describe them.

If we relate this back to Ryle's linguistic dispositions and a mechanistic account then we can see the negation of first personal perspective views of D and E. In the moment we are angry with the person who took our car space. We don't internally shout "sudden stab of anger" rather our anger is directly involved in consciousness and the way we see the world. It is the situation and the way we perceive it that is part of what are making us angry. Anger does not occur in isolation as a linguistic phenomena, our mental log keeper does not say "stab of anger", but rather our anger when it arises is positional, we are angry at the man. That involves the man as a positional objet of consciousness in our world. Later when we reflect and talk about it, when the 'ego' of the prior consciousness appears, we say things like 'he made me angry' or 'I felt a sudden stab of anger'. Here we offer a linguistic description of the anger that is different to the anger itself. Here, as Dummett might put it, we 'encode our thoughts in a transmissible medium, which acts like a telephone and is then decoded by the hearer'. 'We need language, on this view, only because we happen to lack the faculty, that is, of the direct transmission of thoughts.' Simiarly we do not just act from anger the way a car engine runs from fuel, we experience a surge of anger in relation to our conscious experience of the world. The anger is intrinsically a part of, and deeply involved in our conscious experience of the situation or event. Thus we have established the second, and last remaining premise of the original three stated at the start of the paper. Emotions are irreducible to linguistic behaviours.

Discussion

One of the unresolved problems of the paper is the causal relationship between brain chemistry and the positional model of consciousness. The paper deals with this problem in the framework of a thought experiment called the Roid Rager in the Void. For once we realize that consciousness has an effect on brain chemistry: apprehending the man who stole our car park causes hormonal changes in our brain chemistry that causes us to feel rage, and likewise introduction of chemicals that bring about hormonal changes can cause us to feel different emotions, for instance a man who takes an Oxytocin stimulate and suddenly experiences euphoria over a sunset he's seen numerous times before, then a question arises, what happens when we introduce those same hormonal or chemical changes and consciousness doesn't have an object to fixate on? What are the limitations of the positional model for consciousness?

   The paper suggests we imagine the problem in this way. We take a man suffering what the media calls "'roid rage", and we remove all of the typical objects the 'roid rager rages at, we take away his personal trainer, his weights, his protein shakes and his gym equipment. We cast the 'roid rager' adrift in a void in an angry tumbling spiral of flailing fists. What is the anger going to fixate on? Will our roid rager turn introspective and rage at gifts his parents never bought him, or problems he suffered in his life? What if we could remove these from his mind? What does consciousness fixate on when it does not have an object? The paper leaves this issue unresolved.

Conclusion

The paper concludes with the argument for a return to a Pre-Fregeian Psychologism having debunked the underlying tenant of Anti-Psychologism as defined by Dummett, shared by both implicit and explicit language theorists. The paper demonstrates that language is indeed like a code, in which the speaker ciphers his emotions while reflecting on them, using a descriptive vocabulary for describing the 'cortege of the psyche' as he makes consciousness the object of his own introspection in order to communicate them, because he lacks the medium of direct transmission. The mistake of the anti-psychologistic school was to take this reflective vocabulary with its analogical constructs and inherited assumptionds as thought. In their search for a theory of mind they mistook the vocabulary we use to talk about the mind for the mind itself and placed a theory of linguistic meaning as foundational to a theory about the mind.

The paper shows, through the 'feet-of-clay' argument that thought bottoms out in pre, and non-linguistic levels with Brandom's 'game of giving and asking for reasons' and an exploded model of Ryle's linguistic behavioural cartography of the mind. This in turn establishes the second thesis of the paper, introduced at the start of the synopsis, by demonstrating that emotions are not reducible to linguistic behaviours. A 'flash of rage' is a flash of rage about something and we need a theory of consciousness, not a theory of semantic meaning, to give an account for that.

The third thesis of the paper, established by the irreducibility of emotions to linguistic behaviours, called for a theory of consciousness. This call draws attention to causal issues long ignored by anti-psychologistic schools. If apprehending the man who stole our car spot causes changes in our brain chemistry by releasing hormonal changes through the brain cells causing the release of noradrenaline and other chemicals in the brain's limbic system, then we are going to need a theory of consciousness to account for these changes in brain chemistry, since consciously apprehending the man stealing our car spot is part of the process producing that sudden radical change in the brain chemistry.

Lastly the paper also establishes the authority of appeals to introspection for claims about the mind with Sartre's argument that anger is not ultra-joy. If we agree with Sartre, that anger is not a form of ultra joy, no matter how much anger might look like joy either in the physiological descriptions or the activity reported by brain scans, then we acknowledge the authority of introspective claims to individuate on the mind. Likewise, if we turn this insight back on the problem of distinguishing a 'glow of pride' from a 'glow of warmth' we see that the distinction does not depend upon any linguistic behaviour, but depends upon an introspective appeal to what a glow of warmth or pride might be like. The illusion this paper has been at pains to bring out, is that the call for a semantic theory of meaning in talk about the mind, strong enough to decipher instances like that between a 'glow of pride' and a 'glow of anger' is an illusion created from viewing such claims about the mind as a united singular normative source of appeal, like Ryle does with ordinary language. The reason why we have been struggling with a theory of meaning for well over a century is because we continue to see the meaning in language as a unified source of appeal which creates the need for a semantic theory to explain it. We exploded one particular version of this with Ryle's 'agitations' and Wolff's arguments against Ryle's dispositions and found, rather, that underlying this illusion of a unified source in need of a theory of semantics there are rival sources inside language with different types of appeal. What we mistook for a common language was actually based on two different perspectives and a number of shared assumptions built up into analogical constructs between them in which rival position could inherit assumptions. If we agree with Sartre that anger is not ultra joy then we acknowledge the authority of phenomenological insight, and autophenomenological appeal, as so defined by this paper, as the strongest source to individuate on claims about the mind.

The three theses established by the paper lead us to a reconsidered post-anti-psychologistic reconsideration of psychologism. These were (1) autophenomenology is a stronger source of appeal in arguments that involve individuation on the mind and (2) that tells us that language and thought are not the same thing, this was based on the feet of clay argument and final resolution of the 'flash-bang' strain that started with Ryle's Occurrences. (2) argued emotions are irreducible to linguistic behaviours, while some prelinguistic knowledge is necessary for talk about motives and reasons. (1) and (2) point us toward (3). Since firstly autophenomenology is a stronger source for claims about the mind, and secondly, there are some mental states, operations and acts that are irreducible to linguistic accounts because linguistic accounts of such states, operations and acts are like a code for putting thought into, the anti-psychologistic attempt to search for a theory to explain the mind is flawed. It follows that what we need is (3) a theory of consciousness that will allow us to understand and also offer us an account of the non-reductive element of consciousness. The three premises argued together constitute a considered return to psychologism and what the paper argues will be a new chapter in analytic philosophy of mind,

Appendix C

Introductory Paper explaining the scope and aim of the Thesis.

Introductory Paper: *Fun and Phenomenology guest staring Gilbert Ryle, Jean-Paul Sartre and David Hume.*

In this paper I am going to introduce you to a two sided thesis that will serve as a general introduction to my research and larger work. I'll start off by explaining a historical connection between Sartre and Ryle and then progress to how I exploit the philosophical value of this connection, what that means for philosophy in general and why this offers us not only valuable historical insight and perspective, but also, and more importantly a way forward for bridging the gap between the analytic and continental traditions, and bringing them both together in philosophy as part of a general research project, so that each tradition can benefit from the insights and advantages offered by the other.

To do that, and in order to reach the attractive insight on offer from this paper, I'm going to start with Sartre, explain a little of his often overlooked phenomenology of the imagination, discuss what he calls 'the illusion of imminence', how this ties into larger ontological themes in the main body of his work, and offer by way of an introduction, practical steps to unlocking his phenomenology.

From there I'm going to move into Gilbert Ryle's philosophy. Here, I'm going to lay out some of Ryle's linguistic and analytic work, as a general introduction to his work, then I'm going to discuss the historical link between Ryle and Sartre, show the differences and similarities between them, and then reveal how I exploit the historical grounds for philosophical purposes. Most importantly I shall show how the two philosophers serve as the groundwork for bridging the divide between analytic and continental philosophy, and can offer us valuable insight into framing ordinary language philosophy and phenomenology, in terms of the anti-psychologistic thesis of early analytic work and the larger phenomenological project of discovering a theoretical basis for analysis of consciousness.

I will then discuss how I develop a taxonomy of Ryle's arguments to draw out an ambiguity and the implications of this for psychologism. Ryle has been so hugely influential and there are things in Ryle that have not been understood. Going back and re-looking into these foundations opens up new vistas and has far ranging implications throughout contemporary philosophy. Given the length and scope of this paper I will only be able to briefly sketch these out.

Beginning now with Sartre and The Illusion of Imminence, which will be extremely important in a moment, Sartre says, in the Imaginary, and I paraphrase

'At first reflective glance. . . (we think that) an image is in consciousness and that the object of the image (is) in the image. We depict consciousness as a place peopled with small imitations and these imitations are the images. Without any doubt, the origin of this illusion must be sought in our habit of thinking in space and in terms of space. I will call it the Illusion of Imminence. It finds its clearest expression in Hume, who distinguishes ideas and impressions.'[566]

Here Sartre quotes directly from Hume

'The perception, which enters with the most force and violence, we may name impressions. . . By ideas I mean the faint images of these in thinking and reasoning.[567]'

Sartre's strategy is to attack Hume's assumption that ideas and impressions are the same sorts of things, with impressions just being the more lively of the two, and to attack it on phenomenological grounds. To understand his argument, we're going to have to do a small bit of phenomenology, and draw out some of the frameworks Sartre uses in his investigations. This will become important in a moment when we move into Ryle.

In the Imaginary, Sartre discovers three basic structural forms of consciousness. From these three he then moves onto several sub-forms of each, and lays out the groundwork for much of his work in *Being and Nothingness*. The three forms are firstly, perceptual consciousness, secondly both imaged and imaginary forms of consciousness, and thirdly conceptual consciousness. The majority of Sartre's work in *The Imaginary* is centered on a number of exercises that explore each structural form of consciousness, and deriving various steps and procedures from these for guiding one's own consciousness into different states to understand the differences between them. Of the three forms, we only have time in this paper to go into two, and of those two, our investigations are extremely limited by the scope of this paper to just some essential points.

To begin the process of distinguishing imaginary from perceptual consciousness, Sartre uses an exercise involving a piece of paper. Here I quote directly from *The Imaginary*.

Sartre writes,

Let us consider this sheet of paper on the table. The more we look at it, the more it reveals its characteristics to us. . . Each new orientation of my

[566]Jean Paul Sartre *The Imaginary*. Translated by Jonathen Webber. Abington: Routledge, 2004. pg 5
[567]Sartre. *The Imaginary*, 2004 Pg 5

attention, of my analysis, reveals to me a new detail; the upper edge of the sheet is slightly warped, the end of the third line is dotted, etc. But I can keep an image in view as long as I want: I will never find anything there but what I put there. This remark is of the utmost importance in distinguishing the image from perception[568].

So the basic idea is that a mentally imaged consciousness is fundamentally different from a perceptual consciousness. If we take a sheet of paper, look at it, close our eyes and imagine it, then open our eyes we will discover that the sheet of paper we perceive has a plenitude of detail, to which by comparison, the one we imagine doesn't. The perceptual object of consciousness as it presents itself to us contains a higher 'resolution' we might say than its imaged or imagined counterpart and this is a structural difference in the type of consciousness being used itself.

Using these types of exercises, the example with the sheet of paper which I just explained is one. Another is trying to image the Parthenon in the mind and count the pillars from the image is another, or you might try it with the Opera house and count the peaks, or visualize the Harbor bridge and count the struts. Another one is the famous cube example, using these Sartre is able to draw out elements of imaged consciousness and perceptual consciousness that distinguish them from each other. Imaged consciousness, for example, contains what Sartre calls quasi-observation. That is we are able to observe in our mental images only that which we put there and are aware of and know about. Sartre argues that we can not count the columns in the Parthenon from the image we form or recollect because we do not know how many there are. It has a poverty of detail. It exists in what he terms 'irreal' space which is neither continuous nor connected with real space. It lacks spontaneity. This he develops into an extended refutation of George Berkley, beginning in *The Imaginary*, and taking up much of the introduction of *Being and Nothingness*. Perceptual objects contain what Sartre refers to as a plenitude. That is, if we try to observe them in their entirety they possess more information that we can take in. They contain spontaneity. The sheet of paper on the desk, when you pick it up, may have something unexpected written on the other side, where as the one in your mind will only have something written on the other side if you put it there. This is a fundamental difference in kind to the structure of an imaged consciousness as opposed to a perceptual consciousness. Sartre calls this spontaneity. Thus from resolution, spontaneity, plenitude, quasi-observation and irreal space against the continuity of real space, there is a synthetic unity in perceptual objects that points towards what Sartre refers to in *Being and Nothingness* as the transphenomena of being.

Using this approach Sartre is able to reveal structural traits of consciousness that might otherwise be over looked. The structural negations that take place within consciousness and make up much of the mystical sounding 'nothingness' in

---

[568]Sartre. *The Imaginary*, 2004 Pg 9

Being and Nothingness arise, firstly, from the plenitude of reality. To see the pen on the desk, that is, to bring the pen before consciousness as the sole object of consciousness, you must negate the table, the glass of water, the wall behind it, the window, the people playing in the park across the road, and all the rest of reality, in order to see the pen. The object arises from the totality of the world as a negation of the world. The second form of nothingness or negation arises from consciousness apprehending the pen as not the pen. From these Sartre is able to move into the transphenomena of non-being. The imaginary object, however, arises in the mind as a negation of the world as not in the world as not part of the world. These are the sources of irreal space, discontinuity, lack of spontaneity and the poverty of detail which prevents imaginary objects from existing in the real world since they lack the necessary conditions for existing. Between the imaged object and the real object exists an analogon, but the analogon in neither part of the real world nor inside of it. It arises primarily as a structure formed by negating the world in consciousness. Indeed Sartre has a specific exercise for separating imaged objects from the perceptual which involves holding an image in your head while looking at a pen. The moment you can see the image in the mind's eye, Sartre notes, you can no longer see the pen before you. The moment you look at the pen, you cannot see the imaged object.

Sartre then goes on to makes several distinctions between imaginary objects and imaged objects. You may imagine your friend, in your mind, and he might really exist. But his image will contain irreal aspects about him that are different to a perceptual encounter of your friend. Likewise with memory, you might remember and image an object, situation or something that happened. The imaged version will be different from the direct encounter. Sartre uses several examples to draw out elements like co-presence of being, the intimacy and homogeneity of memories, the difference between a reflected and a reflecting consciousness, and the structural elements of a past consciousness and the ego complex, of which, I don't have the time or space to do justice or go into[569]. But what is most important to note are these two fundamental points to Sartre's argument on the imaged consciousness.

Firstly Sartre argues that imaged and imaginary forms of consciousness taken together are structurally different forms of consciousness from perceptual forms of consciousness. Thus, ideas are not the same sorts of things as impressions. This is the key point to his refutation of David Hume. Ideas as images are not the same sorts of things as impressions or perceptions. The two are very different. They contain differences in them like 'resolution' which I just talked about. Quasi-observation spontaneity, synthetic unity and the level of detail in each, as we saw with the example of the piece of paper is fundamentally of a different sort. Sartre's

---

[569] While I agree with the spontaneity of the percept, I think the detail in the imaged consciousness, as Sartre explains it has several problems, for while I can not count the number of vertical support struts on the Harbour Bridge from memory, there are people with photographic memories who can. However, for the purpose of this paper and the relationship with Ryle, I am merely outlaying Sartre's view in regard to the percept and the imaged consciousness.

point is that impressions and ideas are very different sorts of things and we can observe this if we do the phenomenology.

Secondly Sartre argues that the objects of the mental images are not in the images themselves. Nor does he think the things we image are images "in the mind" at all. They are forms of consciousness. To think of the imaged consciousness as photographs, characters or objects in the mind, or as Sartre says to 'depict consciousness as a place peopled with small imitations' is to commit the Illusion of Immanence.

If we turn our attention now to Gilbert Ryle, Ryle writes in *The Concept of Mind*, and I quote

> I want to show that the concept of picturing, visualizing or 'seeing' is a proper and useful concept, but that its use does not entail the existence of pictures which we contemplate or the existence of a gallery in which such pictures are ephemerally suspended. Roughly, imaging occurs, but images are not seen. I do have tunes running in my head, but no tunes are being heard, when I have them running there. True, a person picturing his nursery is, in a certain way, like that person seeing his nursery, but the similarity does not consist in his really looking at a real likeness of his nursery, but in his really seeming to see his nursery itself, when he is not really seeing it. He is not being a spectator of a resemblance of his nursery, but he is resembling a spectator of his nursery.[570]

Ryle's leading argument, on this point, is that the objects of the mind do not necessarily exist. That is, the mind is not a special sort of place populated with small imitations or pictures which we can contemplate in a mental gallery. The mind isn't a place.

Ryle writes, and I quote again

> But, as I shall try to show, the familiar truth that people are constantly seeing things in their minds' eyes and hearing things in their heads is no proof that there exist things which they see and hear, or that the people are seeing or hearing[571].

---

[570]Ryle, Gilbert. *The Concept of Mind*. Middlesex: Peregrine Books, Penguin, 1983. Pg 234
[571] Ryle, *Concept of Mind*, Pg 232

So his argument is much the same as Sartre's. Seeing things with the mind's eye is no proof such things exist, and further.

Ryle writes,

> Hume notoriously thought that there exist both 'impressions' and 'ideas', that is, both sensations and images; and he looked in vain for a clear boundary between the two sorts of 'perceptions'. Ideas, he thought, tend to be fainter than impressions, and in their genesis they are later than impressions, since they are traces, copies or reproductions of impressions. Yet he recognized that impressions can be of any degree of faintness, and that though every idea is a copy, it does not arrive marked 'copy' or 'likeness', any more than impressions arrive marked 'original' or 'sitter'. So, on Hume's showing, simple inspection cannot decide whether a perception is an impression or an idea. Yet the crucial difference remains between what is heard in conversation and what is 'heard' in day-dreams, between the snakes in the Zoo and the snakes 'seen' by the dipsomaniac, between the study that I am in and the nursery in which 'I might be now'. His mistake was to suppose that 'seeing' is a species of seeing, or that 'perception' is the name of a genus of which there are two species, namely impressions and ghosts or echoes of impressions[572].

In a 1959 article originally published in *La Philosophie Analytique* and written in French[573] Ryle admitted that he and Sartre shared the same argument.

Here I quote from the article.

> I shall not repeat the arguments by which Sartre and I exposed the absurdity of Hume's view or of the other view that imagining is witnessing things existing or occurring inside a private chamber. What is more interesting, at least to me, is that after these insidious conceptual mis-constructions had been exposed, . . . I felt conceptual

---

[572] Ryle, *Concept of Mind*, Pg 236
[573] See Ryle, Gilbert. "Phenomenology Vs the Concept of Mind." In *Critical Essays*, edited by Julia Tanney. Oxon: Routledge, 2009., pg 201 for its republication and English translation

embarrassments and these are always a sure sign
that something has gone wrong[574].

Ryle used the same, or a similar argument to Sartre in his arguments on the imagination in *The Concept of Mind*. But what's interesting here is the different treatment between the two philosophers. While the underlying argument is the same, they both take different paths. Ryle's analysis rests upon linguistic elements, specifically Hume's use of the term 'lively', while Sartre's refutation of Hume rests upon phenomenological exercises and distinctions between perception and imaged consciousness, like those I walked you through at the start of this paper.

Ryle writes in The Concept of Mind,

> Hume's attempt to distinguish between ideas and impressions
> by saying that the latter tend to be more lively than the former was
> one of two bad mistakes. Suppose, first, that 'lively' means Vivid'.
> A person may picture vividly, but he cannot see vividly. One 'idea'
> may be more vivid than another 'idea', but impressions cannot
> be described as vivid at all, just as one doll can be more lifelike than
> another, but a baby cannot be lifelike or unlife-like. To say that the
> difference between babies and dolls is that babies are more lifelike
> than dolls is an obvious absurdity. One actor may be more
> convincing than another actor; but a person who is not acting is
> neither convincing nor unconvincing, and cannot therefore be
> described as more convincing than an actor. Alternatively, if Hume
> was using 'vivid' to mean not 'lifelike' but 'intense', 'acute' or
> 'strong', then he was mistaken in the other direction; since, while
> sensations can be compared with other sensations as relatively
> intense, acute or strong, they cannot be so compared with images.
> When I fancy I am hearing a very loud noise, I am not really
> hearing cither a loud or a faint noise; I am not having a mild
> auditory sensation, as I am not having an auditory sensation at all,
> though I am fancying that I am having an intense one. An
> imagined shriek is not ear-splitting, nor yet is it a soothing murmur,
> and an imagined shriek is neither louder nor fainter than a heard
> murmur. It neither drowns it nor is drowned by it.[575]

To put the point like this, the mixed sense of lively can be discovered if one tries to imagine a light so bright and vivid that it blinds us, or try, for example imagining thunder so loud it hurts your ears. A lively image is a different sort of thing to a vivid experience. A life-like painting, is life-like specifically because it isn't the real experience of life. To put Ryle's counter point to Hume precisely, Ryle argues that the difference between idea and impression isn't in the degree of liveliness, but in

---

[574] Ibid, Tanney, 2009. NB. Trans.
[575] Ryle, *Concept of Mind*, Pg 238

the type of liveliness. One type of liveliness refers us to the loudness of actual thunder, which can deafen or hurt our ears, the other type to imagined thunder which cannot hurt the ears.

So Ryle's argument depends upon the way we use and understand the term 'lively', while Sartre's depends upon getting us to perform phenomenological acts and see the difference. One form of argument depends specifically on a linguistic, grammatical and semantic analysis, the other depends upon phenomenological exploration. Now the reason why this argument is so important to Analytic and Continental Philosophy is because it provides us with a clear bridge between the two traditions. We have here the same counter argument to Hume, but treated in two specifically different types of way. One way engages the phenomenological tradition that we receive from Brentano, through Husserl and Bergson, and makes up the body of Sartre's early work. The other embodies a linguistic tradition that arises in the midst of the analytic schools, and sets the foundations for much of the analytic philosophy of mind that follows over the next half a century. The importance of Ryle's work to Functionalists, the Ordinary Language school of philosophy, along with Trans-Atlantic work figures like Wilfrid Sellars, David Lewis, Robert Brandom, D.M. Armstrong, J.J.C. Smart, is hard to overestimate. Indeed Ryle's work can be seen as a corner stone to much of what happens in Analytic philosophy in the second half of the twentieth century, as David Chalmers points out in *The Conscious Mind*.

This bridge between the two schools, between Ryle and Sartre, offers us a unique perspective to do foundational and fundamental work on the two major dominant traditions of philosophy. This is project that Richard Rorty, along with Michael Dummett and even Jaques Derrida, (who's own interest run to the ordinary language school, in particular J. L. Austin) – no doubt would have encouraged. Where Ryle is anti-psychologistic, Sartre is interested in the structures of consciousness, where Ryle eschews consciousness, Sartre is interested in doing phenomenology. Ryle is interested in exploring the ordinary language people use to talk about the mind. It is hard to play down the scope and the possibility for philosophy that this bridge opens up.

The way I tackle that bridge is to produce a taxonomy of Ryle's arguments. The basis of that taxonomy rests upon Ordinary Language arguments that justify certain moves by a special type of claim. An Ordinary Language argument is a special argument about what it makes sense to say within a language. That type of claim is usually grounded by an appeal to ordinary language as a normative source to justify the claim by claiming 'it makes sense to say' whatever is being said. Where the phrase 'it makes sense to say' or its equivalent enters an argument it signals a move to draw on ordinary language as a normative source to support the argument. Ryle both negates other philosophers and forwards many of his own arguments on this basis. The most common of these other arguments are what I refer to as 'Linguistic Behavioral arguments' because they rely upon analysis of the behavior of words to make a philosophical claim about the mind. There are many of these I could offer as an example, but due to the length and scope of the paper I'll offer just two.

Ryle uses a linguistic behavioral argument to draw a distinction between dispositions and episodic verbs. He does this by examining the behavior of individual verbs like 'know', 'run', 'possess', 'tingle', 'aspire', 'wake up', and finds that one set of verbs refer to instances in time, while the second set refer to time frames. 'Reg ran down the road' refers to a specific incidence, an episode, whereas 'Reg aspired to be a bishop' refers to a disposition. It is from the analysis of the behavior of the verbs Ryle is able to ground his distinction between episodes and dispositions. A second Linguistic Behavioral argument might help to make this type of argument a little clearer.

In *The Concept of Mind* Ryle identifies two uses of the verb 'to remember'. The first refers to a specific type of disposition, and is usually followed by how, as in, '*do you remember how* to ride a bicycle'. These refer to the skill and capacity dispositions. '*Do you remember how* to do simultaneous equations?' '*Do you remember how* to formulate a proof in the predicate calculus?' This is the sense in which one has not forgotten. A second sense is used to refer to episodes or instances in time, like, do you remember when Reg ran down the road? Do you remember when you graduated? As far as Ryle depends upon linguistic analysis of the behavour of the verb 'to remember' when coupled with the 'how' and the 'when' to make a make a distinction between retaining capacity dispositions and episodic memories he is using a Linguistic Behavioral style argument. But, as soon as he crosses over into asking you to actually think about the difference between the two, to recall when you graduated, to access those happy memories of your graduation he is crossing over into a phenomenological style argument.

Now there are some arguments which are purely phenomenological. One, for example, is asking you to anticipate your next thought. It's not until you try the exercise that you realize you can't anticipate your next thought. For as soon as you try you discover the thought with which you are trying to anticipate your next thought is the next thought itself. There is no Linguistic Behavioral argument. Ryle tenders no hard currency in the analytic coin of verbs, nouns, epithets or linguistic exchanges to be had. There is no analysis of the behavior of language. These arguments, and others like it, when they appear engage a pure, occult phenomenological stream running through *The Concept of Mind*.

My strategy in the paper is first to break up the normative value of Ordinary Langue arguments by showing that they produce a contradiction. I then show that underneath Ordinary Language style arguments, there are rival normative sources. These rival sources constitute the Linguistic Behavioral and Occult Phenomenological strains in Ryle. The way I break up Ordinary Language arguments is to show that there is an indirect domain that inherits assumptions from two direct, rival domains. These domains are the first personal view, and the third personal view. The first personal view incorporates what I refer to in the paper as auto-phenonomological arguments. The third personal view incorporates what I refer to as hetero-phenomenological arguments. The basis for these fine grain distinctions arises from the Daniel Dennett and David Chalmers debates. The indirect domain contains assumptions taken from the first personal view and the third personal view. The first personal direct view contains assumptions that make up the third personal indirect view and the third personal direct view contains assumptions that make up the first personal indirect view. The indirect view is where first and third personal views meet, and it is here, that Ryle mistakenly thought language was a unified source for intersubjectivley accessing the mind. This is where the normative power of an Ordinary Language claim originated, and it is from here I produce a contradiction that shows the indirect source isn't united at all, but the product of two rival views. Realizing this allows us to assimilate the indirect views into the direct views. The first personal indirect view becomes assimilated into the third personal view. The third personal indirect view becomes assimilated into the first personal view.

The remainder of the thesis is then taken up completing the thread of argument over which of the two is a better authority and the three central theses of psychologism which I advance. I will explain these, before bringing the paper to a close.

Following the break-up of Ordinary Language arguments as a unified normative source, into rival sources of analysis I label the first and third personal points of view, for purely technical reasons, encompassing phenomenology and linguistic behavoural style arguments, as outlined in the paper, I then ask what is the stronger type of argument is it a) phenomenology encompassing the first personal point of view, or is it b) Linguistic Behavioral style arguments which encorporate the third personal point of view? This is the part of the thesis that begins to deal directly with psychologism. Not b) implies not anti-psychologism. The reason for that is anti-psychologism takes it for granted that a theory of language is fundamental to a theory of meaning, and a theory of meaning is essential to a theory of the mind. It starts with Frege, who argues for a type of psycholgism which Dummett later identifies asendorsing explicit meaning. Dummett however, pace Frege, argues for implicit meaning. Both positions are anti-psychologistic in that they think that a theory of meaning begins with language not the mind. Ryle occupies a similar position to Dummett, but he argues one step further, that is while Dummett argues that meaning is implicit in the use of a languag, as opposed to explicit as Dummett identifies in Frege's theory of meaning. Ryle thinks that a theory of the mind is fundamentally based in meaning, and meaning is based in ordinary language implicitly. Ryle is anti-psychologistic about the mind, not just about meaning in language. That is to say, Ryle argues that a theory of the mind must take it that meaning in language is implicit, thus language is fundamental to developing a theory about the mind.

So if we decide in favor of a) and argue that it is the stronger fundamental basis for a claim about the nature of the mind then it tells us that anti-psychologism in the form of Linguistic Behavioral arguments are no good. Since Ordinary Language arguments are no good, this strictly cuts off the two types of arguments Ryle uses that begin from exploring the implicit meaning in everyday talk people use to his essential project of revealing the nature of mind. As such, the effect of supporting a) is to undermine anti-psychologism as a basis for a theory that can tell us something about the mind since we've already dealt with Ordinary Language arguments and divided them into two rival claims, one dealing with phenomenological style arguments and the other with Linguistic Behavioral arguments. If we decide in favor of the former, then it removes language as the foundational basis for understanding the mind, in the strict sense of which I have thus laid out. The reason is that we have assimilated language into the third personal point of view when we broke up Ordinary Language arguments. By supporting auto-phenomenology we indorse a view that argues Linguistic Behavioral arguments, like the difference between sets of verbs that convey episodic structures like 'run', 'wake up' and 'tingle' and another set, the dispositional set, which contains the verbs 'posses', 'aspire' and 'know', are not as strong a source in a fundamental sense, to telling us something about the mind, as 'auto-phenomenological' arguments like in the example I gave from Ryle of trying to anticipate your next thought, and not being able to see the problem merely from the construction or classification of verbs, nouns or other grammatical constructions, but actually attempting to anticipate your next thought, or for instance, like in the example I gave from Sartre of examining a piece of paper, and then visualizing it, then comparing the two forms of consciousness, imaged and perceptual in order to distinguish them from each other.

Let me reiterate the argument thus far. It follows from the breakup of Ordinary Language style arguments, from a unified source as Ryle uses it, into two rival sources following my taxonomy and a contradiction I locate, explained at length in my thesis paper, that we have two rival sources. The first a), deals with phenomenology from the first personal direct view, or what I refer to as 'auto-phenomenology', that is, arguments that involve trying to antipate one's next thought, like trying to imagine lighting so bright it hurts one's eyes, or looking at and then imaging a sheet of paper to distinguish perceptual forms of consciousness from imaged forms consciousness. The second rival source, b) deals with Linguistic Behavioral style arguments and implicates the direct third personal point of view. These are what I refer to as sources of hetero-phenomenology. The next question to ask, once we realize these are rival forms of analysis, that is there is no longer a unified under Ordinary Language body of arguments but two rival forms of arguments, is to question which of the two is the stronger foundational source for telling us about the mind? Support for a) will count against anti-psychologism, since a) starts with auto-phenomenology. Thus a) will imply not b).This is because b) places the authority for foundational claims about the mind inside of Linguistic Behaviorism, which draws upon an implicit theory of meaning. An implicit theory of meaning is pro Anti-Psychologistic since it argues that a theory of meaning begins with language implicitly like Dummett argues for. Anti-psychologism argues for meanining in language either explicitly like Dummett identifies in Frege, or implicitly like Dummett and Ryle both argue for. Thus arguing for a) is also arguing against explicit meaning like Frege argues for. Hence why my paper argues for a

Pre-Fregeian form of psychologism. Indeed the thesis of my paper argues that a) is the stronger foundational source for claims about the nature of mind. The question after that, of course, is why is a) more foundational than b)?

The answer to why is a) a more foundational to telling us about the nature of mind than b), involves a thread I begin building early in the paper, and which is far too complex to give the details here, but from which develops three specific psychologistic theses. This is the type of psychologism I argue for in the paper. These three theses offer the broad outlines for what a theory about the mind might look like. Thus the very first psychologistic thesis covers the second stage of the thesis, which is what happens when you destroy Ordinary Language as a basis for justification for talk about the mind. It does this, but it also takes it a little further. It argues that auto-phenomenology is a stronger basis than hetero-phenomenolgy for telling us fundamental things about the mind. Where it goes a little further is by way of drawing out the implications in the assimilation of the indirect first and third personal views, which formerly contained the assumptions Ryle capitalized on to unify phenomenological and Linguistic Behavioral style arguments as Ordinary Language arguments, prior to the break up of Ordinary Language arguments as a unified normative source to appeal to with *'it makes sense to say'* style of claims, to their assimilation into the first and third personal direct views. Since, by implication, Hetero-phenomenolgy covers all aspects of the third personal view, including the indirect first personal view, which as I mentioned above has been assimilated into the third personal point of view. Hetero-phenomenology thus covers alongside language other domains of direct third personal perspective whichprior to the assimilation, incorporated indirect first personal perspectives like physiology, physical behavior, body language and various forms of brain science, which includes what Sellars would refer to as an 'observational language' as well as the chemistry and physiology. The core of my strategy for arguing in support of the first psychologistic theses involves arguments about ways we come to understand the nature of the emotions.

The essential step in developing this position is arguing that direct first person experience is what grounds our indirect third person assumptions about the mind. But there are other related and competing reasons why direct first person experience of the emotions is stronger than direct third personal views. Behavior is an unreliable indicator for emotion since people can behave differently when they feel similar emotions and may behave similarly when they feel different emotions. Emotions can look identical from the direct third person observation of physiological symptoms. Sartre makes much of this argument. Disbelief looks like disgust in the brain scans. Joy has similar physiology to anger but joy is not a form of anger. Most importantly our brain scans start off with comparative assumptions about the emotions which we use to map them. For example emotion y, whether it is anger, distress, the aesthetic sense of beauty, sadness, etcetera, which can be identified using various techniques like fMIR, as activity x in part z of the brain, will be first mapped off various assumptions that have their grounds in direct first person experience of either the subject or experimenter. I detail a number of these by way of example in my paper.

The second psychologistic thesis is straight forward enough and is the strongest argument in the paper. This is the argument that emotions are irreducible to linguistic behavior. The reason why they are irreducible, I argue, is that meaningful talk about the emotions depends upon non-linguistic elements which are the foundational experiences of the emotions themselves. That is to say, if meaning is essential to a theory of language, then it follows that a theory of meaning must begin with a theory that deals with the emotions. If auto-phenomenology is a stronger foundational source than hetero-phenomenology, then a theory of meaning must start with auto-phenomenological sources as its foundation. This introduces the third psychologistic theses of the paper. The broad general thesis, which is that to understand the nature of mind we need to develop a theory of consciousness. The reason for this, is linguistic descriptions of the emotions like 'flash of anger' or 'thrill of anticipation' fail to encompass the structure of the emotion as the person experiences it, centrally they fail to take into consideration the object to which the emotion is attached or arises in relation to. The attachment or relation implicates a theory of consciousness. A 'flash of anger' is not just a 'flash of anger' but rather, it is a 'flash of anger' about something. A thrill of anticipation is a thrill of anticipation about something. It involves a structural aspect that involves apprehending an object or situation, and which, I argue, implicates the direct need for a theory of consciousness. So if we think about this just in practical terms relating back to the point I just made about inheriting assumptions, a theory of consciousness is highly important for what researchers and people dealing with the mind are actually getting up to in practice, in areas like neuroscience. Since various emotions can be linked to the increase of various chemicals in the mind, and certain situations or circumstances produce various emotions, like these 'flashes of anger' and 'thrills of joy' then there is a missing causal agent in the model that needs something like an account of consciousness to explain it. That is, if the act of conscious apprehension of the object of joy is producing a chemical reaction in the mind, say the flooding of the neurotransmitter oxytocin which is associated with feelings of love, then we need a theory of consciousness to begin explaining what that roll is. That is, the cause of the sudden increase of oxytocin has something to do with the conscious apprehension of the object of joy and love in the patient. Many of the assumptions that neuroscientists make mask the absence of a foundational theory of consciousness which can reveal to us more about the nature of mind.

Thus to conclude this paper, I shall leave you with a very brief sketch of my research and how I have developed a thesis from the bridge that spans Analytic and Continental philosophy and forwarded a position on psychologism and phenomenology. What I argue in my thesis paper is that a theory of the mind cannot begin with language. It must begin with the mind, and only once we have a theory of mind, will we be able to understand language. The way I do that is to open up Ryle's ordinary language model of the mind using Sartre's phenomenology. That is, I develop from Sartre, a way of systematizing Ryle's arguments and drawing out assumptions that rest upon phenomenological acts of the mind. This gives me a taxonomy. The taxonomy allows me to break up Ordinary Language arguments into rival sources, and argue for one of those sources. This then produces the three psychologistic theses I have outlined. Firstly that Auto-phenomenology is a stronger foundational source than hetero-phenomenology. Secondly, the emotions are irreducible to linguistic statements or descriptions. Thirdly to understand the mind we need a theory of consciousness. These arguments are far too extensive to go into details here, they involve elements of emotional, social along with visual and linguistic classification and much seminal and valuable work from the Anglo-American philosopher Wilfred Sellars. But what I've hoped to have done, in writing this short paper, is perhaps to kindle an interest in bridging the gap between analytic and continental philosophy and to offer you a little taste of that process.

Works cited

Ryle, Gilbert. *The Concept of Mind*. Middlesex: Peregrine Books, Penguin, 1983.

Jean Paul Sartre *The Imaginary*. Translated by Jonathen Webber. Abington: Routledge, 2004.

Works Cited

Adams, Fred. "Embodied Cognition." *Phenomenology and the Cognitive Sciences* 9, no. 4 (2010): Pp 619-628

Adams, Fred. Aizawa, Ken. "Defending the Bounds of Cognition." In *The Extended Mind*, edited by Richard Menary, Pp 67-80. Massachusetts: The M.I.T. Press, 2010.

Andy Clark, David J. Chalmers. "The Extended Mind." In *The Extended Mind*, edited by Richard Menary, 27 - 41. Massachusetts: The M.I.T. Press, 2010.

Aquila, Richard E. "Two Problems of Being and Nonbeing in Sartre's Being and Nothingness." *Philosophy and Phenomenological Research* 38, no. 2 (1977): 167 - 189.

Armstrong, David. *A Materialist Theory of Mind* London: Routledge and Kegan Paul, 1968.

Armstrong, David*Sketch for a Systematic Metaphysics*. Oxford: Oxford University Press, 2010.

Austin, J. L. *How to Do Things with Words*. Massachusetts: Harvard University, 1975.

Ayer, A. J. *The Problem of Knowledge*. Middlesex: Penguin, 1956.

Barnes, Hazel. "Sartre's Concept of the Self." *Review of Existential Psychology and Psychiatry* 17, no. 1 (1981): Pp 41-66.

Barthes, Roland. *Elements of Semiology*. Translated by Colin Smith Annette Lavers. New York: Hill and Wang, 1984.

Bennett, Jonathen. *A Philosophical Guide to Conditionals*. Oxford: Oxford University Press, 2006.

Bergson, Henri. *Time and Free Will, an Essay on the Immediate Data of Consciousness*. Translated by M.A. F. L. Pogson. New York: Dover, 2001.

Berkeley, Istvan. "Gilbert Ryle and the Chinese Sceptic: Do Epistermologists Need to Know How To?" *The Electronic Journal of Analytic Philosophy*, no. 7 (2002): http://ejap.louisiana.edu/archives.html.

Berne, Eric. *Games People Play*. New York: Random House, 2004.

Block, Ned. "On a Confusion About the Function of Consciousness." *Behavioral and Brain Sciences* 18, (1995): Pp 227-47.

Branden, Nathaniel. "5. Isn't Everyone Selfish." In *The Virtues of Selfishness*, edited by Ayn Rand, 66-70. New York: New American Library, 2007.

Brandom, Robert. "Study Guide." In *Empiricism & the Philosophy of Mind.* , edited by Richard Rorty. United States: President & Fellows of Harvard University, 1997.

_____. *Articulating Reasons*. Harvard: Harvard University Press, 2001.

Bremer, Manuel. "The Egological Structure of Consciousness: Lessons from Sartre for Analytical Philosophy of Mind." In *A Companion to Phenomenology and Existentialism*, edited by Mark A. Wrathall Hubert L. Dreyfus. http://www.blackwellreference.com/public/tocnode?id=g9781405110778_chun k_g978140511077825#citation: Blackwell Reference Online, 2006.

Brentano, Franz. *Psychology from an Empirical Standpoint*. Translated by Peter Simons. 2nd Revised ed. Abington: Routledge, 1995.

Bruner. "On Perceptual Readiness." *Psychological Review* 64, (1957): 123-152.

Carnap, Rudolf. *Meaning and Necessity: A Study in Semantics and Modal Logic*. London: Phoenix Books; The University of Chicago Press, 1958.

Carter, Rita. *Mapping the Mind*. Los Angeles: University of California Press, 2010.

Castaneda, Hector-Neri. "God and Knowledge: Omniscience and Indexical Reference." In *Thinking, Language & Experience*, 137 - 158. Minneapolis: University of Minnesota Press, 1989.

_____. "The Transparent Subjective Mechanism for Encountering a World." *Nous* 24, no. 5 (1990): 735-749.

_____. "First Person Statements About the Past." In *The Phenomeno-Logic of the I*, edited by James G. Hart Tomis Kapitan. Indianapolis Indiana University Press 1999.

_____. "'He'. A Study in the Logic of Self Consciousness." In *The Phenomeno-Logic of the I* edited by James G. Hart Tomis Kapitan. Indianapolis Indiana University Press
1999.

_____. "I-Structures and the Reflexivity of Self-Consciousness." In *The Phenomeno-Logic of the I* edited by James G. Hart Tomis Kapitan. Indianapolis Indiana University Press 1999.

_____. "Persons, Egos and I's: Their Sameness Relations." In *The Phenomeno-Logic of the I*, edited by Tomis Kapitan James G. Hart. Indianapolis: Indiana University Press, 1999.

Chalmers, David J. *The Conscious Mind*. New York: Oxford, 1996.

_____. "How Can We Construct a Science of Consciousness." In *The Character of Consciousness* 2004.

_____. "The Metaphysics of Consciousness." In *The Character of Consciousness*, Pp 103 - 205. New York: Oxford, 2010.

_____. "The Science of Consciousness." In *The Character of Consciousness*, Pp 47 - 100. New York: Oxford, 2010.

_____."First-Person Methods in the Science of Consciousness." *Arizona Consciousness Bulletin*, (1999).

_____., Bayne, Tim. "What Is the Unity of Consciousness." In *The Unity of Consciousness: Binding, Integration, Dissociation* edited by Chris Frith Axel Cleeremans. Oxford Scholarship Online: March 2012 @ http://www.oxfordscholarship.com/view/10.1093/acprof:oso/9780198508571.00 1.0001/acprof-9780198508571 downloaded 05/06/2012:: Oxford, 2003.

_____.,"The Tyranny of the Subjunctive." http://consc.net/papers/tyranny.html, 1998. Lecture notes resented at Princeton

Crane, Tim. *Aspects of Psychologism*. Massachusetts: Harvard University Press, 2014.

Danto Arthur. "Ad Reinhardt." *The National* 253, no. 6 (1991): Pp 240 - 244.

Danto, Arthur C. *Sartre*. Glasgow: Penguin Books 1975.

Dawkins, Richard. *The Greatest Show on Earth*. Random House: London, 2009.

Dennett, Daniel C. *Consciousness Explained*. London: Penguin, 1993.

———. "Re-Introducing the Concept of Mind." *The Electronic Journal of Analytic Philosophy*, no. 7 (2002): http://ejap.louisiana.edu/archives.html.

Derrida, Jacques. *Writing and Difference*. Translated by Alan Bass. University of Chicago Press. Chicago. 1978.

Derrida, Jacques. *Of Grammatology: "Corrected Edtion"*. Translated by Gayatri Chakravorty Spivak. John Hopkins University Press. Baltimore. 1997.

Desan, Wilfrid. *The Tragic Finale: An Essay on the Philosophy of Jean-Paul Sartre*. Cambridge: Harvard University Press, 1954.

Descartes, Rene. "Discourse on the Method of Rightly Conducting the Reason and Seeking for Truth in the Sciences." In *The Philosophical Works of Descartes*, I. London: Cambridge University Press, 1979. Translated by Elizabeth S. Haldane, G. R. T. Ross

Devitt, Michael. "Realism and Semantics." *Nous* 17, no. 4 (1983): Pp 669-681.

———. "Thoughts and Their Ascription." *Midwest Studies in Philosophy* 9, no. 1 (1984): Pp 385-420.

———. "Realism without Representation." *Philosophical Studies: An International Journal for Philosophy in the Analytic Tradition* Vol. 61, no. 1/2 (1991): 75-77.

———. "The Metaphysics of Nonfactualism." *Nous* 30, no. 10 (1996): Pp 159-176.

———. *Ignorance of Language*. Oxford: Oxford University Press, 2006.

Dretske, Fred. *Explaining Behavior. Reasons in a World of Causes*. Massachusetts: MIT Press, 1988.

Dummett, Michael. "What Do I Know When I Know a Language?" In *The Seas of Language*, Pp 94 - 105. Oxford: Oxford University Press, 1993.

_____. "What Is a Theory of Meaning?" In *The Seas of Language*, Pp 94 - 105. Oxford: Oxford University, 1993.

Ferre, Frederick. "The Logic of Functional Analysis." In *Language, Logic and God*, 58-66. New York: Harper Torchbooks, 1961.

Fodor, Jerry A. *The Modularity of Mind*. Massachusetts: M.I.T. Press, 1983.

_____. *Psychosemantics, the Problem of Meaning in the Philosophy of Mind*. Massachusetts: M.I.T. Press, 1987.

_____. *Lot2*. New York: Oxford University Press, 2008.

Foucault, Michel. *The Archaeology of Knowledge*. Translated by A. M. Sheridan Smith. Oxon: Routledge, 2005.

Frege, Gottlob. "Illustrative Extracts from Frege's Review of Husserl's Philosophie Der Arithmetik." In *Translations from the Philosophical Writings of Gottlob Frege*, Pp 79 - 85. New York: The Philosophical Library, 1952.

_____. "On Concept and Object." In *Translations from the Philosophical Writings of Gottlob Frege*, Pp 42-56. New York: Philosophical Library Inc, 1952.

_____. "On Sense and Reference." In *Translations from the Philosophical Writings of Gottlob Frege*, Pp 59 - 78. New York: The Philosophical Library Inc, 1952.

Freud, Sigmund. *Civilization and Its Discontents*. Translated by David McLintock. London: Penguin, 2004.

_____. "Three Essays on Sexual Theory." In *Psychology of Love*, Pp 111-220. Victoria: Penguin, 2010.

Friedman, Michael. *A Parting of the Ways*. Illinois: Open Court, 2000.

G, Steiner. *After Babel*. Oxford: Oxford University Press, 1975.

G. Lakoff, M. Johnson. *Metaphors We Live By*. Chicago: University of Chicago Press, 1980.

Gensler, Harry G. *Introduction to Logic*. Oxon: Routledge, 2007.

Gilbert Ryle, J. N. Findlay. "Use, Usage and Meaning." *Proceedings of the Aristotelian Society; Supplementary Volumes* 38, (1961): Pp 228-229.

Ginet, Carl. *Knowledge, Perception and Memory* Boston: D. Reidel, 1975.

Goodman, Nelson. "The Problem of Counterfactual Conditionals." *Journal of Philosophy* 44, no. 5 (1947): 113-28.

_____. *Ways of World Making*. Indianapolis: Hackett Publishing, 1954.

Gotterman, Donald. "A Note on Locke's Theory of Self Knowledge." *Journal of The History of Philosophy* 12, no. 2 (1974): 239-242.

Grice, Paul. "Indicative Conditionals." In *Studies in the Way of Words*, Pp 58-87. Harvard: Harvard University Press, 1991.

_____. "Logic and Conversation." In *Studies in the Way of Words*, Pp 22-40. Harvard: Harvard University Press, 1991.

Halliday, M.A.K. "Language Structure and Language Function." In *New Horizons in Linguistics*, edited by John Lyons, Pp140-165. Middlesex: Penguin, 1970.

Hawkes, Terence. *Structuralism and Semiotics*. Suffolk: Routledge, 1988.

Heckman, John. "Introduction." In *Genesis and Structure of Hegel's Phenomenology of Spirit*, Pp xv - xli. Evanston: North-Western University Press, 1974.

Hegel, G. W. F. *The Logic of Hegel: Part I of the Encyclopedia of Philosophical Sciences*. Translated by William Wallace. London: Oxford, The Claredon Press, 1873.

_____. *The Encyclopaedia of Logic: Part 1 of the Encyclopedia of Philosophical Sciences with the Zusatze*. Translated by W. A. Suchting T. F. Geraets, H. S. Harris. Indianapolis: Hackett, 1991.

Heidegger, Martin. *Ontology - the Hermeneutics of Facticity*. Translated by John van Buren. Indiana: Indiana University Press, 2008.

Hemple, Carl G. *Philosophy of Natural Science*. New Jersey: Prentice Hall, 1966.

Hoffe, Otfried. *Immanuel Kant*. Translated by Marshall Farrier. New York, Albany: The State University of New York, 1994.

Hornsby, Jennifer. "Sartre and Action Theory." *Philosophy and Phenomenenological Research* 48, no. 4 (1988): 745-751.

Huffer, Ben. "Actions and Outcomes: Two Aspects of Agency." *Synthese* (2007).
   http://link.springer.com/article/10.1007%2Fs11229-006-9107-z?LI=true#
   [accessed 20/10/2011].

Hume, David. *Enquiries Concerning Human Understanding and Concerning the
   Principles of Morals.* Reprint from the 1777 edition ed. New York: Oxford,
   2005.

Hurley, Susan. "The Varieties of Externalism." In *The Extended Mind,* edited by
   Richard Menary, Pp 101 - 154. Massachusetts: The M.I.T. Press, 2010.

Hyppolite, Jean. *Genesis and Structure of Hegel's Phenomenology of Spirit.*
   Translated by John Heckman Samuel Cherniak Northwestern University
   Studies in Phenomenology & Existential Philosophy, Edited by James M.
   Edie. Evanston: Northwestern University Press, 1976.

Stanley, J. Williamson, T. "Knowing How." *Journal of Philosophy* 98, no. 8 (2001):
   411-444.

Jackson, Frank. "On Assertion and Indicative Conditionals." *The Philosophical
   Review* 88, no. 4 (1976): Pp 565-589.

Jakobson, Roman. *The Science of Language.* London. George Allen & Unwin Ltd.
   1970.

Jean-Paul Sartre, Simone De Beauvoir ed. . *Witness to My Life.* Translated by
   Norman MacAfee Lee Fahnestock. Oxford: Maxwell Macmillian
   International, 1992.

Kant, Immanuel. *The Critique of Pure Reason.* Translated by Vasilis Politis J. M. D.
   Meiklejohn: Orion Publishing Group, 2004.

———. *The Critique of Pure Reason.* Translated by Marcus Weigelt. Max Muller.
   Victoria: Penguin, 2007.

Katz, Fodor. "The Structure of a Semantic Theory." *Language* 39, (1963): Pp 170-
   210.

Keim, Willard D. *Ethics, Morality and International Affairs* Lanham: University
   Press Of America, 2000.

Kojeve, Alexander. *Introduction to the Reading of Hegel, Lectures on the Phenomenology of Spirit*. Translated by Jr. James H. Nichols, Edited by Allan Boom. New York: Cornell, 1969.

Leech, Geoffrey. *Semantics*. Victoria: Penguin, 1974.

Levi-Strauss, Claude. *Structural Anthropology*. Translated by Claire Jacobson, Brooke Grundfest Schoepf. 1963: Basic Books. United States of America.

_____. Translated by Rodney Needham. London. Merlin Press. 1962.

Lewis, David. "An Argument for the Identity Theory." *Journal of Philosophy* 63, no. 1 (1966): 17-25.

_____. *Counterfactuals*. Malden Blackwell, 2001.

_____. "Counterpart Theory and Quantified Modal Logic." In *Philosophy of Logic: An Anthology*, edited by Dale Jacquette. Oxford: Blackwell, 2002.

_____. "Adverbs of Quantification " In *Formal Semantics of Natural Language* edited by Edward L. Keenan. United Kingdom: Cambridge University Press, 2009.

Lyons, John. "Deixis as the Source of Reference." In *Formal Semantics of Natural Language* edited by Edward L. Keenan. United Kingdom: Cambridge University Press, 2009.

Lyotard, Jean-Francois. *The Differend: Phrases in Dispute* Translated by Georges Van Dan Abbeele: University of Minesota Press, 1989.

Mackie, J. L. *Ethics, Inventing Right and Wrong*. Victoria: Penguin, 1986.

Magee, Bryan. *Modern British Philosophy* Oxford: Oxford University Press, 1986.

Mailer, Norman. *The Naked and the Dead*. London: Harperperennial, 2006.

Malcolm, Norman. "Thoughtless Brutes." In *The Nature of Mind*, edited by Rosenthal, Pp 454-461. Oxford: Oxford, 1991.

Mandlebaum, Maurice. *Philosophy, Science and Sense Perception*. Baltimore: John Hopkins University Press, 1964.

Marion, Jean-Luc. *God without Being* Translated by Thomas A. Carlson. Chicago: University of Chicago Press, 1995.

Martin, Thomas. *Oppression and the Human Condition: An Introduction to Sartrean Existentialism*. Lanham: Rowman & Littlefield Publishers 2002.

_____. "Sartre, Sadism and Female Beauty Ideals." *Australian Feminist Studies* 11, no. 24 (1996).

McCosh, James. *Realistic Philosophy*. Vol. II. New York: Scribner, 1900.

McCulloch, Gregory. *Using Sartre*. New York: Routledge, 1994.

Menary, Richard. "Cognitive Intergration and the Extended Mind." In *The Extended Mind*, edited by Richard Menary, Pp 227 - 243. Massachusetts: The M.I.T. Press, 2010.

_____. "Dimensions of Mind." *Phenomenology and the Cognitive Sciences 561-578* 9, no. 4 (2010): Pp 561-578.

_____. "Introduction." In *The Extended Mind*, edited by Richard Menary, Pp 1 - 25. Massachusetts: The M.I.T. Press, 2010.

_____. "Cognitive Practices and Cognitive Character." *Philosophical Explorations* 15 no. 2 (2012): Pp 147 - 164.

Milligan, David. *Reasoning and the Explanation of Human Action*. New Jersey: Humanities Press, 1980.

Mohanty, J. N. "Intentionality." (2011). http://www.blackwellreference.com/public/tocnode?id=g9781405110778_chunk_g97814051107788 [accessed 19/102011].

Morris, Phyllis Sutton. "Sartre on Transcendence of the Ego." *Philosophy and Phenomenenological Research* 46, no. 2 (1985): Pp 179 - 198.

Murphy, Julien S. *Feminist Interpretations of Jean-Paul Sartre* Pennsylvania: Pennsylvania State University 2007.

Myers, G. E. "Motives and Wants." *Mind* Vol. 73, no. 290 (1964): Pp. 173-185.

Negri, Antonio. *Political Descartes: Reason, Ideology and the Bourgeois Project*. Translated by Alberto Toscano Matteo Mandarini Radical Thinkers New York: Verso, 2006.

Nietzsche, Friedrich. *The Genealogy of Morals*. Translated by Horace Samuel. New York: Dover, 2003.

O'Shea, James R. *Wilfrid Sellars* Key Contemporary Thinkers. Cambridge: Polity Press, 2007.

_____. "'The 'Theory Theory' of Mind and the Aims of Sellars' Original Myth of Jones'." *Phenomenology and the Cognitive Sciences* 11, no. 2 (2012): Pp 175-204.

Palmer, A. "Thinking and Performance." In *Knowledge and Necessity: Royal Institute of Philosophy Lectures 1968/9*, edited by Godfrey Vesey, 3, Pp 107-118. London: MacMillian and Co, 1970.

Parkinson, G. H. R. "Translation Theory of Meaning." In *Communication and Understanding*, edited by Godfrey Vesey, 1-19. New Jersey: The Royal Institute of Philosophy; Humanities Press, 1977.

Partee, Barbara Hall. "Deletion and Variable Binding." In *Formal Semantics of Natural Language* edited by Edward L. Keenan. United Kingdom: Cambridge University Press, 2009.

Perry, John. "The Problem of the Essential Indexical " *Nous* 13, no. 1 (1979): 3-21.

Peters, R. S. "Motives and Causes." *Proceedings of the Aristotelian Society, Supplementary Volumes* 26, (1952): pp. 139-194.

_____. *The Concept of Motivation*. London Lowe & Brydone, 1969.

Pinker, Steven. *The Language Instinct*. Victoria: Penguin, 2008.

Plantinga, Alvin. *The Nature of Necessity*. Oxford: Claredon Library of Logic and Philosophy, 1978.

Plato. *The Republic*. Translated by Desmond Lee. Victoria: Penguin, 2003.

Price, Huw. "Metaphysics after Carnap: The Ghost Who Walks." In *Metametaphysics: New Essays on the Foundations of Ontology*, edited by David Manley Daivd J. Chalmers, Ryan Wasserman, Pp 320-346. New York: Oxford University Press, 2009.

Prichard, H. A. "Does Moral Philosophy Rest on a Mistake?" *Mind* 21, no. 81 (1912): Pp 21-37

Quine, W. V. *The Roots of Reference*. Ilinois: Open Court, 1973.

_____. "Reference and Modality." In *From a Logical Point of View*, 139 -159. Massachusetts: Harvard University Press, 1980.

_____.*Word & Object*. Massachusetts: The M.I.T. Press, 1960.

Rachels, James. *The Moral Elements of Moral Philosophy*. 5th ed. Boston: McGraw Hill, 2007.

Rand, Ayn. *The Virtues of Selfishness*. New York: New American Library, 2007.

Ratcliffe, Mathew. "Phenomenology, Neuroscience and Intersubjectivity " In *A Companion to Phenomenology and Existentialism*, edited by Hubert L. Dreyfus and Mark A Wrathall,. http://www.blackwellreference.com/public/tocnode?id=g9781405110778_chun k_g978140511077826, 2006.

Rorty, Amelie. "Enough Already with Theories of Emotion." In *Thinking About Feelings: Philosophers on Emotion*, edited by Robert C. Solomon, 21-33. New York: Oxford University Press.

_____. *Explaining Emotions*. Los Angeles: University of California Press, 1980.

Rorty, Richard. *The Mirror of Nature*. Princeton University Press: Princeton, 2009.

Russell, Bertrand. *The Analysis of Mind*. London: The Muirhead Library of Philosophy, 1951.

Ryle, Gilbert. *The Concept of Mind*. Middlesex: Peregrine Books, Penguin, 1983.

_____. "The Systematic Elusiveness of 'I'." In *The Concept of Mind*. Middlesex: Peregrine Books, 1983.

_____.*On Thinking*. London: Basil Blackwell, 1979.

_____. *Dilemmas*. New York: Press Syndicate of the University of Cambridge, 1987.

_____. "Feelings." In *Collected Essays 1929-1968*, edited by Julia Tanney, II, 284-299. Oxon: Routledge, 2009.

_____. "Ryle's Final Letter to Daniel Dennett." *The Electronic Journal of Analytic Philosophy*, no. 7 (2002): http://ejap.louisiana.edu/archives.html.

_____. "Heidegger's 'Sein Und Zeit'." In *In Critical Essays: Gilbert Ryle*, edited by Julia Tanney, Pp 205-223. Oxon: Routledge, 2009.

_____. "Hume." In *Critical Essays: Gilbert Ryle*, edited by Julia Tanney, Pp 165-174. Oxon: Routledge, 2009.

_____. "John Locke on Human Understanding." In *Critical Essays*, edited by Julia Tanney, I, Pp 132-153. Oxon: Routledge, 2009.

_____. "Phenomenology Vs the Concept of Mind." In *Critical Essays*, edited by Julia Tanney. Oxon: Routledge, 2009.

_____. "Review of Martin Faber: 'The Foundations of Phenomenology'." In *Critical Essays: Gilbert Ryle*, edited by Julia Tanney, I. Oxon: Routledge, 2009.

_____. "Systematically Misleading Expressions." In *Collected Essays 1929-1968*, edited by Julia Tannery, II. New York: Routledge, 2009.

_____. "Thinking and Language." In *Collected Essays 1929-1968*, edited by Julia Tannery, II, 269-283. Oxon: Routledge, 2009.

Sartre, Jean-Paul. "The Wall." In *The Wall and Other Stories*, 1 - 17. New York: New Directions Books, 1948.

_____. *The Ghost of Stalin*. Translated by Martha Fletcher. New York George Braziller, 1968.

_____. "Nausea, Nothingness, Freedom." In *The Existentialists and Jean-Paul Sartre*, edited by Max Charlesworth, 90 - 102. Queensland: Queensland University Press, 1975.

_____. *Life Situations: Essays Written and Spoken*. Translated by Paul Auster. Lydia Davis. New York: Pantheon Books, 1977.

_____. "No Exit." In *No Exit and Three Other Plays*, Pp 1 - 49. New York: Vintage International, 1989.

_____. *Anti-Semite and Jew*. Translated by George J. Becker. New York: Schocken Books, 1995.

_____. *Sketch for a Theory of the Emotions*. Translated by Philip Mairet. New York: Routledge Classics, 2002.

_____. *Being and Nothingness*. Translated by Hazel E. Barnes. Abington: Routledge, 2003.

_____. *The Imaginary*. Translated by Jonathen Webber. Abington: Routledge, 2004.

_____. *Sketch for a Theory of the Emotions*. Translated by Philip Mairet. Abington: Routledge, 2004.

_____. *Transcendence of the Ego*. Abington: Routledge, 2004.

_____. *Existentialism Is a Humanism*. New Haven: Yale University Press, 2007.

_____. "The Itinerary of a Thought." In *Between Existentialism and Marxism*. London: Verso, 2008.

_____. *The Age of Reason*. Translated by Eric Sutton. Victoria: Penguin, 2009.

_____. *Nausea*. Translated by Robert Baldick. Victoria: Penguin, 2010.

Sartre, Jean Paul. *Essays in Existentialism*. Translated by unacknowledged: Citadel Press, 1993.

Saussure, Ferdinand De. *Course in General Linguistics*, Edited by McGaw Hill. New York, 1966.

Schank, Abelson. "Scripts, Plans and Knowledge." In *Proceedings of the 4th international joint conference on Artificial intelligence* 1 Pages 151-157 San Francisco: Morgan Kaufmann Publishers Inc. , 1975.

Scheffler, Israel. *Conditions of Knowledge*. Chicago: Scott, Foresman and Company, 1965.

Sellars, Wilfrid. "The Identity Approach to the Mind-Body Problem." In *In Philosophical Perspectives: Metaphysics and Epistermology*, Pp 190-209. California: Ridgeview, 1967.

_____. "Notes on Intentionality." In *Philosophical Perspectives: Metaphysics and Epistermology*, Pp 128-140: Ridgeview, 1967.

_____. "Actions and Events." *Nous* 7, no. 2 (1973): Pp 179-202.

_____. "Being and Being Known." In *Science, Perception and Reality* Pp 41 - 59. California: Ridgeview, 1991.

_____. "The Language of Theories." In *Science, Perception and Reality*, 106 - 126. California: Ridgeview, 1991.

_____. "Naming and Saying." In *Science, Perception and Reality*, Pp 225 - 246. California: Ridgeview, 1991.

_____. "Philosophy and the Scientific Image of Man." In *Science, Perception and Reality*, Pp 1 - 40. California: Ridgeview, 1991.

_____. "Truth and 'Correspondence'." In *Science, Perception and Reality*, Pp 197 - 224. California: Ridgeview, 1991.

_____. "Appearances and Things in Themselves 2. Persons." In *Science and Metaphysics: Variations on Kantian Themes*, Pp 151-174. California: Ridgeview, 1992.

_____. "Sensibility and Understanding." In *In Science and Metaphysics: Variations on Kantian Themes*, Pp 1-30. California: Ridgeview, 1992.

_____. *Empiricism and the Philosophy of Mind.* . 1963 ed. Electronic Text. 1963 Amendments, edited by Andrew Chrucky. http://www.ditext.com/sellars/epm.html 1995.

_____. "Counterfactuals, Dispositions and the Causal Modalities " *Minnesota Studies in the Philosophy of Science* II, (1958): Pp 225-308.

_____. "Meaning and Ontology." In *Naturalism and Ontology*, Pp 63-96. California, 1996.

_____. *Empiricism & the Philosophy of Mind.* United States: President & Fellows of Harvard University, 1997.

_____. *Pure Pragmatics and Possible Worlds*, Edited by Jeffery F. Sicha. California: Ridgeview, 2005.

Siewert, Charles. "Consciousness." (2011). http://www.blackwellreference.com/public/tocnode?id=g9781405110778_chunk_g97814051107789 [accessed 19/10/2011].

Simpson, David. "Language and Know-How." *Phenomenology and the Cognitive Sciences* 9, no. 5 (2010): 629–643.

Skinner, B. F. *Beyond Human Freedom and Dignity*. Middlesex: Penguin, 1976.

Solomon, Robert. "Emotions in Phenomenology and Existentialism." In *A Companion to Phenomenology and Existentialism*, edited by Hubert L. Dreyfus and Mark A. Wrathall. http://onlinelibrary.wiley.com/doi/10.1002/9780470996508.ch21/: Wiley Online, 2007.

Sousa, Roland de. "Emotion." In *The Stanford Encyclopedia of Philosophy*, edited by Edward Zalta, Spring 2013 Edition. http://plato.stanford.edu/archives/spr2013/entries/emotion/ 2013.

Spade, Paul-Vincent. *Jean-Paul Sartre's Being and Nothingness Class Lecture Notes Fall 1995*. http://pvspade.com/Sartre/sartre.html, 1996.

Stalkner, Robert. "A Theory of Conditionals." In *Studies in Logical Theory*, edited by N. Rescher. Oxford Blackwell, 1968. Pp 98-112

Stalnaker, Robert. Richard Thomason. "A Semantic Analysis of Conditional Logic." *Theoria* 36, no. 1 (1970): Pp 23-42.

Stanley, J. Williamson, T. "Knowing How." *Journal of Philosophy* 98, no. 8 (2001): 411-444.

Stanley, Williamson, *Knowing How*, 2001. Pg 415; Ginet, Carl. *Knowledge, Perception and Memory* Boston: D. Reidel, 1975

Steiner, Claude M. *Scripts People Live: Transactional Analysis of Life Scripts*. New York: Bantam Books, 1982.

Stevenson, Leslie. "Sartre: Radical Freedom." In *Ten Theories of Human Nature*, Pp 181 - 200. New York: Oxford University Press, 2009.

Stitch, S. *From Folk Psychology to Cognitive Science.* . Cambridge: The M.I.T. Press, 1983.

Stitch, Steven. *The Fragmentation of Reason: Preface to a Pragmatic Theory of Cognitive Evaluation* Massachusetts: The M.I.T. Press, 1993.

Stout, Rowland. "What You Know When You Know How Someone Behaves." *The Electronic Journal of Analytic Philosophy*, no. 7 (2002): http://ejap.louisiana.edu/archives.html.

Sutton, John. "Exograms and Interdisciplinararity: History, the Extended Mind, and the Civilizing Process." In *The Extended Mind*, edited by Richard Menary, Pp 189 - 226. Massachusetts: The M.I.T. Press, 2010.

Terry, Winant. "Trans-World Identity of Future Contingents: Sartre on Leibnizian Freedom." *The Southern Journal of Philosophy* 22, no. 4 (1984): 543-564.

"The Fantasy of First-Person Science." In *Daniel C. Dennett, David J. Chalmers.* Transcript at http://ase.tufts.edu/cogstud/papers/chalmersdeb3dft.htm, 2001. Video at http://catcomcon.blogspot.com.au/2012/09/dennett-d-unpublished-fantasy-of-first.html

Triesman, A. M. "Verbal Responses and Contextual Constraints in Language." In *Language*, edited by J. C. Marshall R. C. Oldfield, Pp 276 - 292. Middlesex: Penguin, 1968.

Tsai, Cheng-Hung. "The Metaepistemology of Knowing-How." *Phenomenology and the Cognitive Sciences* 10, no. 4 (2011): Pp 541-556.

Velmans, Max. "Heterophenomenology Versus Critical Phenomenology: A Dialogue with Dan Dennett. 'Unpublished', deposited at " http://cogprints.org/1795/, Febuary 2006, (2001).

Vernon, M. D. *The Psychology of Perception*. Victoria: Penguin, 1962.

Warburton, Nigel. *Philosophy: The Classics*. Oxon: Routledge, 2001.

Webber, Jonathan. "Motivated Aversion: Non-Thetic Awareness in Bad Faith." *Sartre Studies International* 8, no. 1 (2002).

_____. "Introduction." In *The Imaginary*, Pp xiii - xxvi. Oxon: Routledge, 2004.

Weed, Laura. "Philosophy of Mind an Overview." *Philosophy Now* Nov/Dec, no. 87 (2011).

Weitz, Morris. "Professor Ryle's "Logical Behaviorism"." *The Journal of Philosophy* Vol 48, no. 9 (1951): Pp 297 - 301.

Wemin Mo, Wang Wei. "Cogito: From Descartes to Sartre." *Frontiers of Philosophy in China* 2, no. 2 (2007): Pp 247-264.

White, A. R. "The Language of Motives." *Mind* LXVII, no. 266 (1958): Pp 258 - 263.

Willem A. DeVires, Timm Triplett. *Knowledge, Mind and the Given.* Indianapolis: Hackett Publishing Company Inc, 2000.

Wolff, Robert. "Professor Ryle's Discussion of Agitations." *Mind* Vol. 63, no. No. 250 , (1954): Pp. 239-241.

Yule, George. *The Study of Language.* Cambridge University Press. Cambridge. 1985.

www.ingramcontent.com/pod-product-compliance
Lightning Source LLC
Chambersburg PA
CBHW031814170526
45157CB00001B/49